Embedded Robotics

Thomas Bräunl

EMBEDDED ROBOTICS

Mobile Robot Design
and Applications
with Embedded Systems

Second Edition

With 233 Figures and 24 Tables

 Springer

Thomas Bräunl
School of Electrical, Electronic
and Computer Engineering
The University of Western Australia
35 Stirling Highway
Crawley, Perth, WA 6009
Australia

Library of Congress Control Number: 2006925479

ACM Computing Classification (1998): I.2.9, C.3

ISBN-10 3-540-34318-0 Springer Berlin Heidelberg New York
ISBN-13 978-3-540-34318-9 Springer Berlin Heidelberg New York
ISBN-10 3-540-03436-6 1. Edition Springer Berlin Heidelberg New York

Springer is a part of Springer Science+Business Media

springer.com

© Springer-Verlag Berlin Heidelberg 2003, 2006
Printed in Germany

Typesetting: Camera-ready by the author
Production: LE-TEX Jelonek, Schmidt &Vöckler GbR, Leipzig
Cover design: KünkelLopka, Heidelberg

Printed on acid-free paper   45/3180/YL - 5 4 3 2 1   SPIN   12038633

PREFACE

I t all started with a new robot lab course I had developed to accompany my robotics lectures. We already had three large, heavy, and expensive mobile robots for research projects, but nothing simple and safe, which we could give to students to practice on for an introductory course.

We selected a mobile robot kit based on an 8-bit controller, and used it for the first couple of years of this course. This gave students not only the enjoyment of working with real robots but, more importantly, hands-on experience with control systems, real-time systems, concurrency, fault tolerance, sensor and motor technology, etc. It was a very successful lab and was greatly enjoyed by the students. Typical tasks were, for example, driving straight, finding a light source, or following a leading vehicle. Since the robots were rather inexpensive, it was possible to furnish a whole lab with them and to conduct multi-robot experiments as well.

Simplicity, however, had its drawbacks. The robot mechanics were unreliable, the sensors were quite poor, and extendability and processing power were very limited. What we wanted to use was a similar robot at an advanced level. The processing power had to be reasonably fast, it should use precision motors and sensors, and – most challenging – the robot had to be able to do on-board image processing. This had never been accomplished before on a robot of such a small size (about 12cm × 9cm × 14cm). Appropriately, the robot project was called "EyeBot". It consisted of a full 32-bit controller ("EyeCon"), interfacing directly to a digital camera ("EyeCam") and a large graphics display for visual feedback. A row of user buttons below the LCD was included as "soft keys" to allow a simple user interface, which most other mobile robots lack. The processing power of the controller is about 1,000 times faster than for robots based on most 8-bit controllers (25MHz processor speed versus 1MHz, 32-bit data width versus 8-bit, compiled C code versus interpretation) and this does not even take into account special CPU features like the "time processor unit" (TPU).

The EyeBot family includes several driving robots with differential steering, tracked vehicles, omni-directional vehicles, balancing robots, six-legged walkers, biped android walkers, autonomous flying and underwater robots, as well as simulation systems for driving robots ("EyeSim") and underwater robots ("SubSim"). EyeCon controllers are used in several other projects, with and without mobile robots. Numerous universities use EyeCons to drive their own mobile robot creations. We use boxed EyeCons for experiments in a second-year course in Embedded Systems as part of the Electrical Engineering, Information Technology, and Mechatronics curriculums. And one lonely EyeCon controller sits on a pole on Rottnest Island off the coast of Western Australia, taking care of a local weather station.

Acknowledgements

While the controller hardware and robot mechanics were developed commercially, several universities and numerous students contributed to the EyeBot software collection. The universities involved in the EyeBot project are:

- University of Stuttgart, Germany
- University of Kaiserslautern, Germany
- Rochester Institute of Technology, USA
- The University of Auckland, New Zealand
- The University of Manitoba, Winnipeg, Canada
- The University of Western Australia (UWA), Perth, Australia

The author would like to thank the following students, technicians, and colleagues: Gerrit Heitsch, Thomas Lampart, Jörg Henne, Frank Sautter, Elliot Nicholls, Joon Ng, Jesse Pepper, Richard Meager, Gordon Menck, Andrew McCandless, Nathan Scott, Ivan Neubronner, Waldemar Spädt, Petter Reinholdtsen, Birgit Graf, Michael Kasper, Jacky Baltes, Peter Lawrence, Nan Schaller, Walter Bankes, Barb Linn, Jason Foo, Alistair Sutherland, Joshua Petitt, Axel Waggershauser, Alexandra Unkelbach, Martin Wicke, Tee Yee Ng, Tong An, Adrian Boeing, Courtney Smith, Nicholas Stamatiou, Jonathan Purdie, Jippy Jungpakdee, Daniel Venkitachalam, Tommy Cristobal, Sean Ong, and Klaus Schmitt.

Thanks for proofreading the manuscript and numerous suggestions go to Marion Baer, Linda Barbour, Adrian Boeing, Michael Kasper, Joshua Petitt, Klaus Schmitt, Sandra Snook, Anthony Zaknich, and everyone at Springer-Verlag.

Contributions

A number of colleagues and former students contributed to this book. The author would like to thank everyone for their effort in putting the material together.

JACKY BALTES   The University of Manitoba, Winnipeg, contributed to the section on PID control,

ADRIAN BOEING — UWA, coauthored the chapters on the evolution of walking gaits and genetic algorithms, and contributed to the section on SubSim,

CHRISTOPH BRAUNSCHÄDEL — FH Koblenz, contributed data plots to the sections on PID control and on/off control,

MICHAEL DRTIL — FH Koblenz, contributed to the chapter on AUVs,

LOUIS GONZALEZ — UWA, contributed to the chapter on AUVs,

BIRGIT GRAF — Fraunhofer IPA, Stuttgart, coauthored the chapter on robot soccer,

HIROYUKI HARADA — Hokkaido University, Sapporo, contributed the visualization diagrams to the section on biped robot design,

YVES HWANG — UWA, coauthored the chapter on genetic programming,

PHILIPPE LECLERCQ — UWA, contributed to the section on color segmentation,

JAMES NG — UWA, coauthored the sections on probabilistic localization and the DistBug navigation algorithm.

JOSHUA PETITT — UWA, contributed to the section on DC motors,

KLAUS SCHMITT — Univ. Kaiserslautern, coauthored the section on the RoBI-OS operating system,

ALISTAIR SUTHERLAND — UWA, coauthored the chapter on balancing robots,

NICHOLAS TAY — DSTO, Canberra, coauthored the chapter on map generation,

DANIEL VENKITACHALAM — UWA, coauthored the chapters on genetic algorithms and behavior-based systems and contributed to the chapter on neural networks,

EYESIM was implemented by Axel Waggershauser (V5) and Andreas Koestler (V6), UWA, Univ. Kaiserslautern, and FH Giessen.

SUBSIM was implemented by Adrian Boeing, Andreas Koestler, and Joshua Petitt (V1), and Thorsten Rühl and Tobias Bielohlawek (V2), UWA, FH Giessen, and Univ. Kaiserslautern.

Additional Material

Hardware and mechanics of the "EyeCon" controller and various robots of the EyeBot family are available from INROSOFT and various distributors:

`http://inrosoft.com`

All system software discussed in this book, the RoBIOS operating system, C/C++ compilers for Linux and Windows, system tools, image processing tools, simulation system, and a large collection of example programs are available free from:

`http://robotics.ee.uwa.edu.au/eyebot/`

Lecturers who adopt this book for a course can receive a full set of the author's course notes (PowerPoint slides), tutorials, and labs from this website. And finally, if you have developed some robot application programs you would like to share, please feel free to submit them to our website.

Second Edition

Less than three years have passed since this book was first published and I have since used this book successfully in courses on Embedded Systems and on Mobile Robots / Intelligent Systems. Both courses are accompanied by hands-on lab sessions using the EyeBot controllers and robot systems, which the students found most interesting and which I believe contribute significantly to the learning process.

What started as a few minor changes and corrections to the text, turned into a major rework and additional material has been added in several areas. A new chapter on autonomous vessels and underwater vehicles and a new section on AUV simulation have been added, the material on localization and navigation has been extended and moved to a separate chapter, and the kinematics sections for driving and omni-directional robots have been updated, while a couple of chapters have been shifted to the Appendix.

Again, I would like to thank all students and visitors who conducted research and development work in my lab and contributed to this book in one form or another.

All software presented in this book, especially the EyeSim and SubSim simulation systems can be freely downloaded from:

`http://robotics.ee.uwa.edu.au`

Perth, Australia, June 2006

Thomas Bräunl

CONTENTS

PART II: MOBILE ROBOT DESIGN

Contents

Contents

Contents

PART I:
EMBEDDED SYSTEMS

ROBOTS AND CONTROLLERS

1

....

R obotics has come a long way. Especially for mobile robots, a similar trend is happening as we have seen for computer systems: the transition from mainframe computing via workstations to PCs, which will probably continue with handheld devices for many applications. In the past, mobile robots were controlled by heavy, large, and expensive computer systems that could not be carried and had to be linked via cable or wireless devices. Today, however, we can build small mobile robots with numerous actuators and sensors that are controlled by inexpensive, small, and light embedded computer systems that are carried on-board the robot.

There has been a tremendous increase of interest in mobile robots. Not just as interesting toys or inspired by science fiction stories or movies [Asimov 1950], but as a perfect tool for engineering education, mobile robots are used today at almost all universities in undergraduate and graduate courses in Computer Science/Computer Engineering, Information Technology, Cybernetics, Electrical Engineering, Mechanical Engineering, and Mechatronics.

What are the advantages of using mobile robot systems as opposed to traditional ways of education, for example mathematical models or computer simulation?

First of all, a robot is a tangible, self-contained piece of real-world hardware. Students can relate to a robot much better than to a piece of software. Tasks to be solved involving a robot are of a practical nature and directly "make sense" to students, much more so than, for example, the inevitable comparison of sorting algorithms.

Secondly, all problems involving "real-world hardware" such as a robot, are in many ways harder than solving a theoretical problem. The "perfect world" which often is the realm of pure software systems does not exist here. Any actuator can only be positioned to a certain degree of accuracy, and all sensors have intrinsic reading errors and certain limitations. Therefore, a working robot program will be much more than just a logic solution coded in software.

It will be a robust system that takes into account and overcomes inaccuracies and imperfections. In summary: a valid engineering approach to a typical (industrial) problem.

Third and finally, mobile robot programming is enjoyable and an inspiration to students. The fact that there is a moving system whose behavior can be specified by a piece of software is a challenge. This can even be amplified by introducing robot competitions where two teams of robots compete in solving a particular task [Bräunl 1999] – achieving a goal with autonomously operating robots, not remote controlled destructive "robot wars".

1.1 Mobile Robots

Since the foundation of the Mobile Robot Lab by the author at The University of Western Australia in 1998, we have developed a number of mobile robots, including wheeled, tracked, legged, flying, and underwater robots. We call these robots the "EyeBot family" of mobile robots (Figure 1.1), because they are all using the same embedded controller "EyeCon" (EyeBot controller, see the following section).

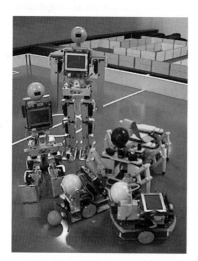

Figure 1.1: Some members of the EyeBot family of mobile robots

The simplest case of mobile robots are wheeled robots, as shown in Figure 1.2. Wheeled robots comprise one or more *driven wheels* (drawn solid in the figure) and have optional passive or *caster wheels* (drawn hollow) and possibly *steered wheels* (drawn inside a circle). Most designs require two motors for driving (and steering) a mobile robot.

The design on the left-hand side of Figure 1.2 has a single driven wheel that is also steered. It requires two motors, one for driving the wheel and one for turning. The advantage of this design is that the driving and turning actions

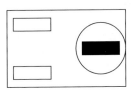 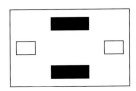 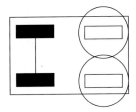

Figure 1.2: Wheeled robots

have been completely separated by using two different motors. Therefore, the control software for driving curves will be very simple. A disadvantage of this design is that the robot cannot turn on the spot, since the driven wheel is not located at its center.

The robot design in the middle of Figure 1.2 is called "differential drive" and is one of the most commonly used mobile robot designs. The combination of two driven wheels allows the robot to be driven straight, in a curve, or to turn on the spot. The translation between driving commands, for example a curve of a given radius, and the corresponding wheel speeds has to be done using software. Another advantage of this design is that motors and wheels are in fixed positions and do not need to be turned as in the previous design. This simplifies the robot mechanics design considerably.

Finally, on the right-hand side of Figure 1.2 is the so-called "Ackermann Steering", which is the standard drive and steering system of a rear-driven passenger car. We have one motor for driving both rear wheels via a differential box and one motor for combined steering of both front wheels.

It is interesting to note that all of these different mobile robot designs require two motors in total for driving and steering.

A special case of a wheeled robot is the omni-directional "Mecanum drive" robot in Figure 1.3, left. It uses four driven wheels with a special wheel design and will be discussed in more detail in a later chapter.

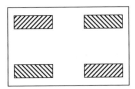 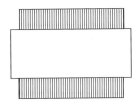 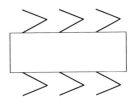

Figure 1.3: Omni-directional, tracked, and walking robots

One disadvantage of all wheeled robots is that they require a street or some sort of flat surface for driving. Tracked robots (see Figure 1.3, middle) are more flexible and can navigate over rough terrain. However, they cannot navigate as accurately as a wheeled robot. Tracked robots also need two motors, one for each track.

Legged robots (see Figure 1.3, right) are the final category of land-based mobile robots. Like tracked robots, they can navigate over rough terrain or climb up and down stairs, for example. There are many different designs for legged robots, depending on their number of legs. The general rule is: the more legs, the easier to balance. For example, the six-legged robot shown in the figure can be operated in such a way that three legs are always on the ground while three legs are in the air. The robot will be stable at all times, resting on a tripod formed from the three legs currently on the ground – provided its center of mass falls in the triangle described by these three legs. The less legs a robot has, the more complex it gets to balance and walk, for example a robot with only four legs needs to be carefully controlled, in order not to fall over. A biped (two-legged) robot cannot play the same trick with a supporting triangle, since that requires at least three legs. So other techniques for balancing need to be employed, as is discussed in greater detail in Chapter 10. Legged robots usually require two or more motors ("degrees of freedom") per leg, so a six-legged robot requires at least 12 motors. Many biped robot designs have five or more motors per leg, which results in a rather large total number of degrees of freedom and also in considerable weight and cost.

Braitenberg vehicles
A very interesting conceptual abstraction of actuators, sensors, and robot control is the vehicles described by Braitenberg [Braitenberg 1984]. In one example, we have a simple interaction between motors and light sensors. If a light sensor is activated by a light source, it will proportionally increase the speed of the motor it is linked to.

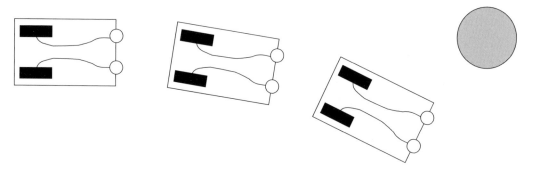

Figure 1.4: Braitenberg vehicles avoiding light (phototroph)

In Figure 1.4 our robot has two light sensors, one on the front left, one on the front right. The left light sensor is linked to the left motor, the right sensor to the right motor. If a light source appears in front of the robot, it will start driving toward it, because both sensors will activate both motors. However, what happens if the robot gets closer to the light source and goes slightly off course? In this case, one of the sensors will be closer to the light source (the left sensor in the figure), and therefore one of the motors (the left motor in the figure) will become faster than the other. This will result in a curve trajectory of our robot and it will miss the light source.

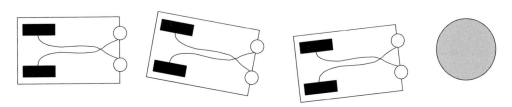

Figure 1.5: Braitenberg vehicles searching light (photovore)

Figure 1.5 shows a very similar scenario of Braitenberg vehicles. However, here we have linked the left sensor to the right motor and the right sensor to the left motor. If we conduct the same experiment as before, again the robot will start driving when encountering a light source. But when it gets closer and also slightly off course (veering to the right in the figure), the left sensor will now receive more light and therefore accelerate the right motor. This will result in a left curve, so the robot is brought back on track to find the light source.

Braitenberg vehicles are only a limited abstraction of robots. However, a number of control concepts can easily be demonstrated by using them.

1.2 Embedded Controllers

The centerpiece of all our robot designs is a small and versatile embedded controller that each robot carries on-board. We called it the "EyeCon" (EyeBot controller, Figure 1.6), since its chief specification was to provide an interface for a digital camera in order to drive a mobile robot using on-board image processing [Bräunl 2001].

Figure 1.6: EyeCon, front and with camera attached

The EyeCon is a small, light, and fully self-contained embedded controller. It combines a 32bit CPU with a number of standard interfaces and drivers for DC motors, servos, several types of sensors, plus of course a digital color camera. Unlike most other controllers, the EyeCon comes with a complete built-in user interface: it comprises a large graphics display for displaying text messages and graphics, as well as four user input buttons. Also, a microphone and a speaker are included. The main characteristics of the EyeCon are:

EyeCon specs

- 25MHz 32bit controller (Motorola M68332)
- 1MB RAM, extendable to 2MB
- 512KB ROM (for system + user programs)
- 1 Parallel port
- 3 Serial ports (1 at V24, 2 at TTL)
- 8 Digital inputs
- 8 Digital outputs
- 16 Timing processor unit inputs/outputs
- 8 Analog inputs
- Single compact PCB
- Interface for color and grayscale camera
- Large graphics LCD (128×64 pixels)
- 4 input buttons
- Reset button
- Power switch
- Audio output
  - Piezo speaker
  - Adapter and volume potentiometer for external speaker
- Microphone for audio input
- Battery level indication
- Connectors for actuators and sensors:
  - Digital camera
  - 2 DC motors with encoders
  - 12 Servos
  - 6 Infrared sensors
  - 6 Free analog inputs

One of the biggest achievements in designing hardware and software for the EyeCon embedded controller was interfacing to a digital camera to allow on-board real-time image processing. We started with grayscale and color Connectix "QuickCam" camera modules for which interface specifications were available. However, this was no longer the case for successor models and it is virtually impossible to interface a camera if the manufacturer does not disclose the protocol. This lead us to develop our own camera module "EyeCam" using low resolution CMOS sensor chips. The current design includes a FIFO hardware buffer to increase the throughput of image data.

A number of simpler robots use only 8bit controllers [Jones, Flynn, Seiger 1999]. However, the major advantage of using a 32bit controller versus an 8bit controller is not just its higher CPU frequency (about 25 times faster) and wider word format (4 times), but the ability to use standard off-the-shelf C and C++ compilers. Compilation makes program execution about 10 times faster than interpretation, so in total this results in a system that is 1,000 times faster. We are using the GNU C/C++ cross-compiler for compiling both the operating system and user application programs under Linux or Windows. This compiler is the industry standard and highly reliable. It is not comparable with any of the C-subset interpreters available.

The EyeCon embedded controller runs our own "RoBIOS" (Robot Basic Input Output System) operating system that resides in the controller's flash-ROM. This allows a very simple upgrade of a controller by simply download-ing a new system file. It only requires a few seconds and no extra equipment, since both the Motorola background debugger circuitry and the writeable flash-ROM are already integrated into the controller.

RoBIOS combines a small monitor program for loading, storing, and exe-cuting programs with a library of user functions that control the operation of all on-board and off-board devices (see Appendix B.5). The library functions include displaying text/graphics on the LCD, reading push-button status, read-ing sensor data, reading digital images, reading robot position data, driving motors, v-omega (vω) driving interface, etc. Included also is a thread-based multitasking system with semaphores for synchronization. The RoBIOS oper-ating system is discussed in more detail in Chapter B.

Another important part of the EyeCon's operating system is the HDT (Hardware Description Table). This is a system table that can be loaded to flash-ROM independent of the RoBIOS version. So it is possible to change the system configuration by changing HDT entries, without touching the RoBIOS operating system. RoBIOS can display the current HDT and allows selection and testing of each system component listed (for example an infrared sensor or a DC motor) by component-specific testing routines.

Figure 1.7 from [InroSoft 2006], the commercial producer of the EyeCon controller, shows hardware schematics. Framed by the address and data buses on the top and the chip-select lines on the bottom are the main system compo-nents ROM, RAM, and latches for digital I/O. The LCD module is memory mapped, and therefore looks like a special RAM chip in the schematics. Optional parts like the RAM extension are shaded in this diagram. The digital camera can be interfaced through the parallel port or the optional FIFO buffer. While the Motorola M68332 CPU on the left already provides one serial port, we are using an ST16C552 to add a parallel port and two further serial ports to the EyeCon system. Serial-1 is converted to V24 level (range +12V to −12V) with the help of a MAX232 chip. This allows us to link this serial port directly to any other device, such as a PC, Macintosh, or workstation for program download. The other two serial ports, Serial-2 and Serial-3, stay at TTL level (+5V) for linking other TTL-level communication hardware, such as the wire-less module for Serial-2 and the IRDA wireless infrared module for Serial-3.

A number of CPU ports are hardwired to EyeCon system components; all others can be freely assigned to sensors or actuators. By using the HDT, these assignments can be defined in a structured way and are transparent to the user

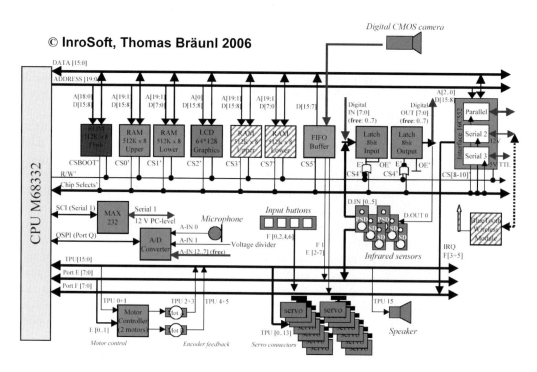

Figure 1.7: EyeCon schematics

program. The on-board motor controllers and feedback encoders utilize the lower TPU channels plus some pins from the CPU port E, while the speaker uses the highest TPU channel. Twelve TPU channels are provided with matching connectors for servos, i.e. model car/plane motors with pulse width modulation (PWM) control, so they can simply be plugged in and immediately operated. The input keys are linked to CPU port F, while infrared distance sensors (PSDs, position sensitive devices) can be linked to either port E or some of the digital inputs.

An eight-line analog to digital (A/D) converter is directly linked to the CPU. One of its channels is used for the microphone, and one is used for the battery status. The remaining six channels are free and can be used for connecting analog sensors.

1.3 Interfaces

A number of interfaces are available on most embedded systems. These are digital inputs, digital outputs, and analog inputs. Analog outputs are not always required and would also need additional amplifiers to drive any actuators. Instead, DC motors are usually driven by using a digital output line and a pulsing technique called "pulse width modulation" (PWM). See Chapter 3 for

video out   camera connector   IR receiver

serial 1   serial 2

graphics LCD

reset button

power switch

speaker microphone   input buttons

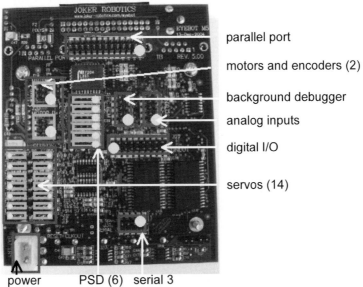

parallel port

motors and encoders (2)

background debugger

analog inputs

digital I/O

servos (14)

power   PSD (6) serial 3

Figure 1.8: EyeCon controller M5, front and back

details. The Motorola M68332 microcontroller already provides a number of digital I/O lines, grouped together in ports. We are utilizing these CPU ports as can be seen in the schematics diagram Figure 1.7, but also provide additional digital I/O pins through latches.

Most important is the M68332's TPU. This is basically a second CPU integrated on the same chip, but specialized to timing tasks. It simplifies tremendously many time-related functions, like periodic signal generation or pulse counting, which are frequently required for robotics applications.

Figure 1.8 shows the EyeCon board with all its components and interface connections from the front and back. Our design objective was to make the construction of a robot around the EyeCon as simple as possible. Most interface connectors allow direct plug-in of hardware components. No adapters or special cables are required to plug servos, DC motors, or PSD sensors into the EyeCon. Only the HDT software needs to be updated by simply downloading the new configuration from a PC; then each user program can access the new hardware.

The parallel port and the three serial ports are standard ports and can be used to link to a host system, other controllers, or complex sensors/actuators. Serial port 1 operates at V24 level, while the other two serial ports operate at TTL level.

The Motorola background debugger (BDM) is a special feature of the M68332 controller. Additional circuitry is included in the EyeCon, so only a cable is required to activate the BDM from a host PC. The BDM can be used to debug an assembly program using breakpoints, single step, and memory or register display. It can also be used to initialize the flash-ROM if a new chip is inserted or the operating system has been wiped by accident.

Figure 1.9: EyeBox units

At The University of Western Australia, we are using a stand-alone, boxed version of the EyeCon controller ("EyeBox" Figure 1.9) for lab experiments in the Embedded Systems course. They are used for the first block of lab experiments until we switch to the EyeBot Labcars (Figure 7.5). See Appendix E for a collection of lab experiments.

1.4 Operating System

Embedded systems can have anything between a complex real-time operating system, such as Linux, or just the application program with no operating system, whatsoever. It all depends on the intended application area. For the EyeCon controller, we developed our own operating system RoBIOS (Robot Basic Input Output System), which is a very lean real-time operating system that provides a monitor program as user interface, system functions (including multithreading, semaphores, timers), plus a comprehensive device driver library for all kinds of robotics and embedded systems applications. This includes serial/parallel communication, DC motors, servos, various sensors, graphics/text output, and input buttons. Details are listed in Appendix B.5.

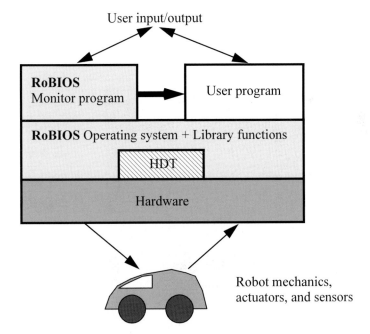

Figure 1.10: RoBIOS structure

The RoBIOS monitor program starts at power-up and provides a comprehensive control interface to download and run programs, load and store programs in flash-ROM, test system components, and to set a number of system parameters. An additional system component, independent of RoBIOS, is the

Hardware Description Table (HDT, see Appendix C), which serves as a user-configurable hardware abstraction layer [Kasper et al. 2000], [Bräunl 2001].

RoBIOS is a software package that resides in the flash-ROM of the controller and acts on the one hand as a basic multithreaded operating system and on the other hand as a large library of user functions and drivers to interface all on-board and off-board devices available for the EyeCon controller. RoBIOS offers a comprehensive user interface which will be displayed on the integrated LCD after start-up. Here the user can download, store, and execute programs, change system settings, and test any connected hardware that has been registered in the HDT (see Table 1.1).

Monitor Program	System Functions	Device Drivers
Flash-ROM management	Hardware setup	LCD output
OS upgrade	Memory manager	Key input
Program download	Interrupt handling	Camera control
Program decompression	Exception handling	Image processing
Program run	Multithreading	Latches
Hardware setup and test	Semaphores	A/D converter
	Timers	RS232, parallel port
	Reset resist. variables	Audio
	HDT management	Servos, motors
		Encoders
		vω driving interface
		Bumper, infrared, PSD
		Compass
		TV remote control
		Radio communication

Table 1.1: RoBIOS features

The RoBIOS structure and its relation to system hardware and the user program are shown in Figure 1.10. Hardware access from both the monitor program and the user program is through RoBIOS library functions. Also, the monitor program deals with downloading of application program files, storing/retrieving programs to/from ROM, etc.

The RoBIOS operating system and the associated HDT both reside in the controller's flash-ROM, but they come from separate binary files and can be downloaded independently. This allows updating of the RoBIOS operating system without having to reconfigure the HDT and vice versa. Together the two binaries occupy the first 128KB of the flash-ROM; the remaining 384KB are used to store up to three user programs with a maximum size of 128KB each (Figure 1.11).

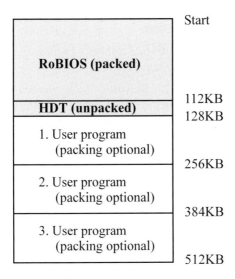

Figure 1.11: Flash-ROM layout

Since RoBIOS is continuously being enhanced and new features and drivers are being added, the growing RoBIOS image is stored in compressed form in ROM. User programs may also be compressed with utility `srec2bin` before downloading. At start-up, a bootstrap loader transfers the compressed RoBIOS from ROM to an uncompressed version in RAM. In a similar way, RoBIOS unpacks each user program when copying from ROM to RAM before execution. User programs and the operating system itself can run faster in RAM than in ROM, because of faster memory access times.

Each operating system comprises machine-independent parts (for example higher-level functions) and machine-dependent parts (for example device drivers for particular hardware components). Care has been taken to keep the machine-dependent part as small as possible, to be able to perform porting to a different hardware in the future at minimal cost.

1.5 References

ASIMOV I. *Robot*, Doubleday, New York NY, 1950

BRAITENBERG, V. *Vehicles – Experiments in Synthetic Psychology*, MIT Press, Cambridge MA, 1984

BRÄUNL, T. *Research Relevance of Mobile Robot Competitions*, IEEE Robotics and Automation Magazine, Dec. 1999, pp. 32-37 (6)

BRÄUNL, T. *Scaling Down Mobile Robots - A Joint Project in Intelligent Mini-Robot Research*, Invited paper, 5th International Heinz Nixdorf Symposium on Autonomous Minirobots for Research and Edutainment, Univ. of Paderborn, Oct. 2001, pp. 3-10 (8)

INROSOFT, http://inrosoft.com, 2006

JONES, J., FLYNN, A., SEIGER, B. *Mobile Robots - From Inspiration to Implementation*, 2nd Ed., AK Peters, Wellesley MA, 1999

KASPER, M., SCHMITT, K., JÖRG, K., BRÄUNL, T. *The EyeBot Microcontroller with On-Board Vision for Small Autonomous Mobile Robots*, Workshop on Edutainment Robots, GMD Sankt Augustin, Sept. 2000, http://www.gmd.de/publications/report/0129/Text.pdf, pp. 15-16 (2)

SENSORS

2

· ·

There are a vast number of different sensors being used in robotics, applying different measurement techniques, and using different interfaces to a controller. This, unfortunately, makes sensors a difficult subject to cover. We will, however, select a number of typical sensor systems and discuss their details in hardware and software. The scope of this chapter is more on interfacing sensors to controllers than on understanding the internal construction of sensors themselves.

What is important is to find the right sensor for a particular application. This involves the right measurement technique, the right size and weight, the right operating temperature range and power consumption, and of course the right price range.

Data transfer from the sensor to the CPU can be either CPU-initiated (*polling*) or sensor-initiated (via *interrupt*). In case it is CPU-initiated, the CPU has to keep checking whether the sensor is ready by reading a status line in a loop. This is much more time consuming than the alternative of a sensor-initiated data transfer, which requires the availability of an interrupt line. The sensor signals via an interrupt that data is ready, and the CPU can react immediately to this request.

Sensor Output	Sample Application
Binary signal (0 or 1)	Tactile sensor
Analog signal (e.g. 0..5V)	Inclinometer
Timing signal (e.g. PWM)	Gyroscope
Serial link (RS232 or USB)	GPS module
Parallel link	Digital camera

Table 2.1: Sensor output

2.1 Sensor Categories

From an engineer's point of view, it makes sense to classify sensors according to their output signals. This will be important for interfacing them to an embedded system. Table 2.1 shows a summary of typical sensor outputs together with sample applications. However, a different classification is required when looking at the application side (see Table 2.2).

		Local	**Global**
Internal	**Passive**	battery sensor, chip-temperature sensor, shaft encoders, accelerometer, gyroscope, inclinometer, compass	**Passive** –
	Active	–	**Active** –
External	**Passive**	on-board camera	**Passive** overhead camera, satellite GPS
	Active	sonar sensor, infrared distance sensor, laser scanner	**Active** sonar (or other) global positioning system

Table 2.2: Sensor classification

From a robot's point of view, it is more important to distinguish:

- Local or on-board sensors
  (sensors mounted on the robot)
- Global sensors
  (sensors mounted outside the robot in its environment and transmitting sensor data back to the robot)

For mobile robot systems it is also important to distinguish:

- Internal or proprioceptive sensors
  (sensors monitoring the robot's internal state)
- External sensors
  (sensors monitoring the robot's environment)

A further distinction is between:

- Passive sensors
  *(sensors that monitor the environment without disturbing it,
  for example digital camera, gyroscope)*
- Active sensors
  *(sensors that stimulate the environment for their measurement,
  for example sonar sensor, laser scanner, infrared sensor)*

Table 2.2 classifies a number of typical sensors for mobile robots according to these categories. A good source for information on sensors is [Everett 1995].

2.2 Binary Sensor

Binary sensors are the simplest type of sensors. They only return a single bit of information, either 0 or 1. A typical example is a tactile sensor on a robot, for example using a microswitch. Interfacing to a microcontroller can be achieved very easily by using a digital input either of the controller or a latch. Figure 2.1 shows how to use a resistor to link to a digital input. In this case, a pull-up resistor will generate a high signal unless the switch is activated. This is called an "active low" setting.

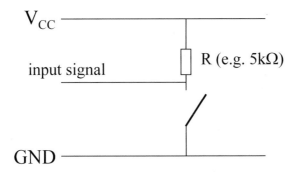

Figure 2.1: Interfacing a tactile sensor

2.3 Analog versus Digital Sensors

A number of sensors produce analog output signals rather than digital signals. This means an A/D converter (analog to digital converter, see Section 2.5) is required to connect such a sensor to a microcontroller. Typical examples of such sensors are:

- Microphone
- Analog infrared distance sensor

- Analog compass
- Barometer sensor

Digital sensors on the other hand are usually more complex than analog sensors and often also more accurate. In some cases the same sensor is available in either analog or digital form, where the latter one is the identical analog sensor packaged with an A/D converter.

The output signal of digital sensors can have different forms. It can be a parallel interface (for example 8 or 16 digital output lines), a serial interface (for example following the RS232 standard) or a "synchronous serial" interface.

The expression "synchronous serial" means that the converted data value is read bit by bit from the sensor. After setting the chip-enable line for the sensor, the CPU sends pulses via the serial clock line and at the same time reads 1 bit of information from the sensor's single bit output line for every pulse (for example on each rising edge). See Figure 2.2 for an example of a sensor with a 6bit wide output word.

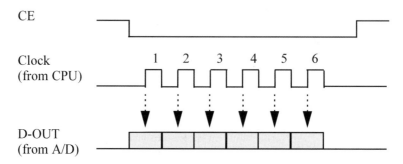

Figure 2.2: Signal timing for synchronous serial interface

2.4 Shaft Encoder

Encoders are required as a fundamental feedback sensor for motor control (Chapters 3 and 4). There are several techniques for building an encoder. The most widely used ones are either magnetic encoders or optical encoders. Magnetic encoders use a Hall-effect sensor and a rotating disk on the motor shaft with a number of magnets (for example 16) mounted in a circle. Every revolution of the motor shaft drives the magnets past the Hall sensor and therefore results in 16 pulses or "ticks" on the encoder line. Standard optical encoders use a sector disk with black and white segments (see Figure 2.3, left) together with an LED and a photo-diode. The photo-diode detects reflected light during a white segment, but not during a black segment. So once again, if this disk has 16 white and 16 black segments, the sensor will receive 16 pulses during one revolution.

Encoders are usually mounted directly on the motor shaft (that is before the gear box), so they have the full resolution compared to the much slower rota-

Encoder ticks

tional speed at the geared-down wheel axle. For example, if we have an encoder which detects 16 ticks per revolution and a gearbox with a ratio of 100:1 between the motor and the vehicle's wheel, then this gives us an encoder resolution of 1,600 ticks per wheel revolution.

Both encoder types described above are called *incremental*, because they can only count the number of segments passed from a certain starting point. They are not sufficient to locate a certain *absolute position* of the motor shaft. If this is required, a Gray-code disk (Figure 2.3, right) can be used in combination with a set of sensors. The number of sensors determines the maximum resolution of this encoder type (in the example there are 3 sensors, giving a resolution of $2^3 = 8$ sectors). Note that for *any* transition between two neighboring sectors of the Gray code disk only a single bit changes (e.g. between $1 = 001$ and $2 = 011$). This would not be the case for a standard binary encoding (e.g. $1 = 001$ and $2 = 010$, which differ by two bits). This is an essential feature of this encoder type, because it will still give a proper reading if the disk just passes between two segments. (For binary encoding the result would be arbitrary when passing between 111 and 000.)

As has been mentioned above, an encoder with only a single magnetic or optical sensor element can only count the number of segments passing by. But it cannot distinguish whether the motor shaft is moving clockwise or counterclockwise. This is especially important for applications such as robot vehicles which should be able to move forward or backward. For this reason most encoders are equipped with two sensors (magnetic or optical) that are positioned with a small phase shift to each other. With this arrangement it is possible to determine the rotation direction of the motor shaft, since it is recorded which of the two sensors first receives the pulse for a new segment. If in Figure 2.3 Enc1 receives the signal first, then the motion is clockwise; if Enc2 receives the signal first, then the motion is counter-clockwise.

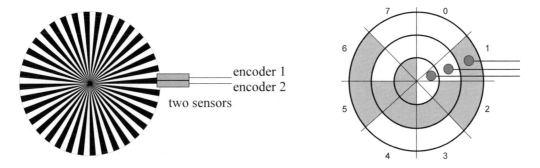

Figure 2.3: Optical encoders, incremental versus absolute (Gray code)

Since each of the two sensors of an encoder is just a binary digital sensor, we could interface them to a microcontroller by using two digital input lines. However, this would not be very efficient, since then the controller would have to constantly poll the sensor data lines in order to record any changes and update the sector count.

Luckily this is not necessary, since most modern microcontrollers (unlike standard microprocessors) have special input hardware for cases like this. They are usually called "pulse counting registers" and can count incoming pulses up to a certain frequency completely independently of the CPU. This means the CPU is not being slowed down and is therefore free to work on higher-level application programs.

Shaft encoders are standard sensors on mobile robots for determining their position and orientation (see Chapter 14).

2.5 A/D Converter

An A/D converter translates an analog signal into a digital value. The characteristics of an A/D converter include:

- Accuracy
  expressed in the number of digits it produces per value
  (for example 10bit A/D converter)
- Speed
  expressed in maximum conversions per second
  (for example 500 conversions per second)
- Measurement range
  expressed in volts
  (for example 0..5V)

A/D converters come in many variations. The output format also varies. Typical are either a parallel interface (for example up to 8 bits of accuracy) or a synchronous serial interface (see Section 2.3). The latter has the advantage that it does not impose any limitations on the number of bits per measurement, for example 10 or 12bits of accuracy. Figure 2.4 shows a typical arrangement of an A/D converter interfaced to a CPU.

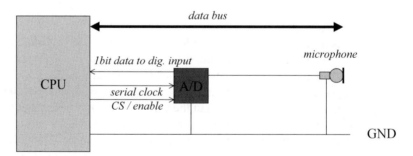

Figure 2.4: A/D converter interfacing

Many A/D converter modules include a multiplexer as well, which allows the connection of several sensors, whose data can be read and converted subsequently. In this case, the A/D converter module also has a 1bit input line, which allows the specification of a particular input line by using the synchronous serial transmission (from the CPU to the A/D converter).

2.6 Position Sensitive Device

Sensors for distance measurements are among the most important ones in robotics. For decades, mobile robots have been equipped with various sensor types for measuring distances to the nearest obstacle around the robot for navigation purposes.

Sonar sensors   In the past, most robots have been equipped with sonar sensors (often Polaroid sensors). Because of the relatively narrow cone of these sensors, a typical configuration to cover the whole circumference of a round robot required 24 sensors, mapping about 15° each. Sonar sensors use the following principle: a short acoustic signal of about 1ms at an ultrasonic frequency of 50kHz to 250kHz is emitted and the time is measured from signal emission until the echo returns to the sensor. The measured time-of-flight is proportional to twice the distance of the nearest obstacle in the sensor cone. If no signal is received within a certain time limit, then no obstacle is detected within the corresponding distance. Measurements are repeated about 20 times per second, which gives this sensor its typical clicking sound (see Figure 2.5).

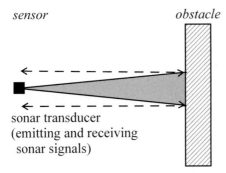

Figure 2.5: Sonar sensor

Sonar sensors have a number of disadvantages but are also a very powerful sensor system, as can be seen in the vast number of published articles dealing with them [Barshan, Ayrulu, Utete 2000], [Kuc 2001]. The most significant problems of sonar sensors are reflections and interference. When the acoustic signal is reflected, for example off a wall at a certain angle, then an obstacle seems to be further away than the actual wall that reflected the signal. Interference occurs when several sonar sensors are operated at once (among the 24 sensors of one robot, or among several independent robots). Here, it can happen that the acoustic signal from one sensor is being picked up by another sensor, resulting in incorrectly assuming a closer than actual obstacle. Coded sonar signals can be used to prevent this, for example using pseudo random codes [Jörg, Berg 1998].

Laser sensors   Today, in many mobile robot systems, sonar sensors have been replaced by either infrared sensors or laser sensors. The current standard for mobile robots is laser sensors (for example Sick Auto Ident [Sick 2006]) that return an almost perfect local 2D map from the viewpoint of the robot, or even a complete 3D distance map. Unfortunately, these sensors are still too large and heavy (and too expensive) for small mobile robot systems. This is why we concentrate on infrared distance sensors.

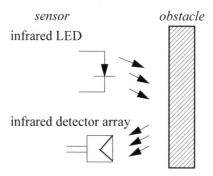

Figure 2.6: Infrared sensor

Infrared sensors   Infrared (IR) distance sensors do not follow the same principle as sonar sensors, since the time-of-flight for a photon would be much too short to measure with a simple and cheap sensor arrangement. Instead, these systems typically use a pulsed infrared LED at about 40kHz together with a detection array (see Figure 2.6). The angle under which the reflected beam is received changes according to the distance to the object and therefore can be used as a measure of the distance. The wavelength used is typically 880nm. Although this is invisible to the human eye, it can be transformed to visible light either by IR detector cards or by recording the light beam with an IR-sensitive camera.

Figure 2.7 shows the Sharp sensor GP2D02 [Sharp 2006] which is built in a similar way as described above. There are two variations of this sensor:

- Sharp GP2D12 with analog output
- Sharp GP2D02 with digital serial output

The analog sensor simply returns a voltage level in relation to the measured distance (unfortunately not proportional, see Figure 2.7, right, and text below). The digital sensor has a digital serial interface. It transmits an 8bit measurement value bit-wise over a single line, triggered by a clock signal from the CPU as shown in Figure 2.2.

In Figure 2.7, right, the relationship between digital sensor read-out (raw data) and actual distance information can be seen. From this diagram it is clear that the sensor does not return a value linear or proportional to the actual distance, so some post-processing of the raw sensor value is necessary. The simplest way of solving this problem is to use a lookup table which can be calibrated for each individual sensor. Since only 8 bits of data are returned, the lookup table will have the reasonable size of 256 entries. Such a lookup table is provided in the hardware description table (HDT) of the RoBIOS operating system (see Section B.3). With this concept, calibration is only required once per sensor and is completely transparent to the application program.

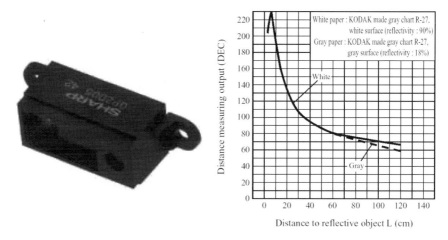

Figure 2.7: Sharp PSD sensor and sensor diagram (source: [Sharp 2006])

Another problem becomes evident when looking at the diagram for actual distances below about 6cm. These distances are below the measurement range of this sensor and will result in an incorrect reading of a *higher* distance. This is a more serious problem, since it cannot be fixed in a simple way. One could, for example, continually monitor the distance of a sensor until it reaches a value in the vicinity of 6cm. However, from then on it is impossible to know whether the obstacle is coming closer or going further away. The safest solution is to mechanically mount the sensor in such a way that an obstacle can never get closer than 6cm, or use an additional (IR) proximity sensor to cover for any obstacles closer than this minimum distance.

IR proximity switches are of a much simpler nature than IR PSDs. IR proximity switches are an electronic equivalent of the tactile binary sensors shown in Section 2.2. These sensors also return only 0 or 1, depending on whether there is free space (for example 1-2cm) in front of the sensor or not. IR proximity switches can be used in lieu of tactile sensors for most applications that involve obstacles with reflective surfaces. They also have the advantage that no moving parts are involved compared to mechanical microswitches.

2.7 Compass

A compass is a very useful sensor in many mobile robot applications, especially self-localization. An autonomous robot has to rely on its on-board sensors in order to keep track of its current position and orientation. The standard method for achieving this in a driving robot is to use shaft encoders on each wheel, then apply a method called "dead reckoning". This method starts with a known initial position and orientation, then adds all driving and turning actions to find the robot's current position and orientation. Unfortunately, due to wheel slippage and other factors, the "dead reckoning" error will grow larger and larger over time. Therefore, it is a good idea to have a compass sensor on-board, to be able to determine the robot's absolute orientation.

A further step in the direction of global sensors would be the interfacing to a receiver module for the satellite-based *global positioning system* (GPS). GPS modules are quite complex and contain a microcontroller themselves. Interfacing usually works through a serial port (see the use of a GPS module in the autonomous plane, Chapter 11). On the other hand, GPS modules only work outdoors in unobstructed areas.

Analog compass   Several compass modules are available for integration with a controller. The simplest modules are analog compasses that can only distinguish eight directions, which are represented by different voltage levels. These are rather cheap sensors, which are, for example, used as directional compass indicators in some four-wheel-drive car models. Such a compass can simply be connected to an analog input of the EyeBot and thresholds can be set to distinguish the eight directions. A suitable analog compass model is:

• Dinsmore Digital Sensor No. 1525 or 1655
   [Dinsmore 1999]

Digital compass   Digital compasses are considerably more complex, but also provide a much higher directional resolution. The sensor we selected for most of our projects has a resolution of 1° and accuracy of 2°, and it can be used indoors:

• Vector 2X
   [Precision Navigation 1998]

This sensor provides control lines for reset, calibration, and mode selection, not all of which have to be used for all applications. The sensor sends data by using the same digital serial interface already described in Section 2.3. The sensor is available in a standard (see Figure 2.8) or gimbaled version that allows accurate measurements up to a banking angle of 15°.

Figure 2.8: Vector 2X compass

2.8 Gyroscope, Accelerometer, Inclinometer

Orientation sensors to determine a robot's orientation in 3D space are required for projects like tracked robots (Figure 7.7), balancing robots (Chapter 9), walking robots (Chapter 10), or autonomous planes (Chapter 11). A variety of sensors are available for this purpose (Figure 2.9), up to complex modules that can determine an object's orientation in all three axes. However, we will concentrate here on simpler sensors, most of them only capable of measuring a single dimension. Two or three sensors of the same model can be combined for measuring two or all three axes of orientation. Sensor categories are:

- **Accelerometer**
  Measuring the acceleration along one axis
  - Analog Devices ADXL05 (single axis, analog output)
  - Analog Devices ADXL202 (dual axis, PWM output)

- **Gyroscope**
  Measuring the rotational change of orientation about one axis
  - HiTec GY 130 Piezo Gyro (PWM input and output)

- **Inclinometer**
  Measuring the absolute orientation angle about one axis
  - Seika N3 (analog output)
  - Seika N3d (PWM output)

Figure 2.9: HiTec piezo gyroscope, Seika inclinometer

2.8.1 Accelerometer

All these simple sensors have a number of drawbacks and restrictions. Most of them cannot handle jitter very well, which frequently occurs in driving or especially walking robots. As a consequence, some software means have to be taken for signal filtering. A promising approach is to combine two different sensor types like a gyroscope and an inclinometer and perform sensor fusion in software (see Figure 7.7).

A number of different accelerometer models are available from Analog Devices, measuring a single or two axes at once. Sensor output is either analog

27

or a PWM signal that needs to be measured and translated back into a binary value by the CPU's timing processing unit.

The acceleration sensors we tested were quite sensitive to positional noise (for example servo jitter in walking robots). For this reason we used additional low-pass filters for the analog sensor output or digital filtering for the digital sensor output.

2.8.2 Gyroscope

The gyroscope we selected from HiTec is just one representative of a product range from several manufacturers of gyroscopes available for model airplanes and helicopters. These modules are meant to be connected between the receiver and a servo actuator, so they have a PWM input and a PWM output. In normal operation, for example in a model helicopter, the PWM input signal from the receiver is modified according to the measured rotation about the gyroscope's axis, and a PWM signal is produced at the sensor's output, in order to compensate for the angular rotation.

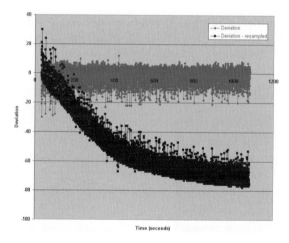

Figure 2.10: Gyroscope drift at rest and correction

Obviously, we want to use the gyroscope only as a sensor. In order to do so, we generate a fixed middle-position PWM signal using the RoBIOS library routine SERVOSet for the input of the gyroscope and read the output PWM signal of the gyroscope with a TPU input of the EyeBot controller. The periodical PWM input signal is translated to a binary value and can then be used as sensor data.

A particular problem observed with the piezo gyroscope used (HiTec GY 130) is drift: even when the sensor is not being moved and its input PWM signal is left unchanged, the sensor output drifts over time as seen in Figure 2.10 [Smith 2002], [Stamatiou 2002]. This may be due to temperature changes in the sensor and requires compensation.

An additional general problem with these types of gyroscopes is that they can only sense the change in orientation (rotation about a single axis), but not the absolute position. In order to keep track of the current orientation, one has to integrate the sensor signal over time, for example using the Runge-Kutta integration method. This is in some sense the equivalent approach to "dead reckoning" for determining the x/y-position of a driving robot. The integration has to be done in regular time intervals, for example 1/100s; however, it suffers from the same drawback as "dead reckoning": the calculated orientation will become more and more imprecise over time.

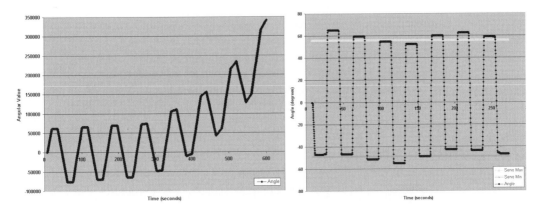

Figure 2.11: Measured gyro in motion (integrated), raw and corrected

Figure 2.11 [Smith 2002], [Stamatiou 2002] shows the integrated sensor signal for a gyro that is continuously moved between two orientations with the help of a servo. As can be seen in Figure 2.11, left, the angle value remains within the correct bounds for a few iterations, and then rapidly drifts outside the range, making the sensor signal useless. The error is due to both sensor drift (see Figure 2.10) and iteration error. The following sensor data processing techniques have been applied:

1. Noise reduction by removal of outlier data values
2. Noise reduction by applying the moving-average method
3. Application of scaling factors to increment/decrement absolute angles
4. Re-calibration of gyroscope rest-average via sampling
5. Re-calibration of minimal and maximal rest-bound via sampling

Two sets of bounds are used for the determination and re-calibration of the gyroscope rest characteristics. The sensor drift has now been eliminated (upper curve in Figure 2.10). The integrated output value for the tilt angle (Figure 2.11, right) shows the corrected noise-free signal. The measured angular value now stays within the correct bounds and is very close to the true angle.

2.8.3 Inclinometer

Inclinometers measure the absolute orientation angle within a specified range, depending on the sensor model. The sensor output is also model-dependent, with either analog signal output or PWM being available. Therefore, interfacing to an embedded system is identical to accelerometers (see Section 2.8.1).

Since inclinometers measure the absolute orientation angle about an axis and not the derivative, they seem to be much better suited for orientation measurement than a gyroscope. However, our measurements with the Seika inclinometer showed that they suffer a time lag when measuring and also are prone to oscillation when subjected to positional noise, for example as caused by servo jitter.

Especially in systems that require immediate response, for example balancing robots in Chapter 9, gyroscopes have an advantage over inclinometers. With the components tested, the ideal solution was a combination of inclinometer and gyroscope.

2.9 Digital Camera

Digital cameras are the most complex sensors used in robotics. They have not been used in embedded systems until recently, because of the processor speed and memory capacity required. The central idea behind the EyeBot development in 1995 was to create a small, compact embedded vision system, and it became the first of its kind. Today, PDAs and electronic toys with cameras are commonplace, and digital cameras with on-board image processing are available on the consumer market.

For mobile robot applications, we are interested in a high frame rate, because our robot is moving and we want updated sensor data as fast as possible. Since there is always a trade-off between high frame rate and high resolution, we are not so much concerned with camera resolution. For most applications for small mobile robots, a resolution of 60×80 pixels is sufficient. Even from such a small resolution we can detect, for example, colored objects or obstacles in the way of a robot (see 60×80 sample images from robot soccer in Figure 2.12). At this resolution, frame rates (reading only) of up to 30 fps (frames per second) are achievable on an EyeBot controller. The frame rate will drop, however, depending on the image processing algorithms applied.

The image resolution must be high enough to detect a desired object from a specified distance. When the object in the distance is reduced to a mere few pixels, then this is not sufficient for a detection algorithm. Many higher-level image processing routines are non-linear in time requirements, but even simple linear filters, for example Sobel edge detectors, have to loop through all pixels, which takes some time [Bräunl 2001]. At 60×80 pixels with 3 bytes of color per pixel this amounts to 14,400 bytes.

Figure 2.12: Sample images with 60×80 resolution

Unfortunately for embedded vision applications, newer camera chips have much higher resolution, for example QVGA (quarter VGA) up to 1,024×1,024, while low-resolution sensor chips are no longer produced. This means that much more image data is being sent, usually at higher transfer rates. This requires additional, faster hardware components for our embedded vision system just to keep up with the camera transfer rate. The achievable frame rate will drop to a few frames per second with no other benefits, since we would not have the memory space to store these high-resolution images, let alone the processor speed to apply typical image processing algorithms to them. Figure 2.13 shows the EyeCam camera module that is used with the EyeBot embedded controller. EyeCam C2 has in addition to the digital output port also an analog grayscale video output port, which can be used for fast camera lens focusing or for analog video recording, for example for demonstration purposes.

Digital + analog camera output

In the following, we will discuss camera hardware interfaces and system software. Image processing routines for user applications are presented in Chapter 17.

2.9.1 Camera Sensor Hardware

In recent years we have experienced a shift in camera sensor technology. The previously dominant CCD (charge coupled device) sensor chips are now being overtaken by the cheaper to produce CMOS (complementary metal oxide semiconductor) sensor chips. The brightness sensitivity range for CMOS sensors is typically larger than that of CCD sensors by several orders of magnitude. For interfacing to an embedded system, however, this does not make a difference. Most sensors provide several different interfacing protocols that can be selected via software. On the one hand, this allows a more versatile hardware design, but on the other hand sensors become as complex as another microcontroller system and therefore software design becomes quite involved.

Typical hardware interfaces for camera sensors are 16bit parallel, 8bit parallel, 4bit parallel, or serial. In addition, a number of control signals have to be provided from the controller. Only a few sensors buffer the image data and allow arbitrarily slow reading from the controller via *handshaking*. This is an

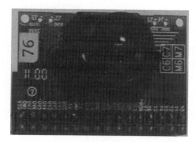

Figure 2.13: EyeCam camera module

ideal solution for slower controllers. However, the standard camera chip provides its own clock signal and sends the full image data as a *stream* with some frame-start signal. This means the controller CPU has to be fast enough to keep up with the data stream.

The parameters that can be set in software vary between sensor chips. Most common are the setting of frame rate, image start in (x,y), image size in (x,y), brightness, contrast, color intensity, and auto-brightness.

The simplest camera interface to a CPU is shown in Figure 2.14. The camera clock is linked to a CPU interrupt, while the parallel camera data output is connected directly to the data bus. Every single image byte from the camera will cause an interrupt at the CPU, which will then enable the camera output and read one image data byte from the data bus.

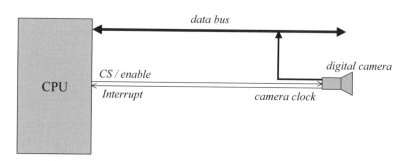

Figure 2.14: Camera interface

Every interrupt creates considerable overhead, since system registers have to be saved and restored on the stack. Starting and returning from an interrupt takes about 10 times the execution time of a normal command, depending on the microcontroller used. Therefore, creating one interrupt per image byte is not the best possible solution. It would be better to buffer a number of bytes and then use an interrupt much less frequently to do a bulk data transfer of image data. Figure 2.15 shows this approach using a FIFO buffer for intermediate storing of image data. The advantage of a FIFO buffer is that it supports unsynchronized read and write in parallel. So while the camera is writing data to the FIFO buffer, the CPU can read data out, with the remaining buffer contents staying undisturbed. The camera output is linked to the FIFO input, with the camera's pixel clock triggering the FIFO write line. From the CPU side, the FIFO data output is connected to the system's data bus, with the chip select triggering the FIFO read line. The FIFO provides three additional status lines:

- Empty flag
- Full flag
- Half full flag

These digital outputs of the FIFO can be used to control the bulk reading of data from the FIFO. Since there is a continuous data stream going into the FIFO, the most important of these lines in our application is the *half full flag*, which we connected to a CPU interrupt line. Whenever the FIFO is half full, we initiate a bulk read operation of 50% of the FIFO's contents. Assuming the CPU responds quickly enough, the *full flag* should never be activated, since this would indicate an imminent loss of image data.

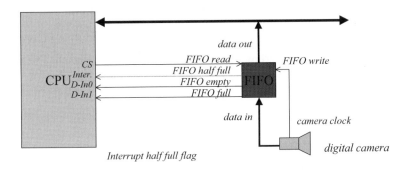

Figure 2.15: Camera interface with FIFO buffer

2.9.2 Camera Sensor Data

We have to distinguish between grayscale and color cameras, although, as we will see, there is only a minor difference between the two. The simplest available sensor chips provide a grayscale image of 120 lines by 160 columns with 1 byte per pixel (for example VLSI Vision VV5301 in grayscale or VV6301 in color). A value of zero represents a black pixel, a value of 255 is a white pixel, everything in between is a shade of gray. Figure 2.16 illustrates such an image. The camera transmits the image data in row-major order, usually after a certain frame-start sequence.

Creating a color camera sensor chip from a grayscale camera sensor chip is very simple. All it needs is a layer of paint over the pixel mask. The standard Bayer pattern technique for pixels arranged in a grid is the *Bayer pattern* (Figure 2.17). Pixels in odd rows (1, 3, 5, etc.) are colored alternately in green and red, while pixels in even rows (2, 4, 6, etc.) are colored alternately in blue and green.

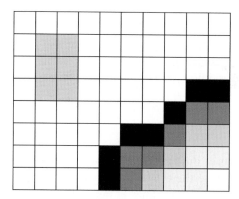

Figure 2.16: Grayscale image

With this colored filter over the pixel array, each pixel only records the intensity of a certain color component. For example, a pixel with a red filter will only record the red intensity at its position. At first glance, this requires 4 bytes per color pixel: green and red from one line, and blue and green (again) from the line below. This would result effectively in a 60×80 color image with an additional, redundant green byte per pixel.

However, there is one thing that is easily overlooked. The four components red, $green_1$, blue, and $green_2$ are not sampled at the same position. For example, the blue sensor pixel is below and to the right of the red pixel. So by treating the four components as one pixel, we have already applied some sort of filtering and lost information.

Bayer Pattern
green, red, green, red, ...
blue, green, blue, green, ...

Figure 2.17: Color image

Demosaicing   A technique called "demosaicing" can be used to restore the image in full 120×160 resolution and in full color. This technique basically recalculates the three color component values (R, G, B) *for each pixel* position, for example by averaging the four closest component neighbors of the same color. Figure 2.18 shows the three times four pixels used for demosaicing the red, green, and blue components of the pixel at position [3,2] (assuming the image starts in the top left corner with [0,0]).

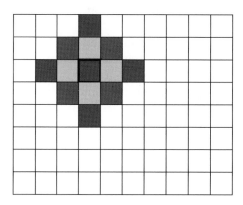

Figure 2.18: Demosaic of single pixel position

Averaging, however, is only the simplest method of image value restoration and does not produce the best results. A number of articles have researched better algorithms for demosaicing [Kimmel 1999], [Muresan, Parks 2002].

2.9.3 Camera Driver

There are three commonly used capture modes available for receiving data from a digital camera:

- **Read mode:** The application requests a frame from the driver and blocks CPU execution. The driver waits for the next complete frame from the camera and captures it. Once a frame has been completely read in, the data is passed to the application and the application continues. In this mode, the driver will first have to wait for the new frame to start. This means that the application will be blocked for up to two frames, one to find the start of a new frame and one to read the current frame.

- **Continuous capture mode:** In this mode, the driver continuously captures a frame from the camera and stores it in one of two buffers. A pointer to the last buffer read in is passed to the application when the application requests a frame.

- **Synchronous continuous capture mode:** In this mode, the driver is working in the background. It receives every frame from the camera and stores it in a buffer. When a frame has been completely read in, a trap signal/software interrupt is sent to the application. The application's signal handler then processes the data. The processing time of the interrupt handler is limited by the acquisition time for one camera image.

Most of these modes may be extended through the use of additional buffers. For example, in the synchronous capture mode, a driver may fill more than a single buffer. Most high-end capture programs running on workstations use the synchronous capture mode when recording video. This of course makes sense, since for recording video, all frames (or as many frames as possible) lead to the best result.

The question is which of these capture modes is best suited for mobile robotics applications on slower microprocessors. There is a significant overhead for the M68332 when reading in a frame from the camera via the parallel port. The camera reads in every byte via the parallel port. Given the low resolution color camera sensor chip VLSI Vision VV6301, 54% of the CPU usage is used to read in a frame, most of which will not actually be used in the application.

Another problem is that the shown image is already outdated (one frame old), which can affect the results. For example, when panning the camera quickly, it may be required to insert delays in the code to wait for the capture driver to catch up to the robot motion.

Therefore, the "read" interface is considered the most suitable one for mobile robotics applications. It provides the least amount of overhead at the cost of a small delay in the processing. This delay can often be eliminated by requesting a frame just before the motion command ends.

2.9.4 Camera RoBIOS Interface

All interaction between camera and CPU occurs in the background through external interrupts from the sensor or via periodic timer interrupts. This makes the camera user interface very simple. The routines listed in Program 2.1 all apply to a number of different cameras and different interfaces (i.e. with or without hardware buffers), for which drivers have been written for the EyeBot.

Program 2.1: Camera interface routines

```
typedef BYTE image    [imagerows][imagecolumns];
typedef BYTE colimage[imagerows][imagecolumns][3];

int CAMInit    (int mode);
int CAMRelease (void);

int CAMGetFrame    (image *buf);
int CAMGetColFrame (colimage *buf, int convert);

int CAMGetFrameMono  (BYTE *buf);
int CAMGetFrameRGB   (BYTE *buf);
int CAMGetFrameBayer (BYTE *buf);

int CAMSet (int  para1, int  para2, int  para3);
int CAMGet (int *para1, int *para2, int *para3);
int CAMMode (int mode);
```

The only mode supported for current EyeCam camera models is NORMAL, while older QuickCam cameras also support zoom modes. CAMInit returns the code number of the camera found or an error code if not successful (see Appendix B.5.4).

The standard image size for grayscale and color images is 62 rows by 82 columns. For grayscale, each pixel uses 1 byte, with values from 0 (black) over 128 (medium-gray) to 255 (white). For color, each pixel comprises 3 bytes in the order red, green, blue. For example, medium green is represented by (0, 128, 0), fully red is (255, 0, 0), bright yellow is (200, 200, 0), black is (0, 0, 0), white is (255, 255, 255).

The standard camera read functions return images of size 62×82 (including a 1-pixel-wide white border) for all camera models, irrespective of their internal resolution:

- `CAMGetFrame`   (read one grayscale image)
- `CAMGetColFrame`   (read one color image)

This originated from the original camera sensor chips (QuickCam and EyeCam C1) supplying 60×80 pixels. A single-pixel-wide border around the image had been added to simplify coding of image operators without having to check image boundaries.

Function `CAMGetColFrame` has a second parameter that allows immediate conversion into a grayscale image in-place. The following call allows grayscale image processing using a color camera:

```
image buffer;
CAMGetColFrame((colimage*)&buffer, 1);
```

Newer cameras like EyeCam C2, however, have up to full VGA resolution. In order to be able to use the full image resolution, three additional camera interface functions have been added for reading images at the camera sensor's resolution (i.e. returning different image sizes for different camera models, see Appendix B.5.4). The functions are:

- `CAMGetFrameMono`   (read one grayscale image)
- `CAMGetFrameColor`   (read one color image in RGB 3byte format)
- `CAMGetFrameBayer`   (read one color image in Bayer 4byte format)

Since the data volume is considerably larger for these functions, they may require considerably more transmission time than the `CAMGetFrame`/`CAMGetColFrame` functions.

Different camera models support different parameter settings and return different camera control values. For this reason, the semantics of the camera routines `CAMSet` and `CAMGet` is not unique among different cameras. For the camera model EyeCam C2, only the first parameter of `CAMSet` is used, allowing the specification of the camera speed (see Appendix B.5.4):

```
FPS60, FPS30, FPS15, FPS7_5, FPS3_75, FPS1_875
```

For cameras EyeCam C2, routine `CAMGet` returns the current frame rate in frames per second (fps), the full supported image width, and image height (see Appendix B.5.4 for details). Function `CAMMode` can be used for switching the camera's auto-brightness mode on or off, if supported by the camera model used (see Appendix B.5.4).

There are a number of shortcomings in this procedural camera interface, especially when dealing with different camera models with different resolutions and different parameters, which can be addressed by an object-oriented approach.

Example camera use   Program 2.2 shows a simple program that continuously reads an image and displays it on the controller's LCD until the rightmost button is pressed (KEY4 being associated with the menu text "End"). The function CAMInit returns the version number of the camera or an error value. This enables the application programmer to distinguish between different camera models in the code by testing this value. In particular, it is possible to distinguish between color and grayscale camera models by comparing with the system constant COLCAM, for example:

```
if (camera<COLCAM) /* then grayscale camera ... */
```

Alternative routines for color image reading and displaying are CAMGet-ColFrame and LCDPutColorGraphic, which could be used in Program 2.2 instead of the grayscale routines.

Program 2.2: Camera application program

```
1   #include "eyebot.h"
2   image    grayimg; /* picture for LCD-output */
3   int      camera;  /* camera version */
4
5   int main()
6   { camera=CAMInit(NORMAL);
7     LCDMenu("","","","End");
8     while (KEYRead()!=KEY4)
9     { CAMGetFrame  (&grayimg);
10      LCDPutGraphic(&grayimg);
11    }
12    return 0;
13  }
```

2.10 References

BARSHAN, B., AYRULU, B., UTETE, S. *Neural network-based target differentiation using sonar for robotics applications*, IEEE Transactions on Robotics and Automation, vol. 16, no. 4, August 2000, pp. 435-442 (8)

BRÄUNL, T. *Parallel Image Processing*, Springer-Verlag, Berlin Heidelberg, 2001

DINSMORE, *Data Sheet Dinsmore Analog Sensor No. 1525*, Dinsmore Instrument Co., http://dinsmoregroup.com/dico, 1999

EVERETT, H.R. *Sensors for Mobile Robots*, AK Peters, Wellesley MA, 1995

JÖRG, K., BERG, M. *Mobile Robot Sonar Sensing with Pseudo-Random Codes*, IEEE International Conference on Robotics and Automation 1998 (ICRA '98), Leuven Belgium, 16-20 May 1998, pp. 2807-2812 (6)

KIMMEL, R. *Demosaicing: Image Reconstruction from Color CCD Samples*, IEEE Transactions on Image Processing, vol. 8, no. 9, Sept. 1999, pp. 1221-1228 (8)

KUC, R. *Pseudoamplitude scan sonar maps*, IEEE Transactions on Robotics and Automation, vol. 17, no. 5, 2001, pp. 767-770

MURESAN, D., PARKS, T. *Optimal Recovery Demosaicing*, IASTED International Conference on Signal and Image Processing, SIP 2002, Kauai Hawaii, http://dsplab.ece.cornell.edu/papers/conference/sip_02_6.pdf, 2002, pp. (6)

PRECISION NAVIGATION, *Vector Electronic Modules*, Application Notes, Precision Navigation Inc., http://www.precisionnav.com, July 1998

SHARP, *Data Sheet GP2D02 - Compact, High Sensitive Distance Measuring Sensor*, Sharp Co., data sheet, http://www.sharp.co.jp/ecg/, 2006

SICK, *Auto Ident Laser-supported sensor systems*, Sick AG, http://www.sick.de/de/products/categories/auto/en.html, 2006

SMITH, C. *Vision Support System for Tracked Vehicle,* B.E. Honours Thesis, The Univ. of Western Australia, Electrical and Computer Eng., supervised by T. Bräunl, 2002

STAMATIOU, N. *Sensor processing for a tracked vehicle*, B.E. Honours Thesis, The Univ. of Western Australia, Electrical and Computer Eng., supervised by T. Bräunl, 2002

ACTUATORS

· ·

T here are many different ways that robotic actuators can be built. Most prominently these are electrical motors or pneumatic actuators with valves. In this chapter we will deal with electrical actuators using direct current (DC) power. These are standard DC motors, stepper motors, and servos, which are DC motors with encapsulated positioning hardware and are not to be confused with servo motors.

3.1 DC Motors

Electrical motors can be:
AC motors
DC motors
Stepper motors
Servos

DC electric motors are arguably the most commonly used method for locomotion in mobile robots. DC motors are clean, quiet, and can produce sufficient power for a variety of tasks. They are much easier to control than pneumatic actuators, which are mainly used if very high torques are required and umbilical cords for external pressure pumps are available – so usually not an option for mobile robots.

Standard DC motors revolve freely, unlike for example stepper motors (see Section 3.4). Motor control therefore requires a feedback mechanism using shaft encoders (see Figure 3.1 and Section 2.4).

Figure 3.1: Motor–encoder combination

The first step when building robot hardware is to select the appropriate motor system. The best choice is an encapsulated motor combination comprising a:

- DC motor

- Gearbox

- Optical or magnetic encoder
  (dual phase-shifted encoders for detection of speed and direction)

Using encapsulated motor systems has the advantage that the solution is much smaller than that using separate modules, plus the system is dust-proof and shielded against stray light (required for optical encoders). The disadvantage of using a fixed assembly like this is that the gear ratio may only be changed with difficulty, or not at all. In the worst case, a new motor/gearbox/encoder combination has to be used.

A magnetic encoder comprises a disk equipped with a number of magnets and one or two Hall-effect sensors. An optical encoder has a disk with black and white sectors, an LED, and a reflective or transmissive light sensor. If two sensors are positioned with a phase shift, it is possible to detect which one is triggered first (using a magnet for magnetic encoders or a bright sector for optical encoders). This information can be used to determine whether the motor shaft is being turned clockwise or counterclockwise.

A number of companies offer small, powerful precision motors with encapsulated gearboxes and encoders:

- Faulhaber   `http://www.faulhaber.de`

- Minimotor   `http://www.minimotor.ch`

- MicroMotor   `http://www.micromo.com`

They all have a variety of motor and gearbox combinations available, so it is important to do some power-requirement calculations first, in order to select the right motor and gearbox for a new robotics project. For example, there is a Faulhaber motor series with a power range from 2W to 4W, with gear ratios available from approximately 3:1 to 1,000,000:1.

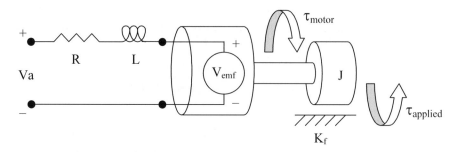

Figure 3.2: Motor model

θ	Angular position of shaft, rad	R	Nominal terminal resistance, Ω
ω	Angular shaft velocity, rad/s	L	Rotor inductance, H
α	Angular shaft accel., rad/s^2	J	Rotor inertia, kg·m^2
i	Current through armature, A	K_f	Frictional const., N·m·s / rad
V_a	Applied terminal voltage, V	K_m	Torque constant, N·m / A
V_e	Back *emf* voltage, V	K_e	Back *emf* constant, V·s / rad
τ_m	Motor torque, N·m	K_s	Speed constant, rad / (V·s)
τ_a	Applied torque (load), N·m	K_r	Regulation constant, (V·s) / rad

Table 3.1: DC motor variables and constant values

Figure 3.2 illustrates an effective linear model for the DC motor, and Table 3.1 contains a list of all relevant variables and constant values. A voltage V_a is applied to the terminals of the motor, which generates a current i in the motor armature. The torque τ_m produced by the motor is proportional to the current, and K_m is the motor's *torque constant*:

$$\tau_m = K_m i$$

It is important to select a motor with the right output power for a desired task. The output power P_o is defined as the rate of work, which for a rotational DC motor equates to the angular velocity of the shaft ω multiplied by the applied torque τ_a (i.e., the torque of the load):

$$P_o = \tau_a \omega$$

The input power P_i, supplied to the motor, is equal to the applied voltage multiplied by the current through the motor:

$$P_i = V_a i$$

The motor also generates heat as an effect of the current flowing through the armature. The power lost to thermal effects P_t is equivalent to:

$$P_t = R i^2$$

The efficiency η of the motor is a measure of how well electrical energy is converted to mechanical energy. This can be defined as the output power produced by the motor divided by the input power required by the motor:

$$\eta = \frac{P_o}{P_i} = \frac{\tau_a \omega}{V_a i}$$

The efficiency is not constant for all speeds, which needs to be kept in mind if the application requires operation at different speed ranges. The electrical system of the motor can be modelled by a resistor-inductor pair in series with a voltage V_{emf}, which corresponds to the *back electromotive force* (see Figure 3.2). This voltage is produced because the coils of the motor are moving through a magnetic field, which is the same principle that allows an electric generator to function. The voltage produced can be approximated as a linear function of the shaft velocity; K_e is referred to as the *back-emf constant*:

$$V_e = K_e\omega$$

Simple motor model
In the simplified DC motor model, motor inductance and motor friction are negligible and set to zero, and the rotor inertia is denoted by J. The formulas for current and angular acceleration can therefore be approximated by:

$$i = \frac{-K_e}{R}\omega + \frac{1}{R}V_a$$

$$\frac{\delta\omega}{\delta t} = \frac{K_m}{J} \cdot i - \frac{\tau_a}{J}$$

Figure 3.3 shows the ideal DC motor performance curves. With increasing torque, the motor velocity is reduced linearly, while the current increases linearly. Maximum output power is achieved at a medium torque level, while the highest efficiency is reached for relatively low torque values. For further reading see [Bolton 1995] and [El-Sharkawi 2000].

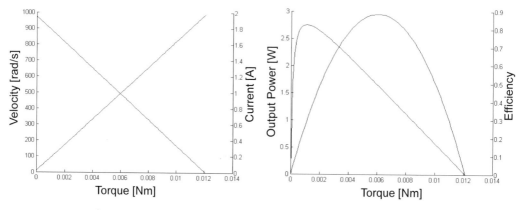

Figure 3.3: Ideal DC motor performance curve

3.2 H-Bridge

H-bridge is needed to run a motor forward and backward
For most applications we want to be able to do two things with a motor:

1. Run it in forward and backward directions.

2. Modify its speed.

An H-bridge is what is needed to enable a motor to run forward/backward. In the next section we will discuss a method called "pulse width modulation" to change the motor speed. Figure 3.4 demonstrates the H-bridge setup, which received its name from its resemblance to the letter "H". We have a motor with two terminals a and b and the power supply with "+" and "–". Closing switches 1 and 2 will connect a with "+" and b with "–": the motor runs forward. In the same way, closing 3 and 4 instead will connect a with "–" and b with "+": the motor runs backward.

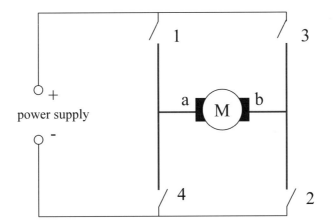

Drive forward: Drive backward:

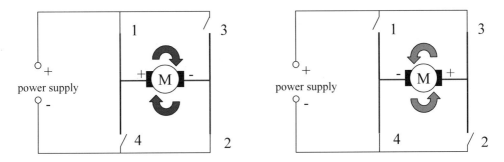

Figure 3.4: H-bridge and operation

The way to implement an H-bridge when using a microcontroller is to use a power amplifier chip in combination with the digital output pins of the controller or an additional latch. This is required because the digital outputs of a microcontroller have very severe output power restrictions. They can only be used to drive other logic chips, but *never* a motor directly. Since a motor can draw a lot of power (for example 1A or more), connecting digital outputs directly to a motor can destroy the microcontroller.

A typical power amplifier chip containing two separate amplifiers is L293D from ST SGS-Thomson. Figure 3.5 demonstrates the schematics. The two inputs x and y are needed to switch the input voltage, so one of them has to be "+", the other has to be "–". Since they are electrically decoupled from the motor, x and y can be directly linked to digital outputs of the microcontroller. So the direction of the motor can then be specified by software, for example setting output x to logic 1 and output y to logic 0. Since x and y are always the opposite of each other, they can also be substituted by a single output port and a negator. The rotation speed can be specified by the "speed" input (see the next section on pulse width modulation).

There are two principal ways of stopping the motor:

- set both x and y to logic 0 (or both to logic 1) or
- set speed to 0

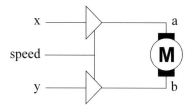

Figure 3.5: Power amplifier

3.3 Pulse Width Modulation

PWM is digital control Pulse width modulation or PWM for short is a smart method for avoiding analog power circuitry by utilizing the fact that mechanical systems have a certain latency. Instead of generating an analog output signal with a voltage proportional to the desired motor speed, it is sufficient to generate digital pulses at the full system voltage level (for example 5V). These pulses are generated at a fixed frequency, for example 20 kHz, so they are beyond the human hearing range.

By varying the pulse width in software (see Figure 3.6, top versus bottom), we also change the equivalent or *effective* analog motor signal and therefore control the motor speed. One could say that the motor system behaves like an integrator of the digital input impulses over a certain time span. The quotient Duty cycle t_{on}/t_{period} is called the "pulse–width ratio" or "duty cycle".

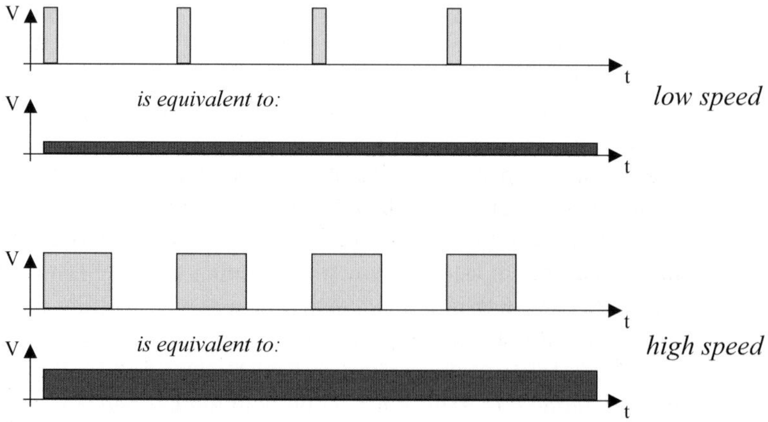

is equivalent to: low speed

is equivalent to: high speed

Figure 3.6: PWM

The PWM can be generated by software. Many microcontrollers like the M68332 have special modes and output ports to support this operation. The digital output port with the PWM signal is then connected to the *speed* pin of the power amplifier in Figure 3.5.

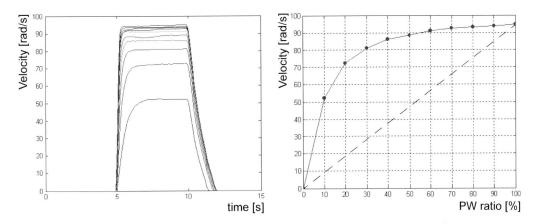

Figure 3.7: Measured motor step response and speed versus PW ratio

Figure 3.7, left, shows the motor speed over time for PWM settings of 10, 20, .., 100. In each case, the velocity builds up at time 5s with some delay, then stays constant, and will slow down with a certain inertia at time 10s. These measurements are called "step response", since the motor input signal jumps in a step function from zero to the desired PWM value.

Unfortunately, the generated motor speed is normally *not* a linear function of the PWM signal ratio, as can be seen when comparing the measurement in Figure 3.7, right, to the dashed line. This shows a typical measurement using a Faulhaber 2230 motor. In order to re-establish an approximately linear speed curve when using the MOTORDrive function (for example MOTORDrive(m1,50) should result in half the speed of MOTORDrive(m1,100)), each motor has to be calibrated.

Motor calibration   Motor calibration is done by measuring the motor speed at various settings between 0 and 100, and then entering the PW ratio required to achieve the desired actual speed in a motor calibration table of the HDT. The motor's maximum speed is about 1,300 rad/s at a PW ratio of 100. It reaches 75% of its maximum speed (975 rad/s) at a PW ratio of 20, so the entry for value 75 in the motor calibration HDT should be 20. Values between the 10 measured points can be interpolated (see Section B.3).

Motor calibration is especially important for robots with differential drive (see Section 4.4 and Section 7.2), because in these configurations normally one motor runs forward and one backward, in order to drive the robot. Many DC motors exhibit some differences in speed versus PW ratio between forward and backward direction. This can be eliminated by using motor calibration.

Open loop control
We are now able to achieve the two goals we set earlier: we can drive a motor forward or backward and we can change its speed. However, we have no way of telling at what speed the motor is actually running. Note that the actual motor speed does depend not only on the PWM signal supplied, but also on external factors such as the load applied (for example the weight of a vehicle or the steepness of its driving area). What we have achieved so far is called open loop control. With the help of feedback sensors, we will achieve closed loop control (often simply called "control"), which is essential to run a motor at a desired speed under varying load (see Chapter 4).

3.4 Stepper Motors

There are two motor designs which are significantly different from standard DC motors. These are stepper motors discussed in this section and servos, introduced in the following section.

Stepper motors differ from standard DC motors in such a way that they have two independent coils which can be independently controlled. As a result, stepper motors can be moved by impulses to proceed exactly a single step forward or backward, instead of a smooth continuous motion in a standard DC motor. A typical number of steps per revolution is 200, resulting in a step size of 1.8°. Some stepper motors allow half steps, resulting in an even finer step size. There is also a maximum number of steps per second, depending on load, which limits a stepper motor's speed.

Figure 3.8 demonstrates the stepper motor schematics. Two coils are independently controlled by two H-bridges (here marked A, $\overline{A}$ and B, $\overline{B}$). Each four-step cycle advances the motor's rotor by a single step if executed in order 1..4. Executing the sequence in reverse order will move the rotor one step back. Note that the switching sequence pattern resembles a gray code. For details on stepper motors and interfacing see [Harman, 1991].

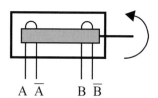

Switching Sequence:

Step	A	B
1	1	1
2	1	0
3	0	0
4	0	1

Figure 3.8: Stepper motor schematics

Stepper motors seem to be a simple choice for building mobile robots, considering the effort required for velocity control and position control of standard DC motors. However, stepper motors are very rarely used for driving mobile robots, since they lack any feedback on load and actual speed (for example a missed step execution). In addition to requiring double the power electronics, stepper motors also have a worse weight/performance ratio than DC motors.

3.5 Servos

Servos are not servo motors! DC motors are sometimes also referred to as "servo motors". This is not what we mean by the term "servo". A servo motor is a high-quality DC motor that qualifies to be used in a "servoing application", i.e. in a closed control loop. Such a motor must be able to handle fast changes in position, speed, and acceleration, and must be rated for high intermittent torque.

Figure 3.9: Servo

A *servo*, on the contrary, is a DC motor with encapsulated electronics for PW control and is mainly used for hobbyist purposes, as in model airplanes, cars, or ships (see Figure 3.9).

A servo has three wires: V_{CC}, ground, and the PW input control signal. Unlike PWM for DC motors, the input pulse signal for servos is not transformed into a velocity. Instead, it is an analog control input to specify the desired position of the servo's rotating disk head. A servo's disk cannot perform a continuous rotation like a DC motor. It only has a range of about ±120° from its middle position. Internally, a servo combines a DC motor with a simple feedback circuit, often using a potentiometer sensing the servo head's current position.

The PW signal used for servos always has a frequency of 50Hz, so pulses are generated every 20ms. The width of each pulse now specifies the desired position of the servo's disk (Figure 3.10). For example, a width of 0.7ms will rotate the disk to the leftmost position (−120°), and a width of 1.7ms will rotate the disk to the rightmost position (+120°). Exact values of pulse duration and angle depend on the servo brand and model.

Like stepper motors, servos seem to be a good and simple solution for robotics tasks. However, servos have the same drawback as stepper motors: they do not provide any feedback to the outside. When applying a certain PW signal to a servo, we do not know when the servo will reach the desired position or whether it will reach it at all, for example because of too high a load or because of an obstruction.

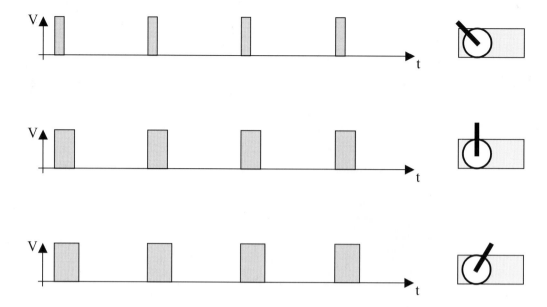

Figure 3.10: Servo control

3.6 References

BOLTON, W. *Mechatronics – Electronic Control Systems in Mechanical Engineering*, Addison Wesley Longman, Harlow UK, 1995

EL-SHARKAWI, M. *Fundamentals of Electric Drives*, Brooks/Cole Thomson Learning, Pacific Grove CA, 2000

HARMAN, T. *The Motorola MC68332 Microcontroller - Product Design, Assembly Language Programming, and Interfacing*, Prentice Hall, Englewood Cliffs NJ, 1991

CONTROL

4

C losed loop control is an essential topic for embedded systems, bringing together actuators and sensors with the control algorithm in software. The central point of this chapter is to use motor feedback via encoders for velocity control and position control of motors. We will exemplify this by a stepwise introduction of PID (Proportional, Integral, Derivative) control.

In Chapter 3, we showed how to drive a motor forward or backward and how to change its speed. However, because of the lack of feedback, the actual motor speed could not be verified. This is important, because supplying the same analog voltage (or equivalent: the same PWM signal) to a motor does not guarantee that the motor will run at the same speed under all circumstances. For example, a motor will run faster when free spinning than under load (for example driving a vehicle) with the same PWM signal. In order to control the motor speed we do need feedback from the motor shaft encoders. Feedback control is called "closed loop control" (simply called "control" in the following), as opposed to "open loop control", which was discussed in Chapter 3.

4.1 On-Off Control

Feedback is everything As we established before, we require feedback on a motor's current speed in order to control it. Setting a certain PWM level alone will not help, since the motor's speed also depends on its load.

The idea behind feedback control is very simple. We have a desired speed, specified by the user or the application program, and we have the current actual speed, measured by the shaft encoders. Measurements and actions according to the measurements can be taken very frequently, for example 100 times per second (EyeBot) or up to 20,000 times per second. The action taken depends on the controller model, several of which are introduced in the following sections. However, in principle the action always looks similar like this:

- In case desired speed is higher than actual speed:
  Increase motor power by a certain degree.
- In case desired speed is lower than actual speed:
  Decrease motor power by a certain degree.

In the simplest case, power to the motor is either switched on (when the speed is too low) or switched off (when the speed is too high). This control law is represented by the formula below, with:

$R(t)$ motor output function over time t

$v_{act}(t)$ actual measured motor speed at time t

$v_{des}(t)$ desired motor speed at time t

K_C constant control value

$$R(t) = \begin{cases} K_C & \text{if } v_{act}(t) < v_{des}(t) \\ 0 & \text{otherwise} \end{cases}$$

Bang-bang controller   What has been defined here, is the concept of an on-off controller, also known as "piecewise constant controller" or "bang-bang controller". The motor input is set to constant value K_C if the measured velocity is too low, otherwise it is set to zero. Note that this controller only works for a positive value of v_{des}. The schematics diagram is shown in Figure 4.1.

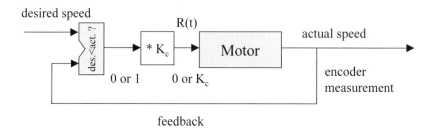

Figure 4.1: On-off controller

The behavior over time of an on-off controller is shown in Figure 4.2. Assuming the motor is at rest in the beginning, the actual speed is less than the desired speed, so the control signal for the motor is a constant voltage. This is kept until at some stage the actual speed becomes larger than the desired speed. Now, the control signal is changed to zero. Over some time, the actual speed will come down again and once it falls below the desired speed, the control signal will again be set to the same constant voltage. This algorithm continues indefinitely and can also accommodate changes in the desired speed. Note that

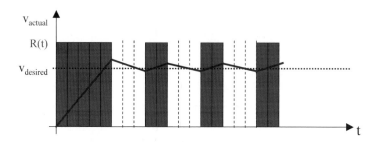

Figure 4.2: On-off control signal

the motor control signal is not continuously updated, but only at fixed time intervals (e.g. every 10ms in Figure 4.2). This delay creates an overshooting or undershooting of the actual motor speed and thereby introduces *hysteresis*.

Hysteresis   The on-off controller is the simplest possible method of control. Many technical systems use it, not limited to controlling a motor. Examples are a refrigerator, heater, thermostat, etc. Most of these technical systems use a hysteresis band, which consists of two *desired* values, one for switching on and one for switching off. This prevents a too high switching frequency near the desired value, in order to avoid excessive wear. The formula for an on-off controller with hysteresis is:

$$R(t + \Delta t) = \begin{cases} K_C & \text{if} & v_{act}(t) < v_{on}(t) \\ 0 & \text{if} & v_{act}(t) > v_{off}(t) \\ R(t) & \text{otherwise} \end{cases}$$

Note that this definition is not a function in the mathematical sense, because the new motor output for an actual speed between the two band limit values is equal to the previous motor value. This means it can be equal to K_C or zero in this case, depending on its history. Figure 4.3 shows the hysteresis curve and the corresponding control signal.

All technical systems have some delay and therefore exhibit some inherent hysteresis, even if it is not explicitly built-in.

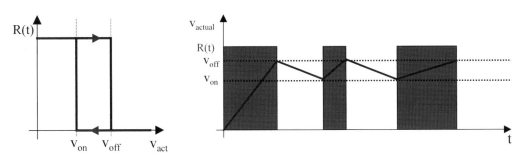

Figure 4.3: On-off control signal with hysteresis band

From theory   Once we understand the theory, we would like to put this knowledge into
to practice practice and implement an on-off controller in software. We will proceed step by step:

1. We need a subroutine for calculating the motor signal as defined in the formula in Figure 4.1. This subroutine has to:
    a. Read encoder data (*input*)
    b. Compute new output value R(t)
    c. Set motor speed (*output*)

2. This subroutine has to be called periodically (for example every 1/100s). Since motor control is a "low-level" task we would like this to run in the background and not interfere with any user program.

Let us take a look at how to implement step 1. In the RoBIOS operating system there are functions available for reading encoder input and setting motor output (see Appendix B.5 and Figure 4.4).

---

1. Write a control subroutine
    a.   Read encoder data (INPUT)
    b.   Compute new output value R(t)
    c.   Set motor speed (OUTPUT)

See library.html:

```
int QUADRead(QuadHandle handle);
      Input:          (handle) ONE decoder-handle
      Output:         32bit counter-value (-2^31 .. 2^31-1)
      Semantics:      Read actual Quadrature-Decoder counter, initially zero.
                      Note: A wrong handle will ALSO result in a 0 counter value!!

int MOTORDrive (MotorHandle handle,int speed);
      Input:              (handle) logical-or of all MotorHandles which should be driven
                          (speed) motor speed in percent
                          Valid values: -100 - 100 (full backward to full forward)
                                        0 for full stop
      Output:         (return code)  0 = ok
                                    -1 = error wrong handle
      Semantics:      Set the given motors to the same given speed
```

Figure 4.4: RoBIOS motor functions

The program code for this subroutine would then look like Program 4.1. Variable r_mot denotes the control parameter R(t). The dotted line has to be replaced by the control function "K_C if $v_{act} < v_{des}$" from Figure 4.1.

Program 4.1: Control subroutine framework

```
1   void controller()
2   { int enc_new, r_mot, err;
3     enc_new = QUADRead(enc1);
4     ...
5     err = MOTORDrive(mot1, r_mot);
6     if (err) printf("error: motor");
7   }
```

Program 4.2 shows the completed control program, assuming this routine is called every 1/100 of a second.

So far we have not considered any potential problems with counter overflow or underflow. However, as the following examples will show, it turns out that overflow/underflow does still result in the correct difference values when using standard signed integers.

Overflow example from positive to negative values:

```
7F FF FF FC        = +2147483644_Dec
80 00 00 06        = -6_Dec
```

Program 4.2: On-off controller

```
 1   int v_des;        /* user input in ticks/s */
 2   #define Kc 75   /* const speed setting    */
 3
 4   void onoff_controller()
 5   { int enc_new, v_act, r_mot, err;
 6     static int enc_old;
 7
 8     enc_new = QUADRead(enc1);
 9     v_act = (enc_new-enc_old) * 100;
10     if (v_act < v_des) r_mot = Kc;
11                 else  r_mot = 0;
12     err = MOTORDrive(mot1, r_mot);
13     if (err) printf("error: motor");
14     enc_old = enc_new;
15   }
```

Overflow and underflow

The difference, second value minus first value, results in:

$$00\ 00\ 00\ 0A \qquad = +10_{Dec}$$

This is the correct number of encoder ticks.

Overflow Example from negative to positive values:

$$FF\ FF\ FF\ FD \qquad = -3_{Dec}$$
$$00\ 00\ 00\ 04 \qquad = +4_{Dec}$$

The difference, second value minus first value, results in +7, which is the correct number of encoder ticks.

---

2. Call control subroutine periodically

> e.g. every 1/100 s

See library.html:

```
TimerHandle OSAttachTimer (int scale, TimerFnc function);
        Input:            (scale) prescale value for 100Hz Timer (1 to ...)
                          (TimerFnc) function to be called periodically
        Output:           (TimerHandle) handle to reference the IRQ-slot
                          A value of 0 indicates an error due to a full list(max. 16).
        Semantics:        Attach a irq-routine (void function(void)) to the irq-list.
                          The scale parameter adjusts the call frequency (100/scale Hz)
                          of this routine to allow many different applications.

int OSDetachTimer (TimerHandle handle)
        Input:            (handle) handle of a previous installed timer irq
        Output:           0 = handle not valid
                          1 = function successfully removed from timer irq list
        Semantics:        Detach a previously installed irq-routine from the irq-list.
```

Figure 4.5: RoBIOS timer functions

Let us take a look at how to implement step 2, using the timer functions in the RoBIOS operating system (see Appendix B.5 and Figure 4.5). There are operating system routines available to initialize a periodic timer function, to be called at certain intervals and to terminate it.

Program 4.3 shows a straightforward implementation using these routines. In the otherwise idle `while`-loop, any top-level user programs should be executed. In that case, the `while`-condition should be changed from:

```
while (1) /* endless loop - never returns */
```

to something than can actually terminate, for example:

```
while (KEYRead() != KEY4)
```

in order to check for a specific end-button to be pressed.

Program 4.3: Timer start

```
1   int main()
2   { TimerHandle t1;
3
4       t1 = OSAttachTimer(1, onoff_controller);
5       while (1) /* endless loop - never returns */
6       { /* other tasks or idle */ }
7       OSDetachTimer(t1); /* free timer, not used */
8       return 0;              /* not used */
9   }
```

Figure 4.6 shows a typical measurement of the step response of an on-off controller. The saw-tooth shape of the velocity curve is clearly visible.

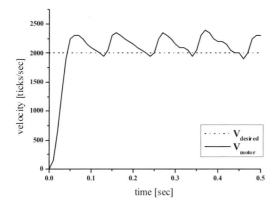

Figure 4.6: Measured step response of on-off controller

4.2 PID Control

PID = P + I + D The simplest method of control is not always the best. A more advanced controller and almost industry standard is the *PID controller*. It comprises a proportional, an integral, and a derivative control part. The controller parts are introduced in the following sections individually and in combined operation.

4.2.1 Proportional Controller

For many control applications, the abrupt change between a fixed motor control value and zero does not result in a smooth control behavior. We can improve this by using a linear or proportional term instead. The formula for the proportional controller (P controller) is:

$$R(t) = K_P \cdot (v_{des}(t) - v_{act}(t))$$

The difference between the desired and actual speed is called the "error function". Figure 4.7 shows the schematics for the P controller, which differs only slightly from the on-off controller. Figure 4.8 shows measurements of characteristic motor speeds over time. Varying the "controller gain" K_P will change the controller behavior. The higher the K_P chosen, the faster the controller responds; however, a too high value will lead to an undesirable oscillating system. Therefore it is important to choose a value for K_P that guarantees a fast response but does not lead the control system to overshoot too much or even oscillate.

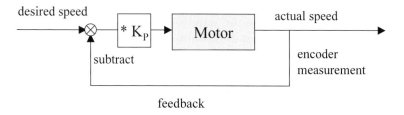

Figure 4.7: Proportional controller

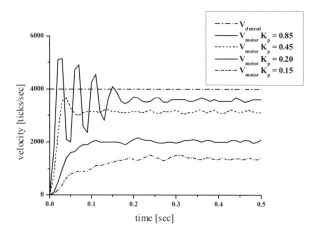

Figure 4.8: Step response for proportional controller

Steady state error   Note that the P controller's equilibrium state is *not* at the desired velocity. If the desired speed is reached exactly, the motor output is reduced to zero, as

57

defined in the P controller formula shown above. Therefore, each P controller will keep a certain "steady-state error" from the desired velocity, depending on the controller gain K_P. As can be seen in Figure 4.8, the higher the gain K_P, the lower the steady-state error. However, the system starts to oscillate if the selected gain is too high.

Program 4.4 shows the brief P controller code that can be inserted into the control frame of Program 4.1, in order to form a complete program.

Program 4.4: P controller code

```
1   e_func = v_des - v_act;   /* error function */
2   r_mot  = Kp*e_func;       /* motor output */
```

4.2.2 Integral Controller

Unlike the P controller, the I controller (integral controller) is rarely used alone, but mostly in combination with the P or PD controller. The idea for the I controller is to reduce the steady-state error of the P controller. With an additional integral term, this steady-state error can be reduced to zero, as seen in the measurements in Figure 4.9. The equilibrium state is reached somewhat later than with a pure P controller, but the steady-state error has been eliminated.

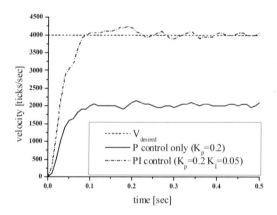

Figure 4.9: Step response for integral controller

When using e(t) as the error function, the formula for the PI controller is:

$$R(t) = K_P \cdot [\, e(t) + 1/T_I \cdot {}_0\!\int^t e(t)dt \,]$$

We rewrite this formula by substituting $Q_I = K_P/T_I$, so we receive independent additive terms for P and I:

$$R(t) = K_P \cdot e(t) + Q_I \cdot {}_0\!\int^t e(t)dt$$

Naive approach The naive way of implementing the I controller part is to transform the integration into a sum of a fixed number (for example 10) of previous error values. These 10 values would then have to be stored in an array and added for every iteration.

Proper PI implementation The proper way of implementing a PI controller starts with discretization, replacing the integral with a sum, using the trapezoidal rule:

$$R_n = K_P \cdot e_n + Q_I \cdot t_{delta} \cdot \sum_{i=1}^{n} \frac{e_i + e_{i-1}}{2}$$

Now we can get rid of the sum by using R_{n-1}, the output value preceding R_n:

$$R_n - R_{n-1} = K_P \cdot (e_n - e_{n-1}) + Q_I \cdot t_{delta} \cdot (e_n + e_{n-1})/2$$

Therefore (substituting K_I for $Q_I \cdot t_{delta}$):

$$R_n = R_{n-1} + K_P \cdot (e_n - e_{n-1}) + K_I \cdot (e_n + e_{n-1})/2$$

Limit controller output values! So we only need to store the previous control value and the previous error value to calculate the PI output in a much simpler formula. Here it is important to *limit the controller output* to the correct value range (for example -100 .. +100 in RoBIOS) and also store the limited value for the subsequent iteration. Otherwise, if a desired speed value cannot be reached, both controller output and error values can become arbitrarily large and invalidate the whole control process [Kasper 2001].

Program 4.5 shows the program fragment for the PI controller, to be inserted into the framework of Program 4.1.

Program 4.5: PI controller code

```
1   static int r_old=0, e_old=0;
2   ...
3   e_func = v_des - v_act;
4   r_mot  = r_old + Kp*(e_func-e_old) + Ki*(e_func+e_old)/2;
5   r_mot = min(r_mot, +100);   /* limit output */
6   r_mot = max(r_mot, -100);   /* limit output */
7   r_old  = r_mot;
8   e_old  = e_func;
```

4.2.3 Derivative Controller

Similar to the I controller, the D controller (derivative controller) is rarely used by itself, but mostly in combination with the P or PI controller. The idea for adding a derivative term is to speed up the P controller's response to a change of input. Figure 4.10 shows the measured differences of a step response between the P and PD controller (left), and the PD and PID controller (right). The PD controller reaches equilibrium faster than the P controller, but still has a steady-state error. The full PID controller combines the advantages of PI and PD. It has a fast response and suffers no steady-state error.

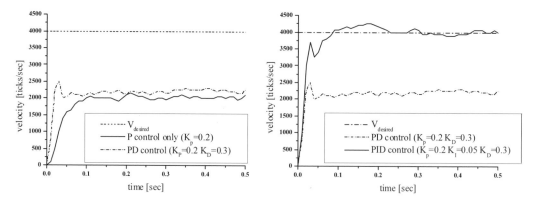

Figure 4.10: Step response for derivative controller and PID controller

When using e(t) as the error function, the formula for a combined PD controller is:

$$R(t) = K_P \cdot [\, e(t) + T_D \cdot de(t)/dt \,]$$

The formula for the full PID controller is:

$$R(t) = K_P \cdot [\, e(t) + 1/T_I \cdot {\textstyle\int_0^t} e(t)dt + T_D \cdot de(t)/dt \,]$$

Again, we rewrite this by substituting T_D and T_I, so we receive independent additive terms for P, I, and D. This is important, in order to experimentally adjust the relative gains of the three terms.

$$R(t) = K_P \cdot e(t) + Q_I \cdot {\textstyle\int_0^t} e(t)dt + Q_D \cdot de(t)/dt$$

Using the same discretization as for the PI controller, we will get:

$$R_n = K_P \cdot e_n + Q_I \cdot t_{delta} \cdot \sum_{i=1}^{n} \frac{e_i + e_{i-1}}{2} + Q_D / t_{delta} \cdot (e_n - e_{n-1})$$

Again, using the difference between subsequent controller outputs, this results in:

$$R_n - R_{n-1} = K_P \cdot (e_n - e_{n-1}) + Q_I \cdot t_{delta} \cdot (e_n + e_{n-1})/2$$
$$+ Q_D / t_{delta} \cdot (e_n - 2 \cdot e_{n-1} + e_{n-2})$$

Finally (substituting K_I for $Q_I \cdot t_{delta}$ and K_D for Q_D / t_{delta}):

Complete PID formula
$$R_n = R_{n-1} + K_P \cdot (e_n - e_{n-1}) + K_I \cdot (e_n + e_{n-1})/2 + K_D \cdot (e_n - 2 \cdot e_{n-1} + e_{n-2})$$

Program 4.6 shows the program fragment for the PD controller, while Program 4.7 shows the full PID controller. Both are to be inserted into the framework of Program 4.1.

Program 4.6: PD controller code

```
1   static int e_old=0;
2   ...
3   e_func = v_des - v_act;             /* error function */
4   deriv  = e_old - e_func;            /* diff. of error fct.   */
5   e_old  = e_func;                    /* store error function */
6   r_mot  = Kp*e_func + Kd*deriv;      /* motor output */
7   r_mot  = min(r_mot, +100);          /* limit output */
8   r_mot  = max(r_mot, -100);          /* limit output */
```

Program 4.7: PID controller code

```
1    static int r_old=0, e_old=0, e_old2=0;
2    ...
3    e_func = v_des - v_act;
4    r_mot  = r_old + Kp*(e_func-e_old) + Ki*(e_func+e_old)/2
5                   + Kd*(e_func - 2* e_old + e_old2);
6    r_mot = min(r_mot, +100);   /* limit output */
7    r_mot = max(r_mot, -100);   /* limit output */
8    r_old  = r_mot;
9    e_old2 = e_old;
10   e_old  = e_func;
```

4.2.4 PID Parameter Tuning

Find parameters experimentally
The tuning of the three PID parameters K_P, K_I, and K_D is an important issue. The following guidelines can be used for experimentally finding suitable values (adapted after [Williams 2006]):

1. Select a typical operating setting for the desired speed, turn off integral and derivative parts, then increase K_P to maximum or until oscillation occurs.

2. If system oscillates, divide K_P by 2.

3. Increase K_D and observe behavior when increasing/decreasing the desired speed by about 5%. Choose a value of K_D which gives a damped response.

4. Slowly increase K_I until oscillation starts. Then divide K_I by 2 or 3.

5. Check whether overall controller performance is satisfactorily under typical system conditions.

Further details on digital control can be found in [Åström, Hägglund 1995] and [Bolton 1995].

4.3 Velocity Control and Position Control

What about starting and stopping? So far, we are able to control a single motor at a certain speed, but we are not yet able to drive a motor at a given speed for a number of revolutions and then come to a stop at exactly the right motor position. The former, maintaining a certain speed, is generally called *velocity control*, while the latter, reaching a specified position, is generally called *position control*.

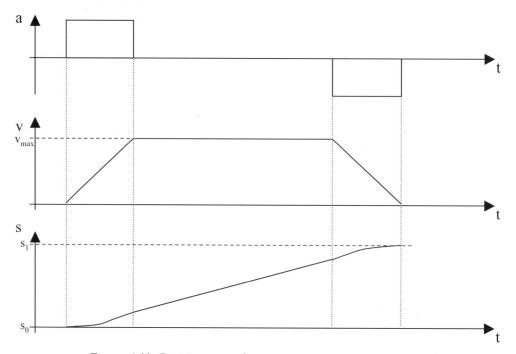

Figure 4.11: Position control

Position control requires an additional controller on top of the previously discussed velocity controller. The position controller sets the desired velocities in all driving phases, especially during the acceleration and deceleration phases (starting and stopping).

Speed ramp   Let us assume a single motor is driving a robot vehicle that is initially at rest and which we would like to stop at a specified position. Figure 4.11 demonstrates the "speed ramp" for the starting phase, constant speed phase, and stopping phase. When ignoring friction, we only need to apply a certain force (here constant) during the starting phase, which will translate into an acceleration of the vehicle. The constant acceleration will linearly increase the vehicle's speed v (integral of a) from 0 to the desired value v_{max}, while the vehicle's position s (integral of v) will increase quadratically.

When the force (acceleration) stops, the vehicle's velocity will remain constant, assuming there is no friction, and its position will increase linearly.

During the stopping phase (deceleration, breaking), a negative force (negative acceleration) is applied to the vehicle. Its speed will be linearly reduced to zero (or may even become negative – the vehicle now driving backwards – if the negative acceleration is applied for too long a time). The vehicle's position will increase slowly, following the square root function.

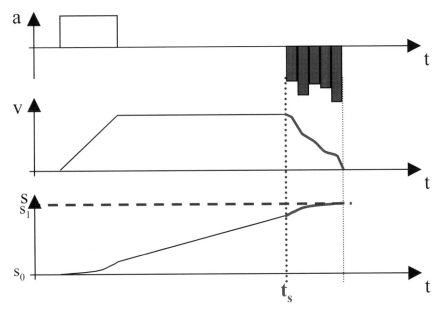

Figure 4.12: Breaking adaptation

The tricky bit now is to control the amount of acceleration in such a way that the vehicle:

a. Comes to rest
    (not moving slowly forward to backward).

b. Comes to rest at exactly the specified position
    (for example we want the vehicle to drive exactly 1 meter and stop within ±1mm).

Figure 4.12 shows a way of achieving this by controlling (continuously updating) the breaking acceleration applied. This control procedure has to take into account not only the current speed as a feedback value, but also the current position, since previous speed changes or inaccuracies may have had already an effect on the vehicle's position.

4.4 Multiple Motors – Driving Straight

Still more tasks to come Unfortunately, this is still not the last of the motor control problems. So far, we have only looked at a single isolated motor with velocity control – and very briefly at position control. The way that a robot vehicle is constructed, how-

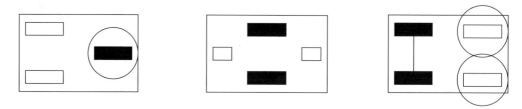

Figure 4.13: Wheeled robots

ever, shows us that a single motor is not enough (see Figure 4.13 repeated from Chapter 1). All these robot constructions require two motors, with the functions of driving and steering either separated or dependent on each other. In the design on the left or the design on the right, the driving and steering functions are separated. It is therefore very easy to drive in a straight line (simply keep the steering fixed at the angle representing "straight") or drive in a circle (keep the steering fixed at the appropriate angle). The situation is quite different in the "differential steering" design shown in the middle of Figure 4.13, which is a very popular design for small mobile robots. Here, one has to constantly monitor and update both motor speeds in order to drive straight. Driving in a circle can be achieved by adding a constant offset to one of the motors. Therefore, a *synchronization* of the two motor speeds is required.

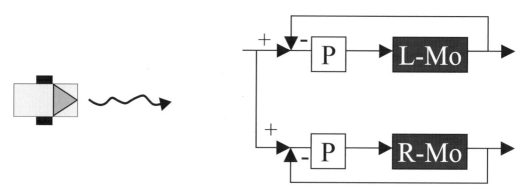

Figure 4.14: Driving straight – first try

There are a number of different approaches for driving straight. The idea presented in the following is from [Jones, Flynn, Seiger 1999]. Figure 4.14 shows the first try for a differential drive vehicle to drive in a straight line. There are two separate control loops for the left and the right motor, each involving feedback control via a P controller. The desired forward speed is supplied to both controllers. Unfortunately, this design will not produce a nice straight line of driving. Although each motor is controlled in itself, there is no control of any slight speed discrepancies between the two motors. Such a setup will most likely result in the robot driving in a wriggly line, but not straight.

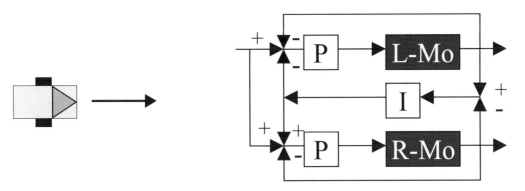

Figure 4.15: Driving straight – second try

An improvement of this control structure is shown in Figure 4.15 as a second try. We now also calculate the difference in motor movements (position, not speed) and feed this back to both P controllers via an additional I controller. The I controller integrates (or sums up) the differences in position, which will subsequently be eliminated by the two P controllers. Note the different signs for the input of this additional value, matching the inverse sign of the corresponding I controller input. Also, this principle relies on the fact that the whole control circuit is very fast compared to the actual motor or vehicle speed. Otherwise, the vehicle might end up in trajectories parallel to the desired straight line.

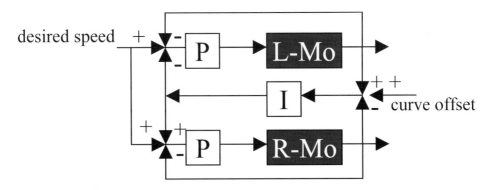

Figure 4.16: Driving straight or in curves

For the final version in Figure 4.16 (adapted from [Jones, Flynn, Seiger, 1999]), we added another user input with the curve offset. With a curve offset of zero. the system behaves exactly like the previous one for driving straight. A positive or negative fixed curve offset, however, will let the robot drive in a counter-clockwise or clockwise circle, respectively. The radius can be calculated from the curve offset amount.

The controller used in the RoBIOS vω library is a bit more complex than the one shown in Figure 4.16. It uses a PI controller for handling the rotational velocity ω in addition to the two PI controllers for left and right motor velocity. More elaborate control models for robots with differential drive wheel arrangements can be found in [Kim, Tsiotras 2002] (conventional controller models) and in [Seraji, Howard 2002] (fuzzy controllers).

4.5 V-Omega Interface

When programming a robot vehicle, we have to abstract all the low-level motor control problems shown in this chapter. Instead, we prefer to have a user-friendly interface with specialized driving functions for a vehicle that automatically takes care of all the feedback control issues discussed.

We have implemented such a driving interface in the RoBIOS operating system, called the "v-omega interface", because it allows us to specify a linear and rotational speed of a robot vehicle. There are lower-level functions for direct control of the motor speed, as well as higher-level functions for driving a complete trajectory in a straight line or a curve (Program 4.8).

Dead reckoning The initialization of the vω interface VWInit specifies the update rate of motor commands, usually 1/100 of a second, but does not activate PI control. The control routine, VWStartControl, will start PI controllers for both driving and steering as a background process with the PI parameters supplied. The corresponding timer interrupt routine contains all the program code for controlling the two driving motors and updating the internal variables for the vehicle's current position and orientation that can be read by using VWGetPosition. This method of determining a vehicle's position by continued addition of driving vectors and maintaining its orientation by adding all rotations is called "dead reckoning". Function VWStalled compares desired and actual motor speeds in order to check whether the robot's wheels have stalled, possibly because it has collided with an obstacle.

Program 4.8: vω interface

```
VWHandle VWInit(DeviceSemantics semantics, int Timescale);
int VWRelease(VWHandle handle);
int VWSetSpeed(VWHandle handle, meterPerSec v, radPerSec w);
int VWGetSpeed(VWHandle handle, SpeedType* vw);
int VWSetPosition(VWHandle handle, meter x, meter y, radians phi);
int VWGetPosition(VWHandle handle, PositionType* pos);
int VWStartControl(VWHandle handle, float Vv,float Tv, float Vw,float Tw);
int VWStopControl(VWHandle handle);
int VWDriveStraight(VWHandle handle, meter delta, meterpersec v);
int VWDriveTurn(VWHandle handle, radians delta, radPerSec w)
int VWDriveCurve(VWHandle handle, meter delta_l, radians delta_phi,
      meterpersec v)
float VWDriveRemain(VWHandle handle)
int VWDriveDone(VWHandle handle)
int VWDriveWait(VWHandle handle)
int VWStalled(VWHandle handle)
```

It is now possible to write application programs using only the vω interface for driving. Program 4.9 shows a simple program for driving a robot 1m straight, then turning 180° on the spot, and driving back in a straight line. The function VWDriveWait pauses user program execution until the driving command has been completed. All driving commands only register the driving parameters and then immediately return to the user program. The actual execution of the PI controller for both wheels is executed completely transparently to the application programmer in the previously described background routines.

Program 4.9: vω application program

```
 1    #include "eyebot.h"
 2    int main()
 3    { VWHandle vw;
 4      vw=VWInit(VW_DRIVE,1);
 5      VWStartControl(vw, 7.0, 0.3 , 10.0, 0.1);/* emp. val.*/
 6
 7      VWDriveStraight(vw, 1.0, 0.5);          /* drive 1m */
 8      VWDriveWait(vw);                  /* wait until done */
 9
10      VWDriveTurn(vw, 3.14, 0.5);      /* turn 180 on spot */
11      VWDriveWait(vw);
12
13      VWDriveStraight(vw, 1.0, 0.5);         /* drive back */
14      VWDriveWait(vw);
15
16      VWStopControl(vw);
17      VWRelease(vw);
18    }
```

With the help of the background execution of driving commands, it is possible to implement high-level control loops when driving. An example could be driving a robot forward in a straight line, until an obstacle is detected (for example by a built-in PSD sensor) and then stop the robot. There is no predetermined distance to be traveled; it will be determined by the robot's environment.

Program 4.10 shows the program code for this example. After initialization of the vω interface and the PSD sensor, we only have a single while-loop. This loop continuously monitors the PSD sensor data (PSDGet) and only if the distance is larger than a certain threshold will the driving command (VWDriveStraight) be called. This loop will be executed many times faster than it takes to execute the driving command for 0.05m in this example; no wait function has been used here. Every newly issued driving command will replace the previous driving command. So in this example, the final driving command will not drive the robot the full 0.05m, since as soon as the PSD sensor registers a shorter distance, the driving control will be stopped and the whole program terminated.

Program 4.10: Typical driving loop with sensors

```
1   #include "eyebot.h"
2   int main()
3   { VWHandle  vw;
4     PSDHandle psd_front;
5
6     vw=VWInit(VW_DRIVE,1);
7     VWStartControl(vw, 7.0, 0.3 , 10.0, 0.1);
8     psd_front = PSDInit(PSD_FRONT);
9     PSDStart(psd_front, TRUE);
10
11    while (PSDGet(psd_front)>100)
12      VWDriveStraight(vw, 0.05, 0.5);
13
14    VWStopControl(vw);
15    VWRelease(vw);
16    PSDStop();
17    return 0;
18  }
```

4.6 References

ÅSTRÖM, K., HÄGGLUND, T. *PID Controllers: Theory, Design, and Tuning*, 2nd Ed., Instrument Society of America, Research Triangle Park NC, 1995

BOLTON, W. *Mechatronics – Electronic Control Systems in Mechanical Engineering*, Addison Wesley Longman, Harlow UK, 1995

JONES, J., FLYNN, A., SEIGER, B. *Mobile Robots - From Inspiration to Implementation*, 2nd Ed., AK Peters, Wellesley MA, 1999

KASPER, M. *Rug Warrior Lab Notes*, Internal report, Univ. Kaiserslautern, Fachbereich Informatik, 2001

KIM, B., TSIOTRAS, P. *Controllers for Unicycle-Type Wheeled Robots: Theoretical Results and Experimental Validation*, IEEE Transactions on Robotics and Automation, vol. 18, no. 3, June 2002, pp. 294-307 (14)

SERAJI, H., HOWARD, A. *Behavior-Based Robot Navigation on Challenging Terrain: A Fuzzy Logic Approach*, IEEE Transactions on Robotics and Automation, vol. 18, no. 3, June 2002, pp. 308-321 (14)

WILLIAMS, C. *Tuning a PID Temperature Controller*, web: http://newton.ex.ac.uk/teaching/CDHW/Feedback/Setup-PID.html, 2006

MULTITASKING

C oncurrency is an essential part of every robot program. A number of more or less independent tasks have to be taken care of, which requires some form of multitasking, even if only a single processor is available on the robot's controller.

Imagine a robot program that should do some image processing and at the same time monitor the robot's infrared sensors in order to avoid hitting an obstacle. Without the ability for multitasking, this program would comprise one large loop for processing one image, then reading infrared data. But if processing one image takes much longer than the time interval required for updating the infrared sensor signals, we have a problem. The solution is to use separate processes or tasks for each activity and let the operating system switch between them.

Threads versus processes The implementation used in RoBIOS is "threads" instead of "processes" for efficiency reasons. Threads are "lightweight processes" in the sense that they share the same memory address range. That way, task switching for threads is much faster than for processes. In this chapter, we will look at cooperative and preemptive multitasking as well as synchronization via semaphores and timer interrupts. We will use the expressions "multitasking" and "process" synonymously for "multithreading" and "thread", since the difference is only in the implementation and transparent to the application program.

5.1 Cooperative Multitasking

The simplest way of multitasking is to use the "cooperative" scheme. Cooperative means that each of the parallel tasks needs to be "well behaved" and does transfer control explicitly to the next waiting thread. If even one routine does not pass on control, it will "hog" the CPU and none of the other tasks will be executed.

The cooperative scheme has less problem potential than the preemptive scheme, since the application program can determine at which point in time it is willing to transfer control. However, not all programs are well suited for it, since there need to be appropriate code sections where a task change fits in.

Program 5.1 shows the simplest version of a program using cooperative multitasking. We are running two tasks using the same code mytask (of course running different code segments in parallel is also possible). A task can recover its own task identification number by using the system function OSGetUID. This is especially useful to distinguish several tasks running the same code in parallel. All our task does in this example is execute a loop, printing one line of text with its id-number and then calling OSReschedule. The system function OSReschedule will transfer control to the next task, so here the two tasks are taking turns in printing lines. After the loop is finished, each task terminates itself by calling OSKill.

Program 5.1: Cooperative multitasking

```
1    #include "eyebot.h"
2    #define SSIZE   4096
3    struct tcb *task1, *task2;
4
5    void mytask()
6    { int id, i;
7      id = OSGetUID(0); /* read slave id no. */
8      for (i=1; i<=100; i++)
9      { LCDPrintf("task %d : %d\n", id, i);
10       OSReschedule();   /* transfer control */
11     }
12     OSKill(0);  /* terminate thread */
13   }
14
15   int main()
16   { OSMTInit(COOP);   /* init multitasking */
17     task1 = OSSpawn("t1", mytask, SSIZE, MIN_PRI, 1);
18     task2 = OSSpawn("t2", mytask, SSIZE, MIN_PRI, 2);
19     if(!task1 || !task2) OSPanic("spawn failed");
20
21     OSReady(task1); /* set state of task1 to READY */
22     OSReady(task2);
23     OSReschedule(); /* start multitasking */
24     /* ------------------------------------------------- */
25     /* processing returns HERE, when no READY thread left */
26     LCDPrintf("back to main");
27     return 0;
28   };
```

The main program has to initialize multitasking by calling OSMTInit; the parameter COOP indicates cooperative multitasking. Activation of processes is done in three steps. Firstly, each task is spawned. This creates a new task structure for a task name (string), a specified function call (here: mytask) with its own local stack with specified size, a certain priority, and an id-number. The required stack size depends on the number of local variables and the calling depth of subroutines (for example recursion) in the task. Secondly, each task is switched to the mode "ready". Thirdly and finally, the main program relinquishes control to one of the parallel tasks by calling OSReschedule itself. This will activate one of the parallel tasks, which will take turns until they both terminate themselves. At that point in time – and also in the case that all parallel processes are blocked, i.e. a "deadlock" has occurred – the main program will be reactivated and continue its flow of control with the next instruction. In this example, it just prints one final message and then terminates the whole program.

The system output will look something like the following:

```
task 2 : 1
task 1 : 1
task 2 : 2
task 1 : 2
task 2 : 3
task 1 : 3
...
task 2 : 100
task 1 : 100
back to main
```

Both tasks are taking turns as expected. Which task goes first is system-dependent.

5.2 Preemptive Multitasking

At first glance, preemptive multitasking does not look much different from cooperative multitasking. Program 5.2 shows a first try at converting Program 5.1 to a preemptive scheme, but unfortunately it is not completely correct. The function mytask is identical as before, except that the call of OSReschedule is missing. This of course is expected, since preemptive multitasking does not require an explicit transfer of control. Instead the task switching is activated by the system timer. The only other two changes are the parameter PREEMPT in the initialization function and the system call OSPermit to enable timer interrupts for task switching. The immediately following call of OSReschedule is optional; it makes sure that the main program immediately relinquishes control.

This approach would work well for two tasks that are not interfering with each other. However, the tasks in this example are interfering by both sending output to the LCD. Since the task switching can occur at any time, it can (and will) occur in the middle of a print operation. This will mix up characters from one line of task1 and one line from task2, for example if task1 is interrupted after printing only the first three characters of its string:

```
task 1 : 1
task 1 : 2
tastask 2 : 1
task 2 : 2
task 2 :k 1: 3
task 1 : 4
. . .
```

But even worse, the task switching can occur in the middle of the system call that writes one character to the screen. This will have all sorts of strange effects on the display and can even lead to a task hanging, because its data area was corrupted.

So quite obviously, synchronization is required whenever two or more tasks are interacting or sharing resources. The corrected version of this preemptive example is shown in the following section, using a semaphore for synchronization.

Program 5.2: Preemptive multitasking – first try (incorrect)

```
1    #include "eyebot.h"
2    #define SSIZE   4096
3    struct tcb *task1, *task2;
4
5    void mytask()
6    { int id, i;
7      id = OSGetUID(0); /* read slave id no. */
8      for (i=1; i<=100; i++)
9        LCDPrintf("task %d : %d\n", id, i);
10     OSKill(0);   /* terminate thread */
11   }
12
13   int main()
14   { OSMTInit(PREEMPT);   /* init multitasking */
15     task1 = OSSpawn("t1", mytask, SSIZE, MIN_PRI, 1);
16     task2 = OSSpawn("t2", mytask, SSIZE, MIN_PRI, 2);
17     if(!task1 || !task2) OSPanic("spawn failed");
18
19     OSReady(task1);  /* set state of task1 to READY */
20     OSReady(task2);
21     OSPermit();       /* start multitasking */
22     OSReschedule();  /* switch to other task */
23     /* ----------------------------------------------- */
24     /* processing returns HERE, when no READY thread left */
25     LCDPrintf("back to main");
26     return 0;
27   };
```

5.3 Synchronization

Semaphores for synchronization In almost every application of preemptive multitasking, some synchronization scheme is required, as was shown in the previous section. Whenever two or more tasks exchange information blocks or share any resources (for example LCD for printing messages, reading data from sensors, or setting actuator values), synchronization is essential. The standard synchronization methods are (see [Bräunl 1993]):

- Semaphores
- Monitors
- Message passing

Here, we will concentrate on synchronization using semaphores. Semaphores are rather low-level synchronization tools and therefore especially useful for embedded controllers.

5.3.1 Semaphores

The concept of semaphores has been around for a long time and was formalized by Dijkstra as a model resembling railroad signals [Dijkstra 1965]. For further historic notes see also [Hoare 1974], [Brinch Hansen 1977], or the more recent collection [Brinch Hansen 2001].

A semaphore is a synchronization object that can be in either of two states: *free* or *occupied*. Each task can perform two different operations on a semaphore: *lock* or *release*. When a task locks a previously "free" semaphore, it will change the semaphore's state to "occupied". While this (the first) task can continue processing, any subsequent tasks trying to lock the now occupied semaphore will be blocked until the first task releases the semaphore. This will only momentarily change the semaphore's state to free – the next waiting task will be unblocked and re-lock the semaphore.

In our implementation, semaphores are declared and initialized with a specified state as an integer value (0: blocked, ≥1: free). The following example defines a semaphore and initializes it to *free*:

```
struct sem  my_sema;
OSSemInit(&my_sema, 1);
```

The calls for locking and releasing a semaphore follow the traditional names coined by Dijkstra: P for locking ("**p**ass") and V for releasing ("**l**eave"). The following example locks and releases a semaphore while executing an exclusive code block:

```
OSSemP(&my_sema);
   /* exclusive block, for example write to screen */
OSSemV(&my_sema);
```

Of course *all* tasks sharing a particular resource or all tasks interacting have to behave using P and V in the way shown above. Missing a P operation can result in a system crash as shown in the previous section. Missing a V operation will result in some or all tasks being blocked and never being released. If tasks share several resources, then one semaphore per resource has to be used, or tasks will be blocked unnecessarily.

Since the semaphores have been implemented using integer counter variables, they are actually "counting semaphores". A counting semaphore initialized with, for example, value 3 allows to perform three subsequent *non-blocking* P operations (decrementing the counter by three down to 0). Initializing a semaphore with value 3 is equivalent to initializing it with 0 and performing three subsequent V operations on it. A semaphore's value can also go below zero, for example if it is initialized with value 1 and two tasks perform a P operation on it. The first P operation will be non-blocking, reducing the semaphore value to 0, while the second P operation will block the calling task and will set the semaphore value to –1.

In the simple examples shown here, we only use the semaphore values 0 (blocked) and 1 (free).

Program 5.3: Preemptive multitasking with synchronization

```
 1   #include "eyebot.h"
 2   #define SSIZE  4096
 3   struct tcb *task1, *task2;
 4   struct sem lcd;
 5
 6   void mytask()
 7   { int id, i;
 8     id = OSGetUID(0); /* read slave id no. */
 9     for (i=1; i<=100; i++)
10     { OSSemP(&lcd);
11         LCDPrintf("task %d : %d\n", id, i);
12       OSSemV(&lcd);
13     }
14     OSKill(0);  /* terminate thread */
15   }
16
17   int main()
18   { OSMTInit(PREEMPT);  /* init multitasking */
19     OSSemInit(&lcd,1);  /* enable semaphore  */
20     task1 = OSSpawn("t1", mytask, SSIZE, MIN_PRI, 1);
21     task2 = OSSpawn("t2", mytask, SSIZE, MIN_PRI, 2);
22     if(!task1 || !task2) OSPanic("spawn failed");
23     OSReady(task1);  /* set state of task1 to READY */
24     OSReady(task2);
25     OSPermit();      /* start multitasking */
26     OSReschedule();  /* switch to other task */
27     /* ---- proc. returns HERE, when no READY thread left */
28     LCDPrintf("back to main");
29     return 0;
30   };
```

5.3.2 Synchronization Example

We will now fix the problems in Program 5.2 by adding a semaphore. Program 5.3 differs from Program 5.2 only by adding the semaphore declaration and initialization in the main program, and by using a bracket of OSSemP and OSSemV around the print statement.

The effect of the semaphore is that only one task is allowed to execute the print statement at a time. If the second task wants to start printing a line, it will be blocked in the P operation and has to wait for the first task to finish printing its line and issue the V operation. As a consequence, there will be no more task changes in the middle of a line or, even worse, in the middle of a character, which can cause the system to hang.

Unlike in cooperative multitasking, task1 and task2 do not necessarily take turns in printing lines in Program 5.3. Depending on the system time slices, task priority settings, and the execution time of the print block enclosed by P and V operations, one or several iterations can occur per task.

5.3.3 Complex Synchronization

In the following, we introduce a more complex example, running tasks with different code blocks and multiple semaphores. The main program is shown in Program 5.4, with slave tasks shown in Program 5.5 and the master task in Program 5.6.

The main program is similar to the previous examples. OSMTInit, OSSpawn, OSReady, and OSPermit operations are required to start multitasking and enable all tasks. We also define a number of semaphores: one for each slave process plus an additional one for printing (as in the previous example). The idea for operation is that one master task controls the operation of three slave tasks. By pressing keys in the master task, individual slave tasks can be either blocked or enabled.

All that is done in the slave tasks is to print a line of text as before, but indented for readability. Each loop iteration has now to pass two semaphore blocks: the first one to make sure the slave is enabled, and the second one to prevent several active slaves from interfering while printing. The loops now run indefinitely, and all slave tasks will be terminated from the master task.

The master task also contains an infinite loop; however, it will kill all slave tasks and terminate itself when KEY4 is pressed. Pressing KEY1 .. KEY3 will either enable or disable the corresponding slave task, depending on its current state, and also update the menu display line.

Program 5.4: Preemptive main

```
1   #include "eyebot.h"
2   #define SLAVES 3
3   #define SSIZE  8192
4   struct tcb *slave_p[SLAVES], *master_p;
5   struct sem sema[SLAVES];
6   struct sem lcd;
7
8   int main()
9   { int i;
10    OSMTInit(PREEMPT); /* init multitasking */
11    for (i=0; i<SLAVES; i++) OSSemInit(&sema[i],0);
12    OSSemInit(&lcd,1);  /* init semaphore */
13    for (i=0; i<SLAVES; i++) {
14      slave_p[i]= OSSpawn("slave-i",slave,SSIZE,MIN_PRI,i);
15      if(!slave_p[i]) OSPanic("spawn for slave failed");
16    }
17    master_p   = OSSpawn("master",master,SSIZE,MIN_PRI,10);
18    if(!master_p) OSPanic("spawn for master failed");
19    for (i=0; i<SLAVES; i++) OSReady(slave_p[i]);
20    OSReady(master_p);
21    OSPermit();      /* activate preemptive multitasking */
22    OSReschedule(); /* start first task */
23    /* -------------------------------------------------- */
24    /* processing returns HERE, when no READY thread left */
25    LCDPrintf("back to main\n");
26  return 0;
27  }
```

Program 5.5: Slave task

```
1   void slave()
2   { int id, i, count = 0;
3
4     /** read slave id no. */
5     id = OSGetUID(0);
6     OSSemP(&lcd);
7       LCDPrintf("slave %d start\n", id);
8     OSSemV(&lcd);
9
10    while(1)
11    { OSSemP(&sema[id]); /* occupy semaphore */
12        OSSemP(&lcd);
13          for (i=0; i<2*id; i++) LCDPrintf("-");
14          LCDPrintf("slave %d:%d\n", id, count);
15        OSSemV(&lcd);
16        count = (count+1) % 100;   /* max count 99 */
17      OSSemV(&sema[id]); /* free semaphore */
18    }
19  } /* end slave */
```

Program 5.6: Master task

```
1   void master()
2   { int i,k;
3     int block[SLAVES] = {1,1,1};   /* slaves blocked */
4     OSSemP(&lcd);   /* optional since slaves blocked */
5       LCDPrintf("master start\n");
6       LCDMenu("V.0", "V.1", "V.2", "END");
7     OSSemV(&lcd);
8
9     while(1)
10    { k = ord(KEYGet());
11      if (k!=3)
12      { block[k] = !block[k];
13        OSSemP(&lcd);
14          if (block[k]) LCDMenuI(k+1,"V");
15            else LCDMenuI(k+1,"P");
16        OSSemV(&lcd);
17        if (block[k])    OSSemP(&sema[k]);
18          else OSSemV(&sema[k]);
19      }
20      else /* kill slaves then exit master */
21      { for (i=0; i<SLAVES; i++) OSKill(slave_p[i]);
22        OSKill(0);
23      }
24    } /* end while */
25  } /* end master */
26
27  int ord(int key)
28  { switch(key)
29    {
30      case KEY1: return 0;
31      case KEY2: return 1;
32      case KEY3: return 2;
33      case KEY4: return 3;
34    }
35    return 0; /* error */
36  }
```

5.4 Scheduling

A scheduler is an operating system component that handles task switching. Task switching occurs in preemptive multitasking when a task's time slice has run out, when the task is being blocked (for example through a P operation on a semaphore), or when the task voluntarily gives up the processor (for example by calling OSReschedule). Explicitly calling OSReschedule is the only possibility for a task switch in cooperative multitasking.

Each task can be in exactly one of three states (Figure 5.1):

- **Ready**
  A task is ready to be executed and waiting for its turn.

- **Running**
  A task is currently being executed.

- **Blocked**
  A task is not ready for execution, for example because it is waiting for a semaphore.

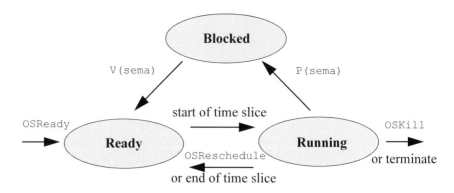

Figure 5.1: Task model

Each task is identified by a task control block that contains all required control data for a task. This includes the task's start address, a task number, a stack start address and size, a text string associated with the task, and an integer priority.

Round robin   Without the use of priorities, i.e. with all tasks being assigned the same priority as in our examples, task scheduling is performed in a "round-robin" fashion. For example, if a system has three "ready" tasks, t1, t2, t3, the execution order would be:

$$t_1, t_2, t_3, t_1, t_2, t_3, t_1, t_2, t_3, ...$$

Indicating the "running" task in bold face and the "ready" waiting list in square brackets, this sequence looks like the following:

t_1 $[t_2, t_3]$
t_2 $[t_3, t_1]$
t_3 $[t_1, t_2]$
...

Each task gets the same time slice duration, in RoBIOS 0.01 seconds. In other words, each task gets its fair share of processor time. If a task is blocked during its running phase and is later unblocked, it will re-enter the list as the *last* "ready" task, so the execution order may be changed. For example:

t_1 (*block t_1*) t_2, t_3, t_2, t_3 (*unblock t_1*) $t_2, t_1, t_3, t_2, t_1, t_3, ...$

Using square brackets to denote the list of ready tasks and curly brackets for the list of all blocked processes, the scheduled tasks look as follows:

t_1 [t_2, t_3] {}   → t_1 *is being blocked*
t_2 [t_3]   {t_1}
t_3 [t_2]   {t_1}
t_2 [t_3]   {t_1}
t_3 [t_2, t_1] {}   → t_3 *unblocks t_1*
t_2 [t_1, t_3] {}
t_1 [t_3, t_2] {}
t_3 [t_2, t_1] {}
...

Whenever a task is put back into the "ready" list (either from running or from blocked), it will be put at the end of the list of all waiting tasks with the same priority. So if all tasks have the same priority, the new "ready" task will go to the end of the complete list.

Priorities   The situation gets more complex if different priorities are involved. Tasks can be started with priorities 1 (lowest) to 8 (highest). The simplest priority model (*not* used in RoBIOS) is static priorities. In this model, a new "ready" task will follow after the last task with the same priority and before all tasks with a lower priority. Scheduling remains simple, since only a single waiting list has to be maintained. However, "starvation" of tasks can occur, as shown in the following example.

Starvation   Assuming tasks t_A and t_B have the higher priority 2, and tasks t_a and t_b have the lower priority 1, then in the following sequence tasks t_a and t_b are being kept from executing (*starvation*), unless t_A and t_B are both blocked by some events.

t_A [t_B, t_a, t_b] {}
t_B [t_A, t_a, t_b] {}
t_A [t_B, t_a, t_b] {}   → t_A *blocked*
t_B [t_A, t_a, t_b] {t_A}   → t_B *blocked*
t_a [t_b]   {t_A, t_B}
...

Dynamic   For these reasons, RoBIOS has implemented the more complex dynamic
priorities priority model. The scheduler maintains eight distinct "ready" waiting lists, one for each priority. After spawning, tasks are entered in the "ready" list matching their priority and each queue for itself implements the "round-robin" principle shown before. So the scheduler only has to determine which queue to select next.

Each queue (not task) is assigned a static priority (1..8) and a dynamic priority, which is initialized with twice the static priority times the number of "ready" tasks in this queue. This factor is required to guarantee fair scheduling for different queue lengths (see below). Whenever a task from a "ready" list is executed, then the dynamic priority of this queue is decremented by 1. Only after the dynamic priorities of all queues have been reduced to zero are the dynamic queue priorities reset to their original values.

The scheduler now simply selects the next task to be executed from the (non-empty) "ready" queue with the highest dynamic priority. If there are no eligible tasks left in any "ready" queue, the multitasking system terminates and continues with the calling `main` program. This model prevents starvation and still gives tasks with higher priorities more frequent time slices for execution. See the example below with three priorities, with static priorities shown on the right, dynamic priorities on the left. The highest dynamic priority after decrementing and the task to be selected for the next time slice are printed in bold type:

$$
\begin{array}{lll}
& -\ 6 & [t_A]_3 \\
& \mathbf{8} & [\mathbf{t_a},\mathbf{t_b}]_2 \\
& 4 & [t_x,t_y]_1 \\[4pt]
t_a & 6 & [t_A]_3 \\
& \mathbf{7} & [\mathbf{t_b}]_2 \\
& 4 & [t_x,t_y]_1 \\[4pt]
t_b & \mathbf{6} & [\mathbf{t_A}]_3 \\
& 6 & [t_a]_2 \\
& 4 & [t_x,t_y]_1 \\[4pt]
t_A & 5 & []_3 \\
& \mathbf{6} & [\mathbf{t_a},\mathbf{t_b}]_2 \\
& 4 & [t_x,t_y]_1 \\[4pt]
t_a & 5 & [\mathbf{t_A}]_3 \\
& 5 & [t_b]_2 \\
& 4 & [t_x,t_y]_1 \\[4pt]
& \ldots \\[4pt]
t_a & 3 & [t_A]_3 \\
& 3 & [t_b]_2 \\
& \mathbf{4} & [\mathbf{t_x},\mathbf{t_y}]_1 \\[4pt]
t_x & \mathbf{3} & [\mathbf{t_A}]_3 \\
& 3 & [t_a,t_b]_2 \\
& 3 & [t_y]_1 \\[4pt]
& \ldots
\end{array}
$$

$(2 \cdot \text{priority} \cdot \text{number\_of\_tasks} = 2 \cdot 3 \cdot 1 = 6)$
$(2 \cdot 2 \cdot 2 = 8)$
$(2 \cdot 1 \cdot 2 = 4)$

5.5 Interrupts and Timer-Activated Tasks

A different way of designing a concurrent application is to use interrupts, which can be triggered by external devices or by a built-in timer. Both are very important techniques; external interrupts can be used for reacting to external sensors, such as counting ticks from a shaft encoder, while timer interrupts can be used for implementing periodically repeating tasks with fixed time frame, such as motor control routines.

The event of an external interrupt signal will stop the currently executing task and instead execute a so-called "interrupt service routine" (ISR). As a general rule, ISRs should have a short duration and are required to clean up any stack changes they performed, in order not to interfere with the foreground task. Initialization of ISRs often requires assembly commands, since interrupt lines are directly linked to the CPU and are therefore machine-dependent (Figure 5.2).

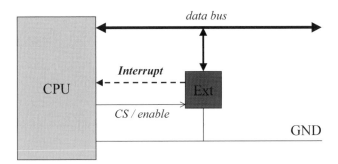

Figure 5.2: Interrupt generation from external device

Somewhat more general are interrupts activated by software instead of external hardware. Most important among software interrupts are timer interrupts, which occur at regular time intervals.

In the RoBIOS operating system, we have implemented a general purpose 100Hz timing interrupt. User programs can attach or detach up to 16 timer interrupt ISRs at a rate between 100Hz (0.01s) and $4.7 \ 10^{-8}$Hz (248.6 days), by specifying an integer frequency divider. This is achieved with the following operations:

```
TimerHandle OSAttachTimer(int scale, TimerFnc function);
int         OSDetachTimer(TimerHandle handle);
```

The timing `scale` parameter (range 1..100) is used to divide the 100Hz timer and thereby specifies the number of timer calls per second (1 for 100Hz, 100 for 1Hz). Parameter `TimerFct` is simply the name of a function without parameters or return value (void).

An application program can now implement a background task, for example a PID motor controller (see Section 4.2), which will be executed several times per second. Although activation of an ISR is handled in the same way as preemptive multitasking (see Section 5.2), an ISR itself will not be preempted, but will keep processor control until it terminates. This is one of the reasons why an ISR should be rather short in time. It is also obvious that the execution time of an ISR (or the sum of all ISR execution times in the case of multiple ISRs) must not exceed the time interval between two timer interrupts, or regular activations will not be possible.

The example in Program 5.7 shows the *timer* routine and the corresponding *main* program. The main program initializes the timer interrupt to once every second. While the foreground task prints consecutive numbers to the screen, the background task generates an acoustic signal once every second.

Program 5.7: Timer-activated example

```
1   void timer()
2   { AUBeep();   /* background task */
3   }
```

```
1   int main()
2   { TimerHandle t; int i=0;
3     t = OSAttachTimer(100, timer);
4       /* foreground task: loop until key press */
5       while (!KEYRead()) LCDPrintf("%d\n", i++);
6     OSDetachTimer(t);
7     return 0;
8   }
```

5.6 References

BRÄUNL, T. *Parallel Programming - An Introduction*, Prentice Hall, Engle-wood Cliffs NJ, 1993

BRINCH HANSEN, P. *The Architecture of Concurrent Programs*, Prentice Hall, Englewood Cliffs NJ, 1977

BRINCH HANSEN, P. (Ed.) *Classic Operating Systems*, Springer-Verlag, Berlin, 2001

DIJKSTRA, E. *Communicating Sequential Processes*, Technical Report EWD-123, Technical University Eindhoven, 1965

HOARE, C.A.R. *Communicating sequential processes*, Communications of the ACM, vol. 17, no. 10, Oct. 1974, pp. 549-557 (9)

WIRELESS COMMUNICATION

6

T here are a number of tasks where a self-configuring network based on wireless communication is helpful for a group of autonomous mobile robots or a single robot and a host computer:

1. To allow robots to communicate with each other
   For example, sharing sensor data or cooperating on a common task or devising a shared plan.

2. To remote-control one or several robots
   For example, giving low-level driving commands or specifying high-level goals to be achieved.

3. To monitor robot sensor data
   For example, displaying camera data from one or more robots or recording a robot's distance sensor data over time.

4. To run a robot with off-line processing
   For example, combining the two previous points, each sensor data packet is sent to a host where all computation takes place, the resulting driving commands being relayed back to the robot.

5. To create a monitoring console for single or multiple robots
   For example, monitoring each robot's position, orientation, and status in a multi-robot scenario in a common environment. This will allow a post-mortem analysis of a robot team's performance and effectiveness for a particular task.

The network needs to be self-configuring. This means there will be no fixed or pre-determined master node. Each agent can take on the role of master. Each agent must be able to sense the presence of another agent and establish a communication link. New incoming agents must be detected and integrated in the network, while exiting agents will be deleted from it. A special error protocol is required because of the high error rate of mobile wireless data exchange. Further details of this project can be found in [Bräunl, Wilke 2001].

6.1 Communication Model

In principle, a wireless network can be considered as a fully connected network, with every node able to access every other node in one hop, i.e. data is always transmitted directly without the need of a third party.

However, if we allowed every node to transmit data at any point in time, we would need some form of collision detection and recovery, similar to CSMA/CD [Wang, Premvuti 94]. Since the number of collisions increases quadratically with the number of agents in the network, we decided to implement a time division network instead. Several agents can form a network using a TDMA (time division multiple access) protocol to utilize the available bandwidth. In TDMA networks, only one transceiver may transmit data at a time, which eliminates data loss from transmission collisions. So at no time may two transceiver modules send data at once. There are basically two techniques available to satisfy this condition (Figure 6.1):

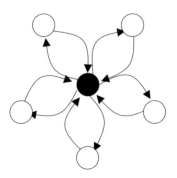 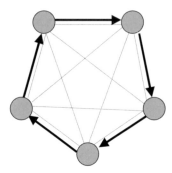

Figure 6.1: Polling versus Virtual Token Ring

- Polling:
  One agent (or a base station) is defined as the "master" node. In a *round-robin* fashion, the master talks to each of the agents subsequently to let it send one message, while each agent is listening to all messages and can filter out only those messages that apply to it.

- Virtual Token Ring:
  A *token* (special data word) is sent from one agent to the next agent in a circular list. Whoever has the token may send its messages and then pass the token on to the next agent. Unlike before, any agent can become master, initiating the ring mechanism and also taking over in case a problem occurs (for example the loss of a token, see Section 6.3).

While both approaches are similar, the token approach has the advantage of less overhead and higher flexibility, yet has the same functionality and safety. Therefore, we chose the "Virtual Token Ring" approach.

1. The master has to create a list of all active robot nodes in the network by a polling process at initialization time. This list has also to be maintained during the operation of the network (see below) and has to be broadcast to all robots in the network (each robot has to know its successor for passing on the token).

2. The master has to monitor the data communication traffic with a time-out watchdog. If no data exchange happens during a fixed amount of time, the master has to assume that the token got lost (for example, the robot that happened to have the token was switched off), and create a new one.

3. If the token is lost several times in a row by the same agent (for example three times), the master decides that this agent is malfunctioning and takes it off the list of active nodes.

4. The master periodically checks for new agents that have become available (for example just switched on or recovered from an earlier malfunction) by sending a "wild card" message. The master will then update the list of active agents, so they will participate in the token ring network.

All agents (and base stations) are identified by a unique id-number, which is used to address an agent. A specific id-number is used for broadcasts, which follows exactly the same communication pattern. In the same way, subgroups of nodes can be implemented, so a group of agents receive a particular message.

The token is passed from one network node to another according to the set of rules that have been defined, allowing the node with the token to access the network medium. At initialization, a token is generated by the master station and passed around the network once to ensure that all stations are functioning correctly.

A node may only transmit data when it is in possession of the token. A node must pass the token after it has transmitted an allotted number of frames. The procedure is as follows:

1. A logical ring is established that links all nodes connected to the wireless network, and the master control station creates a single token.

2. The token is passed from node to node around the ring.

3. If a node that is waiting to send a frame receives the token, it first sends its frame and then passes the token on to the next node.

The major error recovery tool in the network layer is the timer in the master station. The timer monitors the network to ensure that the token is not lost. If the token is not passed across the network in a certain period of time, a new token is generated at the master station and the control loop is effectively reset.

We assume we have a network of about 10 mobile agents (robots), which should be physically close enough to be able to talk directly to each other without the need for an intermediate station. It is also possible to include a base station (for example PC or workstation) as one of the network nodes.

6.2 Messages

Messages are transmitted in a frame structure, comprising the data to be sent plus a number of fields that carry specific information.

Frame structure:

- Start byte
  A specific bit pattern as the frame preamble.

- Message type
  Distinction between user data, system data, etc. (see below).

- Destination ID
  ID of the message's receiver agent or ID of a subgroup of agents or broadcast to all agents.

- Source ID
  ID of the sending agent.

- Next sender's ID
  ID of the next active agent in the token ring.

- Message length
  Number of bytes contained in message (may be 0).

- Message contents
  Actual message contents limited by message length (may be empty).

- Checksum
  Cyclic redundancy checksum (CRC) over whole frame; automatic error handling for bad checksum.

So each frame contains the id-numbers of the transmitter, receiver, and the next-in-line transmitter, together with the data being sent. The message type is used for a number of organizational functions. We have to distinguish three major groups of messages, which are:

1. Messages exchanged at application program level.

2. Messages exchanged to enable remote control at system level (i.e. keypresses, printing to LCD, driving motors).

3. System messages in relation to the wireless network structure that require immediate interpretation.

Distinguishing between the first two message types is required because the network has to serve more than one task. Application programs should be able to freely send messages to any other agent (or group of agents). Also, the network should support remote control or just monitor the behavior of an agent by sending messages, *transparent* to the application program.

This requires the maintenance of two separate receive buffers, one for user messages and one for system messages, plus one send buffer for each agent. That way it can be guaranteed that both application purposes (remote control and data communication between agents) can be executed transparently at the same time.

Message types:

- USER

  A message sent between agents.

- OS

  A message sent from the operating system, for example transparently monitoring an agent's behavior on a base station.

- TOKEN

  No message contents are supplied, i.e. this is an empty message.

- WILD CARD

  The current master node periodically sends a wild card message instead of enabling the next agent in the list for sending. Any new agent that wants to be included in the network structure has to wait for a wild card message.

- ADDNEW

  This is the reply message to the "wild card" of a new agent trying to connect to the network. With this message the new agent tells the current master its id-number.

- SYNC

  If the master has included a new agent in the list or excluded a non-responding agent from the list, it generates an updated list of active agents and broadcasts it to all agents.

6.3 Fault-Tolerant Self-Configuration

The token ring network is a symmetric network, so during normal operation there is no need for a master node. However, a master is required to generate the very first token and in case of a token loss due to a communication problem or an agent malfunction, a master is required to restart the token ring.

There is no need to have the master role fixed to any particular node; in fact the role of master can be assigned temporarily to any of the nodes, if one of the functions mentioned above has to be performed. Since there is no master node, the network has to self-configure at initialization time and whenever a problem arises.

When an agent is activated (i.e. a robot is being switched on), it knows only its own id-number, but not the total number of agents, nor the IDs of other agents, nor who the master is. Therefore, the first round of communication is performed only to find out the following:

- How many agents are communicating?

- What are their id-numbers?

- Who will be master to start the ring or to perform recovery?

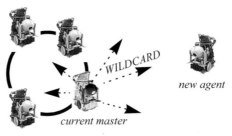

established network

Step1: Master broadcasts WILDCARD

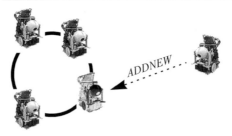

Step2: New agent replies with ADDNEW

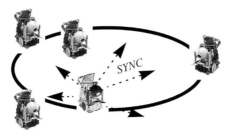

Step3: Master updates network with SYNC broadcast

Figure 6.2: Adding a node to the network

Agent IDs are used to trigger the initial communication. When the wireless network initialization is called for an agent, it first listens to any messages currently being sent. If no messages are picked up within a certain time interval multiplied by the node's own ID, the agent assumes it is the lowest ID and therefore becomes master.

The master will keep sending "wild card" messages at regular time intervals, allowing a new incoming robot to enter and participate in the ring. The master will receive the new agent's ID and assign it a position in the ring. All other agents will be notified about the change in total robot number and the new virtual ring neighborhood structure via a broadcast SYNC message (Figure 6.2). It is assumed that there is only a single new agent waiting at any time.

Each agent has an internal timer process. If, for a certain time, no communication takes place, it is assumed that the token has been lost, for example due to a communication error, an agent malfunction, or simply because one agent has been switched off. In the first case, the message can be repeated, but in the second case (if the token is repeatedly lost by the same agent), that particular agent has to be taken out of the ring.

If the master node is still active, that agent recreates a token and monitors the next round for repeated token loss at the same node. If that is the case, the faulty agent will be decommissioned and a broadcast message will update all agents about the reduced total number and the change in the virtual ring neighborhood structure.

If there is no reply after a certain time period, each agent has to assume that it is the previous master itself which became faulty. In this case, the previous master's successor in the virtual ring structure now takes over its role as master. However, if this process does not succeed after a certain time period, the network start-up process is repeated and a new master is negotiated.

For further reading on related projects in this area, refer to [Balch, Arkin 1995], [Fukuda, Sekiyama 1994], [MacLennan 1991] [Wang, Premvuti 1994], [Werner, Dyer 1990].

6.4 User Interface and Remote Control

As has been mentioned before, the wireless robot communication network *EyeNet* serves two purposes:

- message exchange between robots for application-specific purposes under user program control and

- monitoring and remote controlling one or several robots from a host workstation.

Both communication protocols are implemented as different message types using the same token ring structure, transparent to both the user program and the higher levels of the RoBIOS operating system itself.

In the following, we discuss the design and implementation of a multi-robot console that allows the monitoring of robot behavior, the robots' sensor readings, internal states, as well as current position and orientation in a shared environment. Remote control of individual robots, groups of robots, or all robots is also possible by transmitting input button data, which is interpreted by each robot's application program.

The *optional* host system is a workstation running Unix or Windows, accessing a serial port linked to a wireless module, identical to the ones on the robots. The workstation behaves logically exactly like one of the robots, which make up the other network nodes. That is, the workstation has a unique ID and also receives and sends tokens and messages. System messages from a robot to the host are transparent from the user program and update the robot display window and the robot environment window. All messages being sent in the entire network are monitored by the host without the need to explicitly send them to the host. The EyeBot Master Control Panel has entries for all active robots.

The host only then actively sends a message if a user intervention occurs, for example by pressing an input button in one of the robot's windows. This information will then be sent to the robot in question, once the token is passed to the host. The message is handled via low-level system routines on the robot, so for the upper levels of the operating system it cannot be distinguished whether the robot's physical button has been pressed or whether it was a remote command from the host console. Implementation of the host system largely reuses communication routines from the EyeCon and only makes a few system-dependent changes where necessary.

Remote control One particular application program using the wireless libraries for the Eye-Con and PC under Linux or Windows is the remote control program. A wireless network is established between a number of robots and a host PC, with the application "remote" being run on the PC and the "remote option" being activated on the robots (Hrd / Set / Rmt / On). The remote control can be operated via a serial cable (interface Serial1) or the wireless port (interface Serial2), provided that a wireless key has been entered on the EyeCon (`<I>` / `Reg`).

The remote control protocol runs as part of the wireless communication between all network nodes (robots and PC). However, as mentioned before, the network supports a number of different message types. So the remote control protocol can be run in addition to any inter-robot communication for any application. Switching remote control on or off will not affect the inter-robot communication.

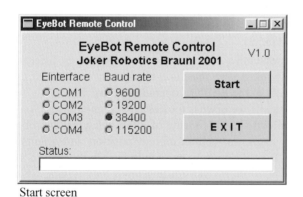
Start screen

Color image transmission

Figure 6.3: Remote control windows

Remote control operates in two directions, which can be enabled independently of each other. All LCD output of a robot is sent to the host PC, where it is displayed in the same way on an EyeCon console window. In the other direction, it is possible to press a button via a mouse-click on the host PC, and this signal is then sent to the appropriate robot, which reacts as if one of its physical buttons had been pressed (see Figure 6.3).

Another advantage of the remote control application is the fact that the host PC supports color, while current EyeCon LCDs are still monochrome for cost

Program 6.1: Wireless "ping" program for controller

```
1    #include "eyebot.h"
2
3    int main()
4    { BYTE myId, nextId, fromId;
5      BYTE mes[20]; /* message buffer */
6      int  len, err;
7
8      LCDPutString("Wireless Network");
9      LCDPutString("----------------");
10     LCDMenu(" "," "," ","END");
11
12     myId = OSMachineID();
13     if (myId==0) { LCDPutString("RadioLib not enabled!\n");
14                    return 1; }
15       else LCDPrintf("I am robot %d\n", myId);
16     switch(myId)
17     { case 1 : nextId = 2; break;
18       case 2 : nextId = 1; break;
19       default: LCDPutString("Set ID 1 or 2\n"); return 1;
20     }
21
22     LCDPutString("Radio");
23     err = RADIOInit();
24     if (err) {LCDPutString("Error Radio Init\n"); return 1;}
25       else LCDPutString("Init\n");
26
27     if (myId == 1)   /* robot 1 gets first to send */
28     { mes[0] = 0;
29       err = RADIOSend(nextId, 1, mes);
30       if (err) { LCDPutString("Error Send\n"); return 1; }
31     }
32
33     while ((KEYRead()) != KEY4)
34     { if (RADIOCheck())   /* check whether mess. is wait. */
35       { RADIORecv(&fromId, &len, mes);   /* wait for mess. */
36         LCDPrintf("Recv %d-%d: %3d\a\n", fromId,len,mes[0]);
37         mes[0]++; /* increment number and send again */
38         err = RADIOSend(nextId, 1, mes);
39         if (err) { LCDPutString("Error Send\n"); return 1; }
40       }
41     }
42     RADIOTerm();
43     return 0;
44   }
```

reasons. If a color image is being displayed on the EyeCon's LCD, the full or a reduced color information of the image is transmitted to and displayed on the host PC (depending on the remote control settings). This way, the processing of color data on the EyeCon can be tested and debugged much more easily.

An interesting extension of the remote control application would be including transmission of all robots' sensor and position data. That way, the movements of a group of robots could be tracked, similar to the simulation environment (see Chapter 13).

6.5 Sample Application Program

Program 6.1 shows a simple application of the wireless library functions. This program allows two EyeCons to communicate with each other by simply exchanging "pings", i.e. a new message is sent as soon as one is received. For reasons of simplicity, the program requires the participating robots' IDs to be 1 and 2, with number 1 starting the communication.

Program 6.2: Wireless host program

```
1   #include "remote.h"
2   #include "eyebot.h"
3
4   int main()
5   { BYTE myId, nextId, fromId;
6     BYTE mes[20]; /* message buffer */
7     int  len, err;
8     RadioIOParameters radioParams;
9
10      RADIOGetIoctl(&radioParams); /* get parameters */
11       radioParams.speed = SER38400;
12       radioParams.interface = SERIAL3; /* COM 3 */
13      RADIOSetIoctl(radioParams);  /* set parameters */
14
15      err = RADIOInit();
16      if (err) { printf("Error Radio Init\n"); return 1; }
17      nextId = 1; /* PC (id 0) will send to EyeBot no. 1 */
18
19      while (1)
20      { if (RADIOCheck())  /* check if message is waiting */
21         { RADIORecv(&fromId, &len, mes);  /* wait next mes. */
22           printf("Recv %d-%d: %3d\a\n", fromId, len, mes[0]);
23           mes[0]++;       /* increment number and send again */
24           err = RADIOSend(nextId, 1, mes);
25           if (err) { printf("Error Send\n"); return 1; }
26         }
27      }
28      RADIOTerm();
29      return 0;
30  }
```

Each EyeCon initializes the wireless communication by using "RADIO-Init", while EyeCon number 1 also sends the first message. In the subsequent while-loop, each EyeCon waits for a message, and then sends another message with a single integer number as contents, which is incremented for every data exchange.

In order to communicate between a host PC and an EyeCon, this example program does not have to be changed much. On the EyeCon side it is only required to adapt the different id-number (the host PC has 0 by default). The program for the host PC is listed in Program 6.2.

It can be seen that the host PC program looks almost identical to the EyeCon program. This has been accomplished by providing a similar EyeBot library for the Linux and Windows environment as for RoBIOS. That way, source programs for a PC host can be developed in the same way and in many cases even with identical source code as for robot application programs.

6.6 References

BALCH, T., ARKIN, R. *Communication in Reactive Multiagent Robotic Systems*, Autonomous Robots, vol. 1, 1995, pp. 27-52 (26)

BRÄUNL, T., WILKE, P. *Flexible Wireless Communication Network for Mobile Robot Agents*, Industrial Robot International Journal, vol. 28, no. 3, 2001, pp. 220-232 (13)

FUKUDA, F., SEKIYAMA, K. *Communication Reduction with Risk Estimate for Multiple Robotic Systems*, IEEE Proceedings of the Conference on Robotics and Automation, 1994, pp. 2864-2869 (6)

MACLENNAN, B. *Synthetic Ecology: An Approach to the Study of Communication*, in C. Langton, D. Farmer, C. Taylor (Eds.), Artificial Life II, Proceedings of the Workshop on Artificial Life, held Feb. 1990 in Santa Fe NM, Addison-Wesley, Reading MA, 1991

WANG, J., PREMVUTI, S. *Resource Sharing in Distributed Robotic Systems based on a Wireless Medium Access Protocol*, Proceedings of the IEEE/RSJ/GI, 1994, pp. 784-791 (8)

WERNER, G., DYER, M. *Evolution of Communication in Artificial Organisms*, Technical Report UCLA-AI-90-06, University of California at Los Angeles, June 1990

PART II:
MOBILE ROBOT DESIGN

DRIVING ROBOTS

7

Using two DC motors and two wheels is the easiest way to build a mobile robot. In this chapter we will discuss several designs such as differential drive, synchro-drive, and Ackermann steering. Omni-directional robot designs are dealt with in Chapter 8. A collection of related research papers can be found in [Rückert, Sitte, Witkowski 2001] and [Cho, Lee 2002]. Introductory textbooks are [Borenstein, Everett, Feng 1998], [Arkin 1998], [Jones, Flynn, Seiger 1999], and [McKerrow 1991].

7.1 Single Wheel Drive

Having a single wheel that is both driven and steered is the simplest conceptual design for a mobile robot. This design also requires two passive caster wheels in the back, since three contact points are always required.

Linear velocity and angular velocity of the robot are completely decoupled. So for driving straight, the front wheel is positioned in the middle position and driven at the desired speed. For driving in a curve, the wheel is positioned at an angle matching the desired curve.

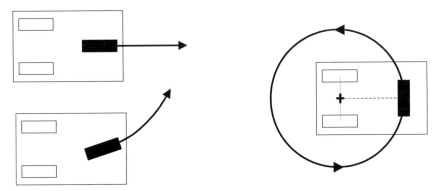

Figure 7.1: Driving and rotation of single wheel drive

Figure 7.1 shows the driving action for different steering settings. Curve driving is following the arc of a circle; however, this robot design cannot turn on the spot. With the front wheel set to 90° the robot will rotate about the midpoint between the two caster wheels (see Figure 7.1, right). So the minimum turning radius is the distance between the front wheel and midpoint of the back wheels.

7.2 Differential Drive

The differential drive design has two motors mounted in fixed positions on the left and right side of the robot, independently driving one wheel each. Since three ground contact points are necessary, this design requires one or two additional passive caster wheels or sliders, depending on the location of the driven wheels. Differential drive is mechanically simpler than the single wheel drive, because it does not require rotation of a driven axis. However, driving control for differential drive is more complex than for single wheel drive, because it requires the coordination of two driven wheels.

The minimal differential drive design with only a single passive wheel cannot have the driving wheels in the middle of the robot, for stability reasons. So when turning on the spot, the robot will rotate about the off-center midpoint between the two driven wheels. The design with two passive wheels or sliders, one each in the front and at the back of the robot, allows rotation about the center of the robot. However, this design can introduce surface contact problems, because it is using four contact points.

Figure 7.2 demonstrates the driving actions of a differential drive robot. If both motors run at the same speed, the robot drives straight forward or backward, if one motor is running faster than the other, the robot drives in a curve along the arc of a circle, and if both motors are run at the same speed in opposite directions, the robot turns on the spot.

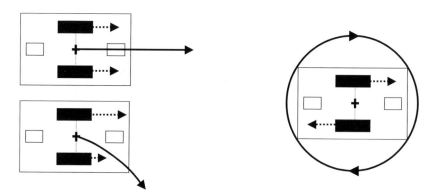

Figure 7.2: Driving and rotation of differential drive

- Driving straight, forward: $v_L = v_R$, $v_L > 0$
- Driving in a right curve: $v_L > v_R$, e.g. $v_L = 2 \cdot v_R$
- Turning on the spot, counter-clockwise: $v_L = -v_R$, $v_L > 0$

Eve   We have built a number of robots using a differential drive. The first one was the *EyeBot Vehicle*, or *Eve* for short. It carried an EyeBot controller (Figure 7.3) and had a custom shaped I/O board to match the robot outline – a design approach that was later dropped in favor of a standard versatile controller.

The robot has a differential drive actuator design, using two Faulhaber motors with encapsulated gearboxes and encapsulated encoders. The robot is equipped with a number of sensors, some of which are experimental setups:

- Shaft encoders (2 units)
- Infrared PSD (1-3 units)
- Infrared proximity sensors (7 units)
- Acoustic bump sensors (2 units)
- QuickCam digital grayscale or color camera (1 unit)

Figure 7.3: Eve

One of the novel ideas is the acoustic bumper, designed as an air-filled tube surrounding the robot chassis. Two microphones are attached to the tube ends. Any collision of the robot will result in an audible bump that can be registered by the microphones. Provided that the microphones can be polled fast enough or generate an interrupt and the bumper is acoustically sufficiently isolated from the rest of the chassis, it is possible to determine the point of impact from the time difference between the two microphone signals.

SoccerBot   Eve was constructed before robot soccer competitions became popular. As it turned out, Eve was about 1cm too wide, according to the RoboCup rules. As a consequence, we came up with a redesigned robot that qualified to compete in the robot soccer events RoboCup [Asada 1998] small size league and FIRA RoboSot [Cho, Lee 2002].

99

The robot has a narrower wheel base, which was accomplished by using gears and placing the motors side by side. Two servos are used as additional actuators, one for panning the camera and one for activating the ball kicking mechanism. Three PSDs are now used (to the left, front, and right), but no infrared proximity sensors or a bumper. However, it is possible to detect a collision by feedback from the driving routines without using any additional sensors (see function VWStalled in Appendix B.5.12).

Figure 7.4: SoccerBot

The digital color camera *EyeCam* is used on the SoccerBot, replacing the obsolete QuickCam. With an optional wireless communication module, the robots can send messages to each other or to a PC host system. The network software uses a Virtual Token Ring structure (see Chapter 6). It is self-organizing and does not require a specific master node.

A team of robots participated in both the RoboCup small size league and FIRA RoboSot. However, only RoboSot is a competition for autonomous mobile robots. The RoboCup small size league does allow the use of an overhead camera as a global sensor and remote computing on a central host system. Therefore, this event is more in the area of real-time image processing than robotics.

Figure 7.4 shows the current third generation of the SoccerBot design. It carries an EyeBot controller and EyeCam camera for on-board image processing and is powered by a lithium-ion rechargeable battery. This robot is commercially available from InroSoft [InroSoft 2006].

LabBot   For our robotics lab course we wanted a simpler and more robust version of the SoccerBot that does not have to comply with any size restrictions. LabBot was designed by going back to the simpler design of Eve, connecting the motors directly to the wheels without the need for gears or additional bearings.

The controller is again flat on the robot top and the two-part chassis can be opened to add sensors or actuators.

Getting away from robot soccer, we had one lab task in mind, namely to simulate foraging behavior. The robot should be able to detect colored cans, collect them, and bring them to a designated location. For this reason, LabBot does not have a kicker. Instead, we designed it with a circular bar in front (Figure 7.5) and equipped it with an electromagnet that can be switched on and off using one of the digital outputs.

Figure 7.5: LabBot with colored band for detection

The typical experiment on the lab course is to have one robot or even two competing robots drive in an enclosed environment and search and collect cans (Figure 7.6). Each robot has to avoid obstacles (walls and other robots) and use image processing to collect a can. The electromagnet has to be switched on after detection and close in on a can, and has to be switched off when the robot has reached the collection area, which also requires on-board localization.

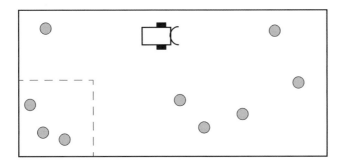

Figure 7.6: Can collection task

7.3 Tracked Robots

A tracked mobile robot can be seen as a special case of a wheeled robot with differential drive. In fact, the only difference is the robot's better maneuverability in rough terrain and its higher friction in turns, due to its tracks and multiple points of contact with the surface.

Figure 7.7 shows EyeTrack, a model snow truck that was modified into a mobile robot. As discussed in Section 7.2, a model car can be simply connected to an EyeBot controller by driving its speed controller and steering servo from the EyeBot instead of a remote control receiver. Normally, a tracked vehicle would have two driving motors, one for each track. In this particular model, however, because of cost reasons there is only a single driving motor plus a servo for steering, which brakes the left or right track.

Figure 7.7: EyeTrack robot and bottom view with sensors attached

EyeTrack is equipped with a number of sensors required for navigating rough terrain. Most of the sensors are mounted on the bottom of the robot. In Figure 7.7, right, the following are visible: top: PSD sensor; middle (left to right): digital compass, braking servo, electronic speed controller; bottom: gyroscope. The sensors used on this robot are:

- Digital color camera
  Like all our robots, EyeTrack is equipped with a camera. It is mounted in the "driver cabin" and can be steered in all three axes by using three servos. This allows the camera to be kept stable when combined with the robot's orientation sensors shown below. The camera will actively stay locked on to a desired target, while the robot chassis is driving over the terrain.

- Digital compass
  The compass allows the determination of the robot's orientation at all

times. This is especially important because this robot does not have two shaft encoders like a differential drive robot.

- Infrared PSDs
  The PSDs on this robot are not just applied to the front and sides in order to avoid obstacles. PSDs are also applied to the front and back at an angle of about 45°, to detect steep slopes that the robot can only descend/ascend at a very slow speed or not at all.

- Piezo gyroscopes
  Two gyroscopes are used to determine to robot's roll and pitch orientation, while yaw is covered by the digital compass. Since the gyroscopes' output is proportional to the rate of change, the data has to be integrated in order to determine the current orientation.

- Digital inclinometers
  Two inclinometers are used to support the two gyroscopes. The inclinometers used are fluid-based and return a value proportional to the robot's orientation. Although the inclinometer data does not require integration, there are problems with time lag and oscillation. The current approach uses a combination of both gyroscopes and inclinometers with sensor fusion in software to obtain better results.

There are numerous application scenarios for tracked robots with local intelligence. A very important one is the use as a "rescue robot" in disaster areas. For example, the robot could still be remote controlled and transmit a video image and sensor data; however, it might automatically adapt the speed according to its on-board orientation sensors, or even refuse to execute a driving command when its local sensors detect a potentially dangerous situation like a steep decline, which could lead to the loss of the robot.

7.4 Synchro-Drive

Synchro-drive is an extension to the robot design with a single driven and steered wheel. Here, however, we have three wheels that are all driven and all being steered. The three wheels are rotated together so they always point in the same driving direction (see Figure 7.8). This can be accomplished, for example, by using a single motor and a chain for steering and a single motor for driving all three wheels. Therefore, overall a synchro-drive robot still has only two degrees of freedom.

A synchro-drive robot is almost a holonomous vehicle, in the sense that it can drive in any desired direction (for this reason it usually has a cylindrical body shape). However, the robot has to stop and realign its wheels when going from driving forward to driving sideways. Nor can it drive and rotate at the same time. Truly holonomous vehicles are introduced in Chapter 8.

An example task that demonstrates the advantages of a synchro-drive is "complete area coverage" of a robot in a given environment. The real-world equivalent of this task is cleaning floors or vacuuming.

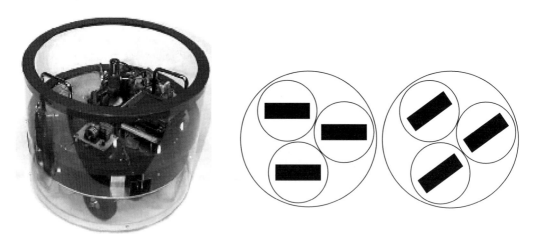

Figure 7.8: Xenia, University of Kaiserslautern, with schematic diagrams

   A behavior-based approach has been developed to perform a goal-oriented complete area coverage task, which has the potential to be the basis for a commercial floor cleaning application. The algorithm was tested in simulation first and thereafter ported to the synchro-drive robot Xenia for validation in a real environment. An inexpensive and easy-to-use external laser positioning system was developed to provide absolute position information for the robot. This helps to eliminate any positioning errors due to local sensing, for example through dead reckoning. By using a simple occupancy-grid representation without any compression, the robot can "clean" a 10m×10m area using less than 1MB of RAM. Figure 7.9 depicts the result of a typical run (without initial wall-following) in an area of 3.3m×2.3m. The photo in Figure 7.9 was taken with an overhead camera, which explains the cushion distortion. For details see [Kamon, Rivlin 1997], [Kasper, Fricke, von Puttkamer 1999], [Peters et al. 2000], and [Univ. Kaiserslautern 2003].

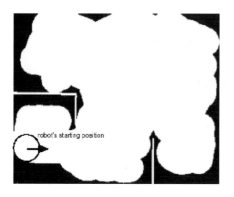
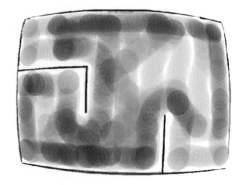

Figure 7.9: Result of a cleaning run, map and photo

7.5 Ackermann Steering

The standard drive and steering system of an automobile are two combined driven rear wheels and two combined steered front wheels. This is known as *Ackermann steering* and has a number of advantages and disadvantages when compared to differential drive:

+   Driving straight is not a problem, since the rear wheels are driven via a common axis.

−   Vehicle cannot turn on the spot, but requires a certain minimum radius.

−   Rear driving wheels experience slippage in curves.

Obviously, a different driving interface is required for Ackermann steering. Linear velocity and angular velocity are completely decoupled since they are generated by independent motors. This makes control a lot easier, especially the problem of driving straight. The driving library contains two independent velocity/position controllers, one for the rear driving wheels and one for the front steering wheels. The steering wheels require a position controller, since they need to be set to a particular angle as opposed to the velocity controller of the driving wheels, in order to maintain a constant rotational speed. An additional sensor is required to indicate the zero steering position for the front wheels.

Figure 7.10 shows the "Four Stooges" robot soccer team from The University of Auckland, which competed in the RoboCup Robot Soccer Worldcup. Each robot has a model car base and is equipped with an EyeBot controller and a digital camera as its only sensor.

Figure 7.10: The Four Stooges, University of Auckland

Model cars   Arguably, the cheapest way of building a mobile robot is to use a model car. We retain the chassis, motors, and servos, add a number of sensors, and replace the remote control receiver with an EyeBot controller. This gives us a ready-to-drive mobile robot in about an hour, as for the example in Figure 7.10.

The driving motor and steering servo of the model car are now directly connected to the controller and not to the receiver. However, we could retain the receiver and connect it to additional EyeBot inputs. This would allow us to transmit "high-level commands" to our controller from the car's remote control.

Model car with servo and speed controller Connecting a model car to an EyeBot is easy. Higher-quality model cars usually have proper servos for steering and either a servo or an electronic power controller for speed. Such a speed controller has the same connector and can be accessed exactly like a servo. Instead of plugging the steering servo and speed controller into the remote control unit, we plug them into two servo outputs on the EyeBot. That is all – the new autonomous vehicle is ready to go.

Driving control for steering and speed is achieved by using the command SERVOSet. One servo channel is used for setting the driving speed (–100 .. +100, fast backward .. stop .. fast forward), and one servo channel is used for setting the steering angle (–100 .. +100, full left .. straight .. full right).

Model car with integrated electronics The situation is a bit more complex for small, cheap model cars. These sometimes do not have proper servos, but for cost reasons contain a single electronic box that comprises receiver and motor controller in a single unit. This is still not a problem, since the EyeBot controller has two motor drivers already built in. We just connect the motors directly to the EyeBot DC motor drivers and read the steering sensor (usually a potentiometer) through an analog input. We can then program the software equivalent of a servo by having the EyeBot in the control loop for the steering motor.

Figure 7.11 shows the wiring details. The driving motor has two wires, which need to be connected to the pins Motor+ and Motor– of the "Motor A" connector of the EyeBot. The steering motor has five wires, two for the motor and three for the position feedback. The two motor wires need to be connected to Motor+ and Motor– of the EyeBot's "Motor B" connector. The connectors of the feedback potentiometer need to be connected to V_{CC} (5V) and Ground on the analog connector, while the slider of the potentiometer is connected to a free analog input pin. Note that some servos are only rated for 4.8V, while others are rated for 6.0V. This has to be observed, otherwise severe motor damage may be the consequence.

Driving such a model car is a bit more complex than in the servo case. We can use the library routine MOTORDrive for setting the linear speed of the driving motors. However, we need to implement a simple PID or bang-bang controller for the steering motor, using the analog input from the potentiometer as feedback, as described in Chapter 4.

The coding of the timing interrupt routine for a simple bang-bang controller is shown in Program 7.1. Routine IRQSteer needs to be attached to the timer interrupt and called 100 times per second in the background. This routine allows accurate setting of the steering angle between the values –100 and +100. However, most cheap model cars cannot position the steering that accurately, probably because of substandard potentiometers. In this case, a much

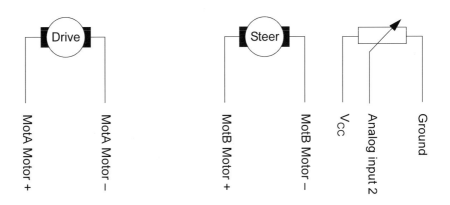

Figure 7.11: Model car connection diagram with pin numbers

reduced steering setting with only five or three values (left, straight, right) is sufficient.

Program 7.1: Model car steering control

```
 1   #include "eyebot.h"
 2   #define STEER_CHANNEL 2
 3   MotorHandle MSteer;
 4   int steer_angle; /* set by application program */
 5
 6   void IRQSteer()
 7   { int steer_current,ad_current;
 8     ad_current=OSGetAD(STEER_CHANNEL);
 9     steer_current=(ad_current-400)/3-100;
10     if (steer_angle-steer_current >  10)
11               MOTORDrive(MSteer,  75);
12       else if (steer_angle-steer_current < -10)
13               MOTORDrive(MSteer, -75);
14           else MOTORDrive(MSteer,   0);
15   }
```

7.6 Drive Kinematics

In order to obtain the vehicle's current trajectory, we need to constantly monitor both shaft encoders (for example for a vehicle with differential drive). Figure 7.12 shows the distance traveled by a robot with differential drive.

We know:

- r wheel radius
- d distance between driven wheels
- ticks_per_rev number of encoder ticks for one full wheel revolution
- $ticks_L$ number of ticks during measurement in left encoder
- $ticks_R$ number of ticks during measurement in right encoder

First we determine the values of s_L and s_R in meters, which are the distances traveled by the left and right wheel, respectively. Dividing the measured ticks by the number of ticks per revolution gives us the number of wheel revolutions. Multiplying this by the wheel circumference gives the traveled distance in meters:

$$s_L = 2\pi{\cdot}r \cdot ticks_L \, / \, ticks\_per\_rev$$
$$s_R = 2\pi{\cdot}r \cdot ticks_R \, / \, ticks\_per\_rev$$

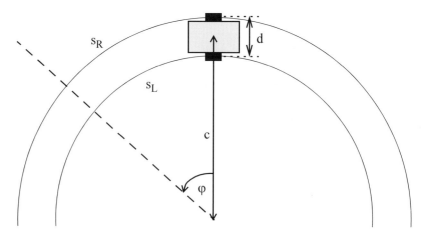

Figure 7.12: Trajectory calculation for differential drive

So we already know the distance the vehicle has traveled, i.e.:

$$s = (s_L + s_R) \, / \, 2$$

This formula works for a robot driving forward, backward, or turning on the spot. We still need to know the vehicle's rotation φ over the distance traveled. Assuming the vehicle follows a circular segment, we can define s_L and s_R as the traveled part of a full circle (φ in radians) multiplied by each wheel's turning radius. If the turning radius of the vehicle's center is c, then during a left turn the turning radius of the right wheel is c + d/2, while the turning radius of the left wheel is c − d/2. Both circles have the same center.

$$s_R = \varphi \cdot (c + d/2)$$
$$s_L = \varphi \cdot (c - d/2)$$

Subtracting both equations eliminates c:

$$s_R - s_L = \varphi \cdot d$$

And finally solving for φ:

$$\varphi = (s_R - s_L) \, / \, d$$

Using wheel velocities $v_{L,R}$ instead of driving distances $s_{L,R}$ and using $\dot{\theta}_{L,\,R}$ as wheel rotations per second with radius r for left and right wheel, we get:

$$v_R = 2\pi r \cdot \dot{\theta}_R$$
$$v_L = 2\pi r \cdot \dot{\theta}_L$$

Kinematics differential drive   The formula specifying the velocities of a differential drive vehicle can now be expressed as a matrix. This is called the forward kinematics:

$$\begin{bmatrix} v \\ \omega \end{bmatrix} = 2\pi r \begin{bmatrix} \dfrac{1}{2} & \dfrac{1}{2} \\[2mm] -\dfrac{1}{d} & \dfrac{1}{d} \end{bmatrix} \begin{bmatrix} \dot{\theta}_L \\ \dot{\theta}_R \end{bmatrix}$$

where:

     v is the vehicle's linear speed (equals ds/dt or $\dot{s}$),
     ω is the vehicle's rotational speed (equals $d\varphi/dt$ or $\dot{\varphi}$),
     $\dot{\theta}_{L, R}$ are the individual wheel speeds in revolutions per second,
     r is the wheel radius,
     d is the distance between the two wheels.

Inverse kinematics   The inverse kinematics is derived from the previous formula, solving for the individual wheel speeds. It tells us the required wheel speeds for a desired vehicle motion (linear and rotational speed). We can find the inverse kinematics by inverting the 2×2 matrix of the forward kinematics:

$$\begin{bmatrix} \dot{\theta}_L \\ \dot{\theta}_R \end{bmatrix} = \frac{1}{2\pi r} \begin{bmatrix} 1 & -\dfrac{d}{2} \\[2mm] 1 & \dfrac{d}{2} \end{bmatrix} \begin{bmatrix} v \\ \omega \end{bmatrix}$$

Kinematics Ackermann drive   If we consider the motion in a vehicle with Ackermann steering, then its front wheel motion is identical with the vehicle's forward motion s in the direction of the wheels. It is also easy to see (Figure 7.13) that the vehicle's overall forward and downward motion (resulting in its rotation) is given by:

     forward = $s \cdot \cos \alpha$
     down = $s \cdot \sin \alpha$

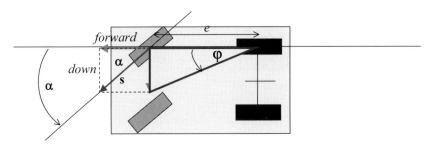

Figure 7.13: Motion of vehicle with Ackermann steering

If e denotes the distance between front and back wheels, then the overall vehicle rotation angle is $\varphi = $ down $/ e$ since the front wheels follow the arc of a circle when turning.

The calculation for the traveled distance and angle of a vehicle with Ackermann drive vehicle is shown in Figure 7.14, with:

α	steering angle,
e	distance between front and back wheels,
s_{front}	distance driven, measured at front wheels,
$\dot\theta$	driving wheel speed in revolutions per second,
s	total driven distance along arc,
φ	total vehicle rotation angle

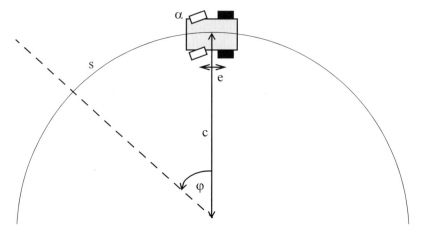

Figure 7.14: Trajectory calculation for Ackermann steering

The trigonometric relationship between the vehicle's steering angle and overall movement is:

$$s = s_{front}$$
$$\varphi = s_{front} \cdot \sin \alpha / e$$

Expressing this relationship as velocities, we get:

$$v_{forward} = v_{motor} = 2\pi r \cdot \dot\theta$$
$$\omega = v_{motor} \cdot \sin \alpha / e$$

Therefore, the kinematics formula becomes relatively simple:

$$\begin{bmatrix} v \\ \omega \end{bmatrix} = 2\pi r \cdot \dot\theta \cdot \begin{bmatrix} 1 \\ \dfrac{\sin \alpha}{e} \end{bmatrix}$$

Note that this formula changes if the vehicle is rear-wheel driven and the wheel velocity is measured there. In this case the *sin* function has to be replaced by the *tan* function.

7.7 References

ARKIN, R. *Behavior-Based Robotics*, MIT Press, Cambridge MA, 1998

ASADA, M., *RoboCup-98: Robot Soccer World Cup II*, Proceedings of the Second RoboCup Workshop, Paris, 1998

BORENSTEIN, J., EVERETT, H., FENG, L. *Navigating Mobile Robots: Sensors and Techniques*, AK Peters, Wellesley MA, 1998

CHO, H., LEE, J.-J. (Eds.) *Proceedings of the 2002 FIRA World Congress*, Seoul, Korea, May 2002

INROSOFT, http://inrosoft.com, 2006

JONES, J., FLYNN, A., SEIGER, B. *Mobile Robots - From Inspiration to Implementation*, 2nd Ed., AK Peters, Wellesley MA, 1999

KAMON, I., RIVLIN, E. *Sensory-Based Motion Planning with Global Proofs*, IEEE Transactions on Robotics and Automation, vol. 13, no. 6, Dec. 1997, pp. 814-822 (9)

KASPER, M. FRICKE, G. VON PUTTKAMER, E. *A Behavior-Based Architecture for Teaching More than Reactive Behaviors to Mobile Robots*, 3rd European Workshop on Advanced Mobile Robots, EUROBOT '99, Zürich, Switzerland, September 1999, IEEE Press, pp. 203-210 (8)

MCKERROW, P., *Introduction to Robotics*, Addison-Wesley, Reading MA, 1991

PETERS, F., KASPER, M., ESSLING, M., VON PUTTKAMER, E. *Flächendeckendes Explorieren und Navigieren in a priori unbekannter Umgebung mit low-cost Robotern*, 16. Fachgespräch Autonome Mobile Systeme AMS 2000, Karlsruhe, Germany, Nov. 2000

PUTTKAMER, E. VON. *Autonome Mobile Roboter*, Lecture notes, Univ. Kaiserslautern, Fachbereich Informatik, 2000

RÜCKERT, U., SITTE, J., WITKOWSKI, U. (Eds.) *Autonomous Minirobots for Research and Edutainment – AMiRE2001*, Proceedings of the 5th International Heinz Nixdorf Symposium, HNI-Verlagsschriftenreihe, no. 97, Univ. Paderborn, Oct. 2001

UNIV. KAISERSLAUTERN, http://ag-vp-www.informatik.uni-kl.de/ Research.English.html, 2003

OMNI-DIRECTIONAL ROBOTS

8

All the robots introduced in Chapter 7, with the exception of syncro-drive vehicles, have the same deficiency: they cannot drive in all possible directions. For this reason, these robots are called "non-holonomic". In contrast, a "holonomic" or omni-directional robot is capable of driving in any direction. Most non-holonomic robots cannot drive in a direction perpendicular to their driven wheels. For example, a differential drive robot can drive forward/backward, in a curve, or turn on the spot, but it cannot drive sideways. The omni-directional robots introduced in this chapter, however, are capable of driving in any direction in a 2D plane.

8.1 Mecanum Wheels

The marvel behind the omni-directional drive design presented in this chapter are Mecanum wheels. This wheel design has been developed and patented by the Swedish company Mecanum AB with Bengt Ilon in 1973 [Jonsson 1987], so it has been around for quite a while. Further details on Mecanum wheels and omni-directional drives can be found in [Carlisle 1983], [Agullo, Cardona, Vivancos 1987], and [Dickerson, Lapin 1991].

Figure 8.1: Mecanum wheel designs with rollers at 45°

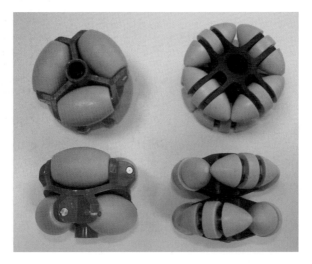

Figure 8.2: Mecanum wheel designs with rollers at 90°

There are a number of different Mecanum wheel variations; Figure 8.1 shows two of our designs. Each wheel's surface is covered with a number of *free rolling* cylinders. It is important to stress that the wheel hub is driven by a motor, but the rollers on the wheel surface are not. These are held in place by ball-bearings and can freely rotate about their axis. While the wheels in Figure 8.1 have the rollers at +/– 45° and there is a left-hand and a right-hand version of this wheel type, there are also Mecanum wheels with rollers set at 90° (Figure 8.2), and these do not require left-hand/right-hand versions.

A Mecanum-based robot can be constructed with either three or four independently driven Mecanum wheels. Vehicle designs with three Mecanum wheels require wheels with rollers set at 90° to the wheel axis, while the design we are following here is based on four Mecanum wheels and requires the rollers to be at an angle of 45° to the wheel axis. For the construction of a robot with four Mecanum wheels, two left-handed wheels (rollers at +45° to the wheel axis) and two right-handed wheels (rollers at –45° to the wheel axis) are required (see Figure 8.3).

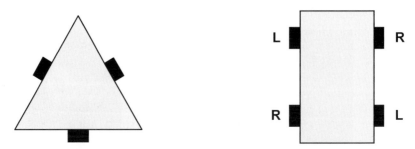

Figure 8.3: 3-wheel and 4-wheel omni-directional vehicles

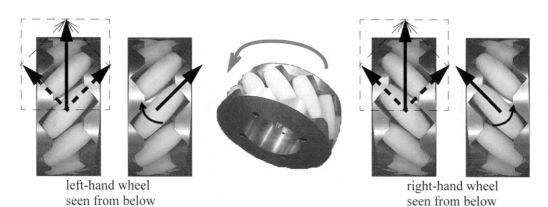

left-hand wheel
seen from below

right-hand wheel
seen from below

Figure 8.4: Mecanum principle, vector decomposition

Although the rollers are freely rotating, this does not mean the robot is spinning its wheels and not moving. This would only be the case if the rollers were placed parallel to the wheel axis. However, our Mecanum wheels have the rollers placed at an angle (45° in Figure 8.1). Looking at an individual wheel (Figure 8.4, view from the bottom through a "glass floor"), the force generated by the wheel rotation acts on the ground through the one roller that has ground contact. At this roller, the force can be split in a vector parallel to the roller axis and a vector perpendicular to the roller axis. The force perpendicular to the roller axis will result in a small roller rotation, while the force parallel to the roller axis will exert a force on the wheel and thereby on the vehicle.

Since Mecanum wheels do not appear individually, but e.g. in a four wheel assembly, the resulting wheel forces at 45° from each wheel have to be combined to determine the overall vehicle motion. If the two wheels shown in Figure 8.4 are the robot's front wheels and both are rotated forward, then each of the two resulting 45° force vectors can be split into a forward and a sideways force. The two forward forces add up, while the two sideways forces (one to the left and one to the right) cancel each other out.

8.2 Omni-Directional Drive

Figure 8.5, left, shows the situation for the full robot with four independently driven Mecanum wheels. In the same situation as before, i.e. all four wheels being driven forward, we now have four vectors pointing forward that are added up and four vectors pointing sideways, two to the left and two to the right, that cancel each other out. Therefore, although the vehicle's chassis is subjected to additional perpendicular forces, the vehicle will simply drive straight forward.

In Figure 8.5, right, assume wheels 1 and 4 are driven backward, and wheels 2 and 4 are driven forward. In this case, all forward/backward veloci-

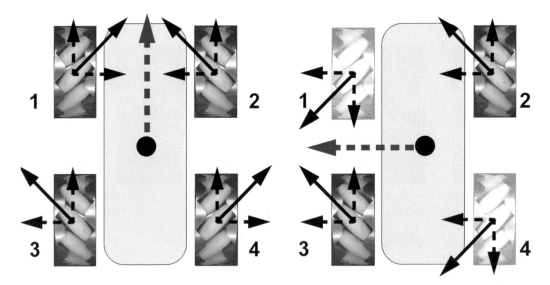

*Figure 8.5: Mecanum principle, driving forward and sliding sideways;
dark wheels rotate forward, bright wheels backward (seen from below)*

ties cancel each other out, but the four vector components to the left add up and let the vehicle slide to the left.

The third case is shown in Figure 8.6. No vector decomposition is necessary in this case to reveal the overall vehicle motion. It can be clearly seen that the robot motion will be a clockwise rotation about its center.

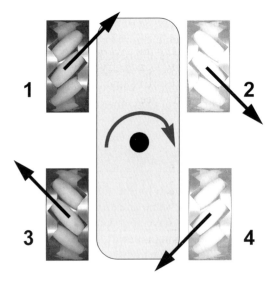

Figure 8.6: Mecanum principle, turning clockwise (seen from below)

The following list shows the basic motions, driving forward, driving sideways, and turning on the spot, with their corresponding wheel directions (see Figure 8.7).

- Driving forward: all four wheels forward
- Driving backward: all four wheels backward
- Sliding left: 1, 4: backward;   2, 3: forward
- Sliding right: 1, 4: forward;   2. 3: backward
- Turning clockwise on the spot: 1, 3: forward;   2, 4: backward
- Turning counter-clockwise: 1, 3: backward;   2, 4: forward

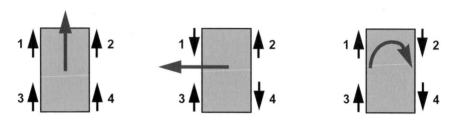

Figure 8.7: Kinematics of omni-directional robot

So far, we have only considered a Mecanum wheel spinning at full speed forward or backward. However, by varying the individual wheel speeds and by adding linear interpolations of basic movements, we can achieve driving directions along any vector in the 2D plane.

8.3 Kinematics

Forward kinematics The *forward kinematics* is a matrix formula that specifies which direction the robot will drive in (linear velocity v_x along the robot's center axis, v_y perpendicular to it) and what its rotational velocity ω will be for *given* individual wheel speeds $\dot{\theta}_{FL}$, .., $\dot{\theta}_{BR}$ and wheels distances d (left/right) and e (front/back):

$$
\begin{bmatrix} v_x \\ v_y \\ \omega \end{bmatrix} = 2\pi r \begin{bmatrix} \dfrac{1}{4} & \dfrac{1}{4} & \dfrac{1}{4} & \dfrac{1}{4} \\[2mm] -\dfrac{1}{4} & \dfrac{1}{4} & \dfrac{1}{4} & -\dfrac{1}{4} \\[2mm] -\dfrac{1}{2(d+e)} & \dfrac{1}{2(d+e)} & -\dfrac{1}{2(d+e)} & \dfrac{1}{2(d+e)} \end{bmatrix} \cdot \begin{bmatrix} \dot{\theta}_{FL} \\ \dot{\theta}_{FR} \\ \dot{\theta}_{BL} \\ \dot{\theta}_{BR} \end{bmatrix}
$$

with:

$\dot{\theta}_{FL}$, etc.   four individual wheel speeds in revolutions per second,

r   wheel radius,

d   distance between left and right wheel pairs,

e	distance between front and back wheel pairs,
v_x	vehicle velocity in forward direction,
v_y	vehicle velocity in sideways direction,
ω	vehicle rotational velocity.

Inverse
kinematics
    The *inverse kinematics* is a matrix formula that specifies the required individual wheel speeds for *given* desired linear and angular velocity (v_x, v_y, ω) and can be derived by inverting the matrix of the forward kinematics [Viboonchaicheep, Shimada, Kosaka 2003].

$$\begin{bmatrix} \dot{\theta}_{FL} \\ \dot{\theta}_{FR} \\ \dot{\theta}_{BL} \\ \dot{\theta}_{BR} \end{bmatrix} = \frac{1}{2\pi r} \begin{bmatrix} 1 & -1 & -(d+e)/2 \\ 1 & 1 & (d+e)/2 \\ 1 & 1 & -(d+e)/2 \\ 1 & -1 & (d+e)/2 \end{bmatrix} \cdot \begin{bmatrix} v_x \\ v_y \\ \omega \end{bmatrix}$$

8.4 Omni-Directional Robot Design

We have so far developed three different Mecanum-based omni-directional robots, the demonstrator models Omni-1 (Figure 8.8, left), Omni-2 (Figure 8.8, right), and the full size robot Omni-3 (Figure 8.9).

The first design, Omni-1, has the motor/wheel assembly tightly attached to the robot's chassis. Its Mecanum wheel design has rims that only leave a few millimeters clearance for the rollers. As a consequence, the robot can drive very well on hard surfaces, but it loses its omni-directional capabilities on softer surfaces like carpet. Here, the wheels will sink in a bit and the robot will then drive on the wheel rims, losing its capability to drive sideways.

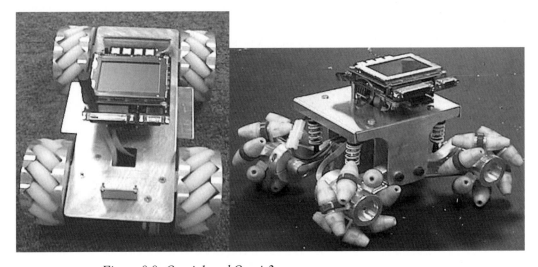

Figure 8.8: Omni-1 and Omni-2

The deficiencies of Omni-1 led to the development of Omni-2. This robot first of all has individual cantilever wheel suspensions with shock absorbers. This helps to navigate rougher terrain, since it will keep all wheels on the ground. Secondly, the robot has a completely rimless Mecanum wheel design, which avoids sinking in and allows omni-directional driving on softer surfaces.

Omni-3 uses a scaled-up version of the Mecanum wheels used for Omni-1 and has been constructed for a payload of 100kg. We used old wheelchair motors and an EyeBot controller with external power amplifiers as the onboard embedded system. The robot has been equipped with infrared sensors, wheel encoders and an emergency switch. Current projects with this robot include navigation and handling support systems for wheelchair-bound handicapped people.

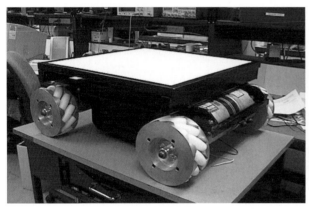 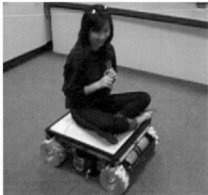

Figure 8.9: Omni-3

8.5 Driving Program

Operating the omni-directional robots obviously requires an extended driving interface. The vω routines for differential drive or Ackermann-steering robots are not sufficient, since we also need to specify a vector for the driving direction in addition to a possible rotation direction. Also, for an omni-directional robot it is possible to drive along a vector and rotate at the same time, which has to be reflected by the software interface. The extended library routines are:

Extending the
vω interface

```
int OMNIDriveStraight(VWHandle handle, meter distance,
    meterPerSec v, radians direction);

int OMNIDriveTurn(VWHandle handle, meter delta1,
    radians direction, radians delta_phi,
    meterPerSec v, radPerSec w);

int OMNITurnSpot(VWHandle handele, radians delta_phi,
    radPerSec w);
```

The code example in Program 8.1, however, does not use this high-level driving interface. Instead it shows as an example how to set individual wheel speeds to achieve the basic omni-directional driving actions: forward/backward, sideways, and turning on the spot.

Program 8.1: Omni-directional driving (excerpt)

```
 1   LCDPutString("Forward\n");
 2   MOTORDrive (motor_fl,  60);
 3   MOTORDrive (motor_fr,  60);
 4   MOTORDrive (motor_bl,  60);
 5   MOTORDrive (motor_br,  60);
 6   OSWait(300);
 7   LCDPutString("Reverse\n");
 8   MOTORDrive (motor_fl,-60);
 9   MOTORDrive (motor_fr,-60);
10   MOTORDrive (motor_bl,-60);
11   MOTORDrive (motor_br,-60);
12   OSWait(300);
13   LCDPutString("Slide-L\n");
14   MOTORDrive (motor_fl,-60);
15   MOTORDrive (motor_fr,  60);
16   MOTORDrive (motor_bl,  60);
17   MOTORDrive (motor_br,-60);
18   OSWait(300);
19   LCDPutString("Turn-Clock\n");
20   MOTORDrive (motor_fl,  60);
21   MOTORDrive (motor_fr,-60);
22   MOTORDrive (motor_bl,  60);
23   MOTORDrive (motor_br,-60);
24   OSWait(300);
```

8.6 References

AGULLO, J., CARDONA, S., VIVANCOS, J. *Kinematics of vehicles with directional sliding wheels*, Mechanical Machine Theory, vol. 22, 1987, pp. 295-301 (7)

CARLISLE, B. *An omni-directional mobile robot*, in B. Rooks (Ed.): Developments in Robotics 1983, IFS Publications, North-Holland, Amsterdam, 1983, pp. 79-87 (9)

DICKERSON, S., LAPIN, B. *Control of an omni-directional robotic vehicle with Mecanum wheels*, Proceedings of the National Telesystems Conference 1991, NTC'91, vol. 1, 1991, pp. 323-328 (6)

JONSSON, S. *New AGV with revolutionary movement*, in R. Hollier (Ed.), Automated Guided Vehicle Systems, IFS Publications, Bedford, 1987, pp. 345-353 (9)

VIBOONCHAICHEEP, P., SHIMADA, A., KOSAKA,Y. *Position rectification control for Mecanum wheeled omni-directional vehicles*, 29th Annual Conference of the IEEE Industrial Electronics Society, IECON'03, vol. 1, Nov. 2003, pp. 854-859 (6)

BALANCING ROBOTS

Balancing robots have recently gained popularity with the introduction of the commercial Segway vehicle [Segway 2006]; however, many similar vehicles have been developed before. Most balancing robots are based on the inverted pendulum principle and have either wheels or legs. They can be studied in their own right or as a precursor for biped walking robots (see Chapter 10), for example to experiment with individual sensors or actuators. Inverted pendulum models have been used as the basis of a number of bipedal walking strategies: [Caux, Mateo, Zapata 1998], [Kajita, Tani 1996], [Ogasawara, Kawaji 1999], and [Park, Kim 1998]. The dynamics can be constrained to two dimensions and the cost of producing an inverted pendulum robot is relatively low, since it has a minimal number of moving parts.

9.1 Simulation

A software simulation of a balancing robot is used as a tool for testing control strategies under known conditions of simulated sensor noise and accuracy. The model has been implemented as an ActiveX control, a software architecture that is designed to facilitate binary code reuse. Implementing the system model in this way means that we have a simple-to-use component providing a real-time visual representation of the system's state (Figure 9.1).

The system model driving the simulation can cope with alternative robot structures. For example, the effects of changing the robot's length or its weight structure by moving the position of the controller can be studied. These will impact on both the robot's center of mass and its moment of inertia.

Software simulation can be used to investigate techniques for control systems that balance inverted pendulums. The first method investigated was an adaptive control system, based on a backpropagation neural network, which learns to balance the simulation with feedback limited to a single failure signal when the robot falls over. Disadvantages of this approach include the requirement for a large number of training cycles before satisfactory performance is obtained. Additionally, once the network has been trained, it is not possible to

Figure 9.1: Simulation system

make quick manual changes to the operation of the controller. For these reasons, we selected a different control strategy for the physical robot.

An alternative approach is to use a simple PD control loop, of the form:

$$u(k) = [W] \cdot [X(k)]$$

where:

u(k)   Horizontal force applied by motors to the ground.

X(k)   k-th measurement of the system state.

W   Weight vector applied to measured robot state.

Tuning of the control loop was performed manually, using the software simulation to observe the effect of modifying loop parameters. This approach quickly yielded a satisfactory solution in the software model, and was selected for implementation on the physical robot.

9.2 Inverted Pendulum Robot

Inverted pendulum

The physical balancing robot is an inverted pendulum with two independently driven motors, to allow for balancing, as well as driving straight and turning (Figure 9.2). Tilt sensors, inclinometers, accelerometers, gyroscopes, and digital cameras are used for experimenting with this robot and are discussed below.

- Gyroscope (Hitec GY-130)
  This is a piezo-electric gyroscope designed for use in remote controlled vehicles, such as model helicopters. The gyroscope modifies a servo control signal by an amount proportional to its measure of angular velocity. Instead of using the gyro to control a servo, we read back the modified servo signal to obtain a measurement of angular velocity. An estimate of angular displacement is obtained by integrating the velocity signal over time.

Figure 9.2: BallyBot balancing robot

- Acceleration sensors (Analog Devices ADXL05)
  These sensors output an analog signal, proportional to the acceleration in the direction of the sensor's axis of sensitivity. Mounting two acceleration sensors at 90° angles means that we can measure the translational acceleration experienced by the sensors in the plane through which the robot moves. Since gravity provides a significant component of this acceleration, we are able to estimate the orientation of the robot.

- Inclinometer (Seika N3)
  An inclinometer is used to support the gyroscope. Although the inclinometer cannot be used alone because of its time lag, it can be used to reset the software integration of the gyroscope data when the robot is close to resting in an upright position.

- Digital camera (EyeCam C2)
  Experiments have been conducted in using an artificial horizon or, more generally, the optical flow of the visual field to determine the robot's trajectory and use this for balancing (see also Chapter 10).

Variable	Description	Sensor
x	Position	Shaft encoders
v	Velocity	Differentiated encoder reading
Θ	Angle	Integrated gyroscope reading
ω	Angular velocity	Gyroscope

Table 9.1: State variables

The PD control strategy selected for implementation on the physical robot requires the measurement of four state variables: $\{x, v, \Theta, \omega\}$, see Table 9.1.

An implementation relying on the gyroscope alone does not completely solve the problem of balancing the physical robot, remaining balanced on average for 5–15 seconds before falling over. This is an encouraging initial result, but it is still not a robust system. The system's balancing was greatly improved by adding an inclinometer to the robot. Although the robot was not able to balance with the inclinometer alone, because of inaccuracies and the time lag of the sensor, the combination of inclinometer and gyroscope proved to be the best solution. While the integrated data of the gyroscope gives accurate short-term orientation data, the inclinometer is used to recalibrate the robot's orientation value as well as the gyroscope's zero position at certain time intervals when the robot is moving at a low speed.

Gyro drift   A number of problems have been encountered with the sensors used. Over time, and especially in the first 15 minutes of operation, the observed "zero velocity" signal received from the gyroscope can deviate (Figure 9.3). This means that not only does our estimate of the angular velocity become inaccurate, but since our estimate of the angle is the integrated signal, it becomes inaccurate as well.

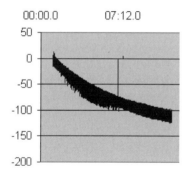

Figure 9.3: Measurement data revealing gyro drift

Motor force   The control system assumes that it is possible to accurately generate a horizontal force using the robot's motors. The force produced by the motors is related to the voltage applied, as well as the current shaft speed and friction. This relationship was experimentally determined and includes some simplification and generalization.

Wheel slippage   In certain situations, the robot needs to generate considerable horizontal force to maintain balance. On some surfaces this force can exceed the frictional force between the robot tires and the ground. When this happens, the robot loses track of its displacement, and the control loop no longer generates the correct output. This can be observed by sudden, unexpected changes in the robot displacement measurements.

Program 9.1 is an excerpt from the balancing program. It shows the periodic timer routine for reading sensor values and updating the system state. Details of this control approach are described in [Sutherland, Bräunl 2001] and [Sutherland, Bräunl 2002].

Program 9.1: Balance timer routine

```
1    void CGyro::TimerSample()
2    { ...
3     iAngVel = accreadX();
4     if (iAngVel > -1)
5     {
6      iAngVel = iAngVel;
7      // Get the elapsed time
8      iTimeNow = OSGetCount();
9      iElapsed = iTimeNow - g_iSampleTime;
10     // Correct elapsed time if rolled over!
11     if (iElapsed < 0) iElapsed += 0xFFFFFFFF; // ROLL OVER
12     // Correct the angular velocity
13     iAngVel -= g_iZeroVelocity;
14     // Calculate angular displacement
15     g_iAngle += (g_iAngularVelocity * iElapsed);
16     g_iAngularVelocity = -iAngVel;
17     g_iSampleTime = iTimeNow;
18     // Read inclinometer (drain residual values)
19     iRawADReading = OSGetAD(INCLINE_CHANNEL);
20     iRawADReading = OSGetAD(INCLINE_CHANNEL);
21     // If recording, and we have started...store data
22     if (g_iTimeLastCalibrated > 0)
23       { ... /* re-calibrate sensor */
24       }
25    }
26    // If correction factor remaining to apply, apply it!
27    if (g_iGyroAngleCorrection > 0)
28    { g_iGyroAngleCorrection -= g_iGyroAngleCorrectionDelta;
29      g_iAngle -= g_iGyroAngleCorrectionDelta;
30    }
31  }
```

Second balancing robot   A second two-wheel balancing robot had been built in a later project [Ooi 2003], Figure 9.4. Like the first robot it uses a gyroscope and inclinometer as sensors, but it employs a Kalman filter method for balancing [Kalman 1960], [Del Gobbo, Napolitano, Famouri, Innocenti 2001]. A number of Kalman-based control algorithms have been implemented and compared with each other, including a pole-placement controller and a Linear Quadratic Regulator (LQR) [Nakajima, Tsubouchi, Yuta, Koyanagi 1997], [Takahashi, Ishikawa, Hagiwara 2001]. An overview of the robot's control system from [Ooi 2003] is shown in Figure 9.5.

The robot also accepts driving commands from an infrared remote control, which are interpreted as a bias by the balance control system. They are used to drive the robot forward/backward or turn left/right on the spot.

Figure 9.4: Second balancing robot design

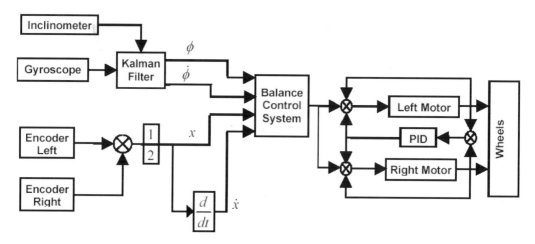

Figure 9.5: Kalman-based control system

9.3 Double Inverted Pendulum

Another design is taking the inverted pendulum approach one step further by replacing the two wheels with four independent leg joints. This gives us the equivalent of a double inverted pendulum; however, with two independent legs controlled by two motors each, we can do more than balancing – we can walk.

Dingo   The double inverted pendulum robot Dingo is very close to a walking robot, but its movements are constrained in a 2D plane. All sideways motions can be ignored, since the robot has long, bar-shaped feet, which it must lift over each other. Since each foot has only a minimal contact area with the ground, the robot has to be constantly in motion to maintain balance.

Figure 9.6 shows the robot schematics and the physical robot. The robot uses the same sensor equipment as BallyBot, namely an inclinometer and a gyroscope.

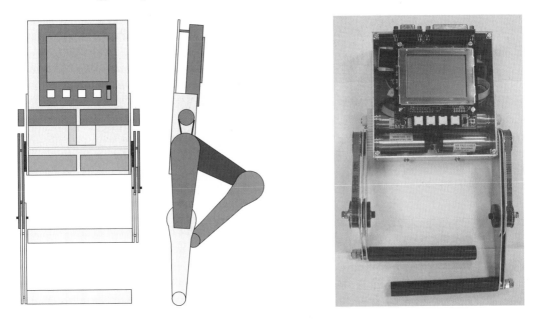

Figure 9.6: Double inverted pendulum robot

9.4 References

CAUX, S., MATEO, E., ZAPATA, R. *Balance of biped robots: special double-inverted pendulum*, IEEE International Conference on Systems, Man, and Cybernetics, 1998, pp. 3691-3696 (6)

DEL GOBBO, D., NAPOLITANO, M., FAMOURI, P., INNOCENTI, M., *Experimental application of extended Kalman filtering for sensor validation*, IEEE Transactions on Control Systems Technology, vol. 9, no. 2, 2001, pp. 376-380 (5)

KAJITA, S., TANI, K. *Experimental Study of Biped Dynamic Walking in the Linear Inverted Pendulum Mode*, IEEE Control Systems Magazine, vol. 16, no. 1, Feb. 1996, pp. 13-19 (7)

KALMAN R.E, *A New Approach to Linear Filtering and Prediction Problems*, Transactions of the ASME - Journal of Basic Engineering, Series D, vol. 82, 1960, pp. 35-45

NAKAJIMA, R., TSUBOUCHI, T., YUTA, S., KOYANAGI, E., *A Development of a New Mechanism of an Autonomous Unicycle*, IEEE International Conference on Intelligent Robots and Systems, IROS '97, vol. 2, 1997, pp. 906-912 (7)

OGASAWARA, K., KAWAJI, S. *Cooperative motion control for biped locomotion robots*, IEEE International Conference on Systems, Man, and Cybernetics, 1999, pp. 966-971 (6)

OOI, R., *Balancing a Two-Wheeled Autonomous Robot*, B.E. Honours Thesis, The Univ. of Western Australia, Mechanical Eng., supervised by T. Bräunl, 2003, pp. (56)

PARK, J.H., KIM, K.D. *Bipedal Robot Walking Using Gravity-Compensated Inverted Pendulum Mode and Computed Torque Control*, IEEE International Conference on Robotics and Automation, 1998, pp. 3528-3533 (6)

SEGWAY, *Welcome to the evolution in mobility*, http://www.segway.com, 2006

SUTHERLAND, A., BRÄUNL, T. *Learning to Balance an Unknown System*, Proceedings of the IEEE-RAS International Conference on Humanoid Robots, Humanoids 2001, Waseda University, Tokyo, Nov. 2001, pp. 385-391 (7)

SUTHERLAND, A., BRÄUNL, T. *An Experimental Platform for Researching Robot Balance*, 2002 FIRA Robot World Congress, Seoul, May 2002, pp. 14-19 (6)

TAKAHASHI, Y., ISHIKAWA, N., HAGIWARA, T. *Inverse pendulum controlled two wheel drive system*, Proceedings of the 40th SICE Annual Conference, International Session Papers, SICE 2001, 2001, pp. 112 -115 (4)

WALKING ROBOTS

W alking robots are an important alternative to driving robots, since the majority of the world's land area is unpaved. Although driving robots are more specialized and better adapted to flat surfaces – they can drive faster and navigate with higher precision – walking robots can be employed in more general environments. Walking robots follow nature by being able to navigate rough terrain, or even climb stairs or over obstacles in a standard household situation, which would rule out most driving robots.

Robots with six or more legs have the advantage of stability. In a typical walking pattern of a six-legged robot, three legs are on the ground at all times, while three legs are moving. This gives static balance while walking, provided the robot's center of mass is within the triangle formed by the three legs on the ground. Four-legged robots are considerably harder to balance, but are still fairly simple when compared to the dynamics of biped robots. Biped robots are the most difficult to balance, with only one leg on the ground and one leg in the air during walking. Static balance for biped robots can be achieved if the robot's feet are relatively large and the ground contact areas of both feet are overlapping. However, this is not the case in human-like "android" robots, which require dynamic balance for walking.

A collection of related research papers can be found in [Rückert, Sitte, Witkowski 2001] and [Cho, Lee 2002].

10.1 Six-Legged Robot Design

Figure 10.1 shows two different six-legged robot designs. The "Crab" robot was built from scratch, while "Hexapod" utilizes walking mechanics from Lynxmotion in combination with an EyeBot controller and additional sensors.

The two robots differ in their mechanical designs, which might not be recognized from the photos. Both robots are using two servos (see Section 3.5) per leg, to achieve leg lift (up/down) and leg swing (forward/backward) motion. However, Crab uses a mechanism that allows all servos to be firmly mounted on the robot's main chassis, while Hexapod only has the swing servos mounted to the robot body; the lift servos are mounted on small sub-assemblies, which are moved with each leg.

The second major difference is in sensor equipment. While Crab uses sonar sensors with a considerable amount of purpose-built electronics, Hexapod uses infrared PSD sensors for obstacle detection. These can be directly interfaced to the EyeBot without any additional electronic circuitry.

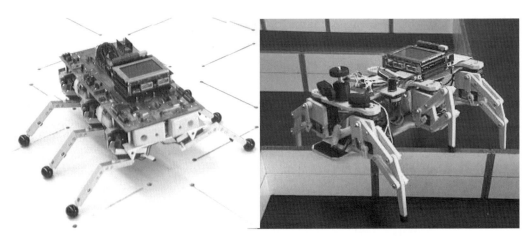

Figure 10.1: Crab six-legged walking robot, Univ. Stuttgart,
and Lynxmotion Hexapod base with EyeCon, Univ. Stuttgart

Program 10.1 shows a very simple program generating a walking pattern for a six-legged robot. Since the same EyeCon controller and the same RoBIOS operating system are used for driving and walking robots, the robot's HDT (Hardware Description Table) has to be adapted to match the robot's physical appearance with corresponding actuator and sensor equipment.

Data structures like `GaitForward` contain the actual positioning data for a gait. In this case it is six key frames for moving one full cycle for *all* legs. Function `gait` (see Program 10.2) then uses this data structure to "step through" these six individual key frame positions by subsequent calls of `move_joint`.

Function `move_joint` moves all the robot's 12 joints from one position to the next position using key frame averaging. For each of these iterations, new positions for all 12 leg joints are calculated and sent to the servos. Then a certain delay time is waited before the routine proceeds, in order to give the servos time to assume the specified positions.

Program 10.1: Six-legged gait settings

```
 1   #include "eyebot.h"
 2   ServoHandle servohandles[12];
 3   int semas[12]= {SER_LFUD, SER_LFFB, SER_RFUD, SER_RFFB,
 4                   SER_LMUD, SER_LMFB, SER_RMUD, SER_RMFB,
 5                   SER_LRUD, SER_LRFB, SER_RRUD, SER_RRFB};
 6   #define MAXF  50
 7   #define MAXU  60
 8   #define CNTR 128
 9   #define UP (CNTR+MAXU)
10   #define DN (CNTR-MAXU)
11   #define FW (CNTR-MAXF)
12   #define BK (CNTR+MAXF)
13   #define GaitForwardSize 6
14   int GaitForward[GaitForwardSize][12]= {
15    {DN,FW,  UP,BK,  UP,BK,  DN,FW,  DN,FW,  UP,BK},
16    {DN,FW,  DN,BK,  DN,BK,  DN,FW,  DN,FW,  DN,BK},
17    {UD,FW,  DN,BK,  DN,BK,  UP,FW,  UP,FW,  DN,BK},
18    {UP,BK,  DN,FW,  DN,FW,  UP,BK,  UP,BK,  DN,FW},
19    {DN,BK,  DN,FW,  DN,FW,  DN,BK,  DN,BK,  DN,FW},
20    {DN,BK,  UP,FW,  UP,FW,  DN,BK,  DN,BK,  UP,FW},
21   };
22   #define GaitTurnRightSize 6
23   int GaitRight[GaitTurnRightSize][12]= { ...};
24   #define GaitTurnLeftSize 6
25   int GaitLeft[GaitTurnLeftSize][12]= { ...};
26   int PosInit[12]=
27    {CT,CT,  CT,CT,  CT,CT,  CT,CT,  CT,CT,  CT,CT};
```

Program 10.2: Walking routines

```
 1   void move_joint(int pos1[12], int pos2[12], int speed)
 2   { int i, servo, steps = 50;
 3     float size[12];
 4     for (servo=0; servo<NumServos; servo++)
 5       size[servo] = (float) (pos2[servo]-pos1[servo]) /
 6                     (float) steps;
 7     for (i=0;i<steps;i++)
 8     { for(servo=0; servo<NumServos; servo++)
 9         SERVOSet(servohandles[servo], pos1[servo]+
10                     (int)((float) i *size[servo]));
11       OSWait(10/speed);
12     }
13   }
```

```
 1   void gait(int g[][12], int size, int speed)
 2   { int i;
 3     for (i=0; i<size; i++)
 4       move_joint(g[i], g[(i+1)%size], speed);
 5   }
```

10.2 Biped Robot Design

Finally, we get to robots that resemble what most people think of when hearing the term "robot". These are biped walking robots, often also called "humanoid robots" or "android robots" because of their resemblance to human beings.

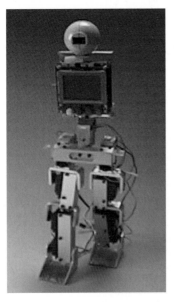 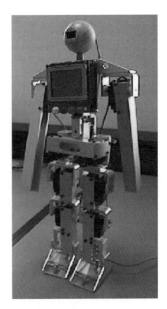

Figure 10.2: Johnny and Jack humanoid robots, UWA

Our first attempts at humanoid robot design were the two robots *Johnny Walker* and *Jack Daniels*, built in 1998 and named because of their struggle to maintain balance during walking (see Figure 10.2, [Nicholls 1998]). Our goal was humanoid robot design and control with limited funds. We used servos as actuators, linked in an aluminum U-profile. Although servos are very easily interfaced to the EyeBot controller and do not require an explicit feedback loop, it is exactly this feature (their built-in hardwired feedback loop and lack of external feedback) which causes most control problems. Without feedback sensors from the joints it is not possible to measure joint positions or joint torques.

These first-generation robots were equipped with foot switches (microswitches and binary infrared distance switches) and two-axes accelerometers in the hips. Like all of our other robots, both Johnny and Jack are completely autonomous robots, not requiring any umbilical cords or "remote brains". Each robot carries an EyeBot controller as on-board intelligence and a set of rechargeable batteries for power.

The mechanical structure of Johnny has nine degrees of freedom (*dof*), four per leg plus one in the torso. Jack has eleven *dof*, two more for its arms. Each of the robots has four *dof* per leg. Three servos are used to bend the leg at the ankle, knee, and hip joints, all in the same plane. One servo is used to turn the leg in the hip, in order to allow the robot to turn. Both robots have an additional *dof* for bending the torso sideways as a counterweight. Jack is also equipped with arms, a single *dof* per arm enabling it to swing its arms, either for balance or for touching any objects.

A second-generation humanoid robot is *Andy Droid*, developed by InroSoft (see Figure 10.3). This robot differs from the first-generation design in a number of ways [Bräunl, Sutherland, Unkelbach 2002]:

- Five *dof* per leg
  Allowing the robot to bend the leg and also to lean sideways.

- Lightweight design
  Using the minimum amount of aluminum and steel to reduce weight.

- Separate power supplies for controller and motors
  To eliminate incorrect sensor readings due to high currents and voltage fluctuations during walking.

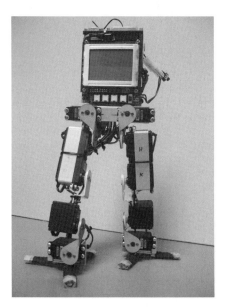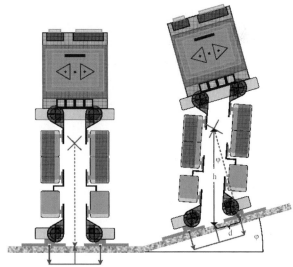

Figure 10.3: Andy Droid humanoid robot

Figure 10.3, left, shows Andy without its arms and head, but with its second generation foot design. Each foot consists of three adjustable toes, each equipped with a strain gauge. With this sensor feedback, the on-board controller can directly determine the robot's pressure point over each foot's support area and therefore immediately counteract to an imbalance or adjust the walking gait parameters (Figure 10.3, right, [Zimmermann 2004]).

Andy has a total of 13 *dof*, five per leg, one per arm, and one optional *dof* for the camera head. The robot's five *dof* per leg comprise three servos for bending the leg at the ankle, knee, and hips joints, all in the same plane (same as for Johnny). Two additional servos per leg are used to bend each leg sideways in the ankle and hip position, allowing the robot to lean sideways while keeping the torso level. There is no servo for turning a leg in the hip. The turning motion can still be achieved by different step lengths of left and right leg. An additional *dof* per arm allows swinging of the arms. Andy is 39cm tall and weighs approximately 1kg without batteries [Bräunl 2000], [Montgomery 2001], [Sutherland, Bräunl 2001], [Bräunl, Sutherland, Unkelbach 2002].

Digital servos   Andy 2 from InroSoft (Figure 10.4) is the successor robot of Andy Droid. Instead of analog servos it uses digital servos that serve as both actuators and sensors. These digital servos are connected via RS232 and are daisy-chained, so a single serial port is sufficient to control all servos. Instead of pulse width modulation (PWM) signals, the digital servos receive commands as an ASCII sequence, including the individual servo number or a broadcast command. This further reduces the load on the controller and generally simplifies operation. The digital servos can act as sensors, returning position and electrical current data when being sent the appropriate command sequence.

Figure 10.4: Andy 2 humanoid robot

Figure 10.5 visualizes feedback uploaded from the digital servos, showing the robot's joint positions and electrical currents (directly related to joint torque) during a walking gait [Harada 2006]. High current (torque) joints are color-coded, so problem areas like the robot's right hip servo in the figure can be detected.

A sample robot motion program without using any sensor feedback is shown in Program 10.3 (main) and Program 10.4 (move subroutine). This program for the Johnny/Jack servo arrangement demonstrates the robot's movements by letting it execute some *squat* exercises, continuously bending both knees and standing up straight again.

Figure 10.5: Visualization of servo sensor data

The main program comprises four steps:

- Initializing all servos.
- Setting all servos to the "up" position.
- Looping between "up" and "down" until keypress.
- Releasing all servos.

Robot configurations like "up" and "down" are stored as arrays with one integer value per *dof* of the robot (nine in this example). They are passed as parameters to the subroutine "move", which drives the robot servos to the desired positions by incrementally setting the servos to a number of intermediate positions.

Subroutine "move" uses local arrays for the current position (now) and the individual servo increment (diff). These values are calculated once. Then for a pre-determined constant number of steps, all servos are set to their next incremental position. An OSWait statement between loop iterations gives the servos some time to actually drive to their new positions.

Program 10.3: Robot gymnastics – main

```
1   int main()
2   { int i, delay=2;
3     typedef enum{rHipT,rHipB,rKnee,rAnkle, torso,
4                   lAnkle,lKnee,lHipB,lHipT}    link;
5     int up  [9] = {127,127,127,127,127,127,127,127,127};
6     int down[9] = {127, 80,200, 80,127,200, 80,200,127};
7
8     /* init servos */
9     serv[0]=SERVOInit(RHipT);  serv[1]=SERVOInit(RHipB);
10    serv[2]=SERVOInit(RKnee);  serv[3]=SERVOInit(RAnkle);
11    serv[4]=SERVOInit(Torso);
12    serv[5]=SERVOInit(LAnkle); serv[6]=SERVOInit(LKnee);
13    serv[7]=SERVOInit(LHipB);  serv[8]=SERVOInit(LHipT);
14    /* put servos in up position */
15    LCDPutString("Servos up-pos..\n");
16    for(i=0;i<9;i++) SERVOSet(serv[i],up[i]);
17    LCDMenu(""," "," ","END");
18
19    while (KEYRead() != KEY4) /* exercise until key press */
20    { move(up,down, delay);   /* move legs in bent pos.*/
21      move(down,up, delay);   /* move legs straight    */
22    }
23
24    /* release servo handles */
25    SERVORelease(RHipT);   SERVORelease(RHipB);
26    SERVORelease(RKnee);   SERVORelease(RAnkle);
27    SERVORelease(Torso);
28    SERVORelease(LAnkle);  SERVORelease(LKnee);
29    SERVORelease(LHipB);   SERVORelease(LHipT);
30    return 0;
31  }
```

Program 10.4: Robot gymnastics – move

```
1   void move(int old[], int new[], int delay)
2   { int i,j; /* using int constant STEPS */
3     float now[9], diff[9];
4
5     for (j=0; j<9; j++) /* update all servo positions */
6     { now[j]  = (float) old[j];
7       diff[j] = (float) (new[j]-old[j]) / (float) STEPS;
8     }
9     for (i=0; i<STEPS; i++) /* move servos to new pos.*/
10    { for (j=0; j<9; j++)
11      { now[j] +=  diff[j];
12        SERVOSet(serv[j], (int) now[j]);
13      }
14      OSWait(delay);
15    }
16  }
```

10.3 Sensors for Walking Robots

Sensor feedback is the foundation of dynamic balance and biped walking in general. This is why we put quite some emphasis on the selection of suitable sensors and their use in controlling a robot. On the other hand, there are a number of biped robot developments made elsewhere, which do not use any sensors at all and rely solely on a stable mechanical design with a large support area for balancing and walking.

Our humanoid robot design, *Andy*, uses the following sensors for balancing:

- Infrared proximity sensors in feet
  Using two sensors per foot, these give feedback on whether the heel or the toe has contact with the ground.

- Strain gauges in feet
  Each foot comprises three toes of variable length with exactly one contact point. This allows experimenting with different foot sizes. One strain gauge per toe allows calculation of foot torques including the "zero moment point" (see Section 10.5).

- Acceleration sensors
  Using acceleration sensors in two axes to measure dynamic forces on the robot, in order to balance it.

- Piezo gyroscopes
  Two gyroscopes are used as an alternative to acceleration sensors, which are subject to high-frequency servo noise. Since gyroscopes only return the change in acceleration, their values have to be integrated to maintain the overall orientation.

- Inclinometer
  Two inclinometers are used to support the gyroscopes. Although inclinometers cannot be used alone because of their time lag, they can be used to eliminate sensor drift and integration errors of the gyroscopes.

In addition, Andy uses the following sensors for navigation:

- Infrared PSDs
  With these sensors placed on the robot's hips in the directions forward, left, and right, the robot can sense surrounding obstacles.

- Digital camera
  The camera in the robot's head can be used in two ways, either to support balancing (see "artificial horizon" approach in Section 10.5) or for detecting objects, walking paths, etc.

Acceleration sensors for two axes are the main sensors for balancing and walking. These sensors return values depending on the current acceleration in one of two axes, depending on the mounting angle. When a robot is moving only slowly, the sensor readings correspond to the robot's relative attitude, i.e. leaning forward/backward or left/right. Sensor input from the infrared sensors

139

and acceleration sensors is used as feedback for the control algorithm for balancing and walking.

10.4 Static Balance

There are two types of walking, using:

- Static balance
  The robot's center of mass is at all times within the support area of its foot on the ground – or the combined support area of its two feet (convex hull), if both feet are on the ground.

- Dynamic balance
  The robot's center of mass may be outside the support area of its feet during a phase of its gait.

In this section we will concentrate on static balance, while dynamic balance is the topic of the following section.

Our approach is to start with a semi-stable pre-programmed, but parameterized gait for a humanoid robot. Gait parameters are:

1. Step length
2. Height of leg lift
3. Walking speed
4. Leaning angle of torso in forward direction (constant)
5. Maximal leaning angle of torso sideways (variable, in sync with gait)

We will then update the gait parameters in real time depending on the robot's sensors. Current sensor readings are compared with desired sensor readings at each time point in the gait. Differences between current and desired sensor readings will result in immediate parameter adaptations to the gait pattern. In order to get the right model parameters, we are conducting experiments with the BallyBot balancing robot as a testbed for acceleration, inclination, and gyro sensors (see Chapter 9).

We constructed a gait generation tool [Nicholls 1998], which is being used to generate gait sequences off-line, that can subsequently be downloaded to the robot. This tool allows the independent setting of each *dof* for each time step and graphically displays the robot's attitude in three orthogonal views from the major axes (Figure 10.6). The gait generation tool also allows the playback of an entered gait sequence. However, it does not perform any analysis of the mechanics for the viability of a gait.

The first step toward walking is to achieve static balance. For this, we have the robot standing still, but use the acceleration sensors as a feedback with a software PI controller to the two hip joints. The robot is now actively standing straight. If pushed back, it will bend forward to counterbalance, and vice versa. Solving this isolated problem is similar to the inverted pendulum problem.

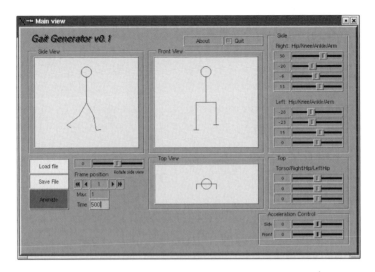

Figure 10.6: Gait generation tool

Unfortunately, typical sensor data is not as clean as one would like. Figure 10.7 shows typical sensor readings from an inclinometer and foot switches for a walking experiment [Unkelbach 2002]:

- The top curve shows the inclinometer data for the torso's side swing. The measured data is the absolute angle and does not require integration like the gyroscope.

- The two curves on the bottom show the foot switches for the right and left foot. First both feet are on the ground, then the left foot is lifted up and down, then the right foot is lifted up and down, and so on.

Program 10.5 demonstrates the use of sensor feedback for balancing a standing biped robot. In this example we control only a single axis (here forward/backward); however, the program could easily be extended to balance left/right as well as forward/backward by including a second sensor with another PID controller in the control loop.

The program's endless while-loop starts with the reading of a new acceleration sensor value in the forward/backward direction. For a robot at rest, this value should be zero. Therefore, we can treat this value directly as an error value for our PID controller. The PID controller used is the simplest possible. For the integral component, a number (e.g. 10) of previous error values should be added. In the example we only use two: the last plus the current error value. The derivative part uses the difference between the previous and current error value. All PID parameter values have to be determined experimentally.

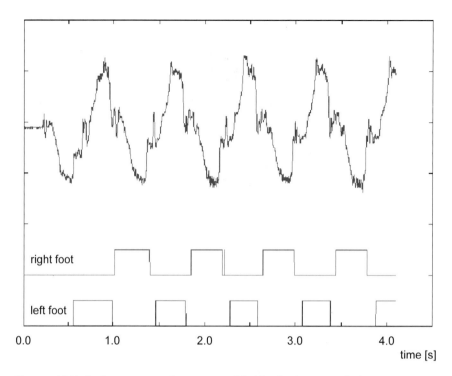

Figure 10.7: Inclinometer side swing and left/right foot switch data

Program 10.5: Balancing a biped robot

```
1   void balance( void ) /* balance forward/backward */
2   { int posture[9]= {127,127,127,127,127,127,127,127,127};
3     float err, lastErr =0.0, adjustment;
4     /* PID controller constants */
5     float kP = 0.1, kI = 0.10, kD = 0.05, time = 0.1;
6     int i, delay = 1;
7
8     while (1)   /* endless loop */
9     { /* read derivative. sensor signal = error */
10      err = GetAccFB();
11      /* adjust hip angles using PID */
12      adjustment =   kP*(err)
13                   + kI*(lastErr + err)*time
14                   + kD*(lastErr - err);
15      posture[lHipB] += adjustment;
16      posture[rHipB] += adjustment;
17      SetPosture(posture);
18      lastErr = err;
19      OSWait(delay);
20    }
21  }
```

10.5 Dynamic Balance

Walking gait patterns relying on static balance are not very efficient. They require large foot areas and only relatively slow gaits are possible, in order to keep dynamic forces low. Walking mechanisms with dynamic balance, on the other hand, allow the construction of robots with smaller feet, even feet that only have a single contact point, and can be used for much faster walking gaits or even running.

As has been defined in the previous section, dynamic balance means that at least during some phases of a robot's gait, its center of mass is not supported by its foot area. Ignoring any dynamic forces and moments, this means that the robot would fall over if no counteraction is taken in real time. There are a number of different approaches to dynamic walking, which are discussed in the following.

10.5.1 Dynamic Walking Methods

In this section, we will discuss a number of different techniques for dynamic walking together with their sensor requirements.

1. **Zero moment point (ZMP)**
   [Fujimoto, Kawamura 1998], [Goddard, Zheng, Hemami 1992], [Kajita, Yamaura, Kobayashi 1992], [Takanishi et al. 1985]
   This is one of the standard methods for dynamic balance and is published in a number of articles. The implementation of this method requires the knowledge of all dynamic forces on the robot's body plus all torques between the robot's foot and ankle. This data can be determined by using accelerometers or gyroscopes on the robot's body plus pressure sensors in the robot's feet or torque sensors in the robot's ankles.

   With all contact forces and all dynamic forces on the robot known, it is possible to calculate the "zero moment point" (ZMP), which is the dynamic equivalent to the static center of mass. If the ZMP lies within the support area of the robot's foot (or both feet) on the ground, then the robot is in dynamic balance. Otherwise, corrective action has to be taken by changing the robot's body posture to avoid it falling over.

2. **Inverted pendulum**
   [Caux, Mateo, Zapata 1998], [Park, Kim 1998], [Sutherland, Bräunl 2001]
   A biped walking robot can be modeled as an inverted pendulum (see also the balancing robot in Chapter 9). Dynamic balance can be achieved by constantly monitoring the robot's acceleration and adapting the corresponding leg movements.

3. **Neural networks**
   [Miller 1994], [Doerschuk, Nguyen, Li 1995], [Kun, Miller 1996]
   As for a number of other control problems, neural networks can be used to

achieve dynamic balance. Of course, this approach still needs all the sensor feedback as in the other approaches.

4. **Genetic algorithms**
[Boeing, Bräunl 2002], [Boeing, Bräunl 2003]
A population of virtual robots is generated with initially random control settings. The best performing robots are reproduced using genetic algorithms for the next generation.

This approach in practice requires a mechanics simulation system to evaluate each individual robot's performance and even then requires several CPU-days to evolve a good walking performance. The major issue here is the transferability of the simulation results back to the physical robot.

5. **PID control**
[Bräunl 2000], [Bräunl, Sutherland, Unkelbach 2002]
Classic PID control is used to control the robot's leaning front/back and left/right, similar to the case of static balance. However, here we do not intend to make the robot stand up straight. Instead, in a teaching stage, we record the *desired front and side lean* of the robot's body during all phases of its gait. Later, when controlling the walking gait, we try to achieve this *offset* of front and side lean by using a PID controller. The following parameters can be set in a standard walking gate to achieve this leaning:

- Step length
- Height of leg lift
- Walking speed
- Amount of forward lean of torso
- Maximal amount of side swing

6. **Fuzzy control**
[Unpublished]
We are working on an adaptation of the PID control, replacing the classic PID control by fuzzy logic for dynamic balance.

7. **Artificial horizon**
[Wicke 2001]
This innovative approach does not use any of the kinetics sensors of the other approaches, but a monocular grayscale camera. In the simple version, a black line on white ground (an "artificial horizon") is placed in the visual field of the robot. We can then measure the robot's orientation by changes of the line's position and orientation in the image. For example, the line will move to the top if the robot is falling forward, it will be slanted at an angle if the robot is leaning left, and so on (Figure 10.8).

With a more powerful controller for image processing, the same principle can be applied even without the need for an artificial horizon. As long as there is enough texture in the background, general optical flow can be used to determine the robot's movements.

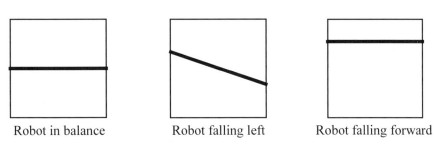

Robot in balance	Robot falling left	Robot falling forward

Figure 10.8: Artificial horizon

Figure 10.9 shows Johnny Walker during a walking cycle. Note the typical side-swing of the torso to counterbalance the leg-lifting movement. This creates a large momentum around the robot's center of mass, which can cause problems with stability due to the limited accuracy of the servos used as actuators.

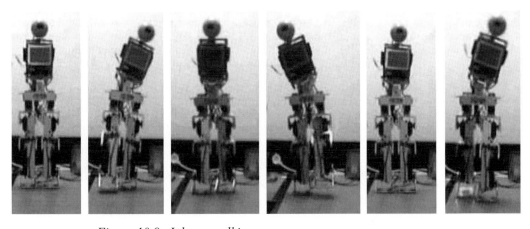

Figure 10.9: Johnny walking sequence

Figure 10.10 shows a similar walking sequence with Andy Droid. Here, the robot performs a much smoother and better controlled walking gait, since the mechanical design of the hip area allows a smoother shift of weight toward the side than in Johnny's case.

10.5.2 Alternative Biped Designs

All the biped robots we have discussed so far are using servos as actuators. This allows an efficient mechanical and electronic design of a robot and therefore is a frequent design approach in many research groups, as can be seen from the group photo of FIRA HuroSot World Cup Competition in 2002 [Baltes, Bräunl 2002]. With the exception of one robot, all robots were using servos (see Figure 10.11).

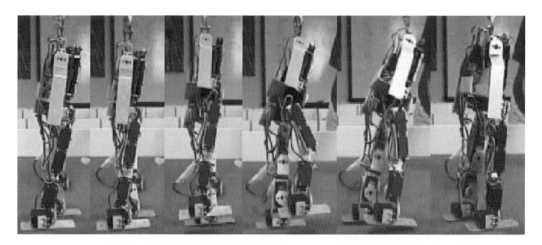

Figure 10.10: Andy walking sequence

*Figure 10.11: Humanoid robots at FIRA HuroSot 2002 with robots from
(left to right): Korea, Australia, Singapore, New Zealand, and Korea*

Other biped robot designs also using the EyeCon controller are *Tao Pie Pie*
from University of Auckland, New Zealand, and University of Manitoba, Canada, [Lam, Baltes 2002] and *ZORC* from Universität Dortmund, Germany
[Ziegler et al. 2001].

As has been mentioned before, servos have severe disadvantages for a
number of reasons, most importantly because of their lack of external feedback. The construction of a biped robot with DC motors, encoders, and end-
switches, however, is much more expensive, requires additional motor driver
electronics, and is considerably more demanding in software development. So
instead of redesigning a biped robot by replacing servos with DC motors and
keeping the same number of degrees of freedom, we decided to go for a minimal approach. Although Andy has 10 *dof* in both legs, it utilizes only three

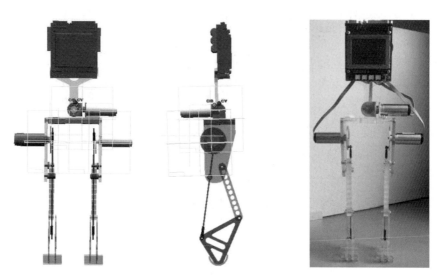

Figure 10.12: Minimal biped design Rock Steady

independent *dof*: bending each leg up and down, and leaning the whole body left or right. Therefore, it should be possible to build a robot that uses only three motors and uses mechanical gears or pulleys to achieve the articulated joint motion.

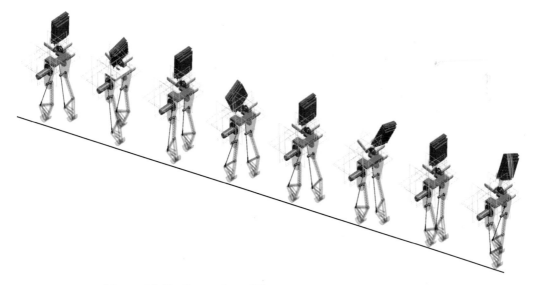

Figure 10.13: Dynamic walking sequence

The CAD designs following this approach and the finished robot are shown in Figure 10.12 [Jungpakdee 2002]. Each leg is driven by only one motor, while the mechanical arrangement lets the foot perform an ellipsoid curve for each motor revolution. The feet are only point contacts, so the robot has to keep moving continuously, in order to maintain dynamic balance. Only one motor is used for shifting a counterweight in the robot's torso sideways (the original drawing in Figure 10.12 specified two motors). Figure 10.13 shows the simulation of a dynamic walking sequence [Jungpakdee 2002].

10.6 References

BALTES. J., BRÄUNL, T. *HuroSot - Laws of the Game*, FIRA 1st Humanoid Robot Soccer Workshop (HuroSot), Daejeon Korea, Jan. 2002, pp. 43-68 (26)

BOEING, A., BRÄUNL, T. *Evolving Splines: An alternative locomotion controller for a bipedal robot*, Seventh International Conference on Control, Automation, Robotics and Vision, ICARV 2002, CD-ROM, Singapore, Dec. 2002, pp. 1-5 (5)

BOEING, A., BRÄUNL, T. *Evolving a Controller for Bipedal Locomotion*, Proceedings of the Second International Symposium on Autonomous Minirobots for Research and Edutainment, AMiRE 2003, Brisbane, Feb. 2003, pp. 43-52 (10)

BRÄUNL, T. *Design of Low-Cost Android Robots*, Proceedings of the First IEEE-RAS International Conference on Humanoid Robots, Humanoids 2000, MIT, Boston, Sept. 2000, pp. 1-6 (6)

BRÄUNL, T., SUTHERLAND, A., UNKELBACH, A. *Dynamic Balancing of a Humanoid Robot*, FIRA 1st Humanoid Robot Soccer Workshop (HuroSot), Daejeon Korea, Jan. 2002, pp. 19-23 (5)

CAUX, S., MATEO, E., ZAPATA, R. *Balance of biped robots: special double-inverted pendulum*, IEEE International Conference on Systems, Man, and Cybernetics, 1998, pp. 3691-3696 (6)

CHO, H., LEE, J.-J. (Eds.) *Proceedings 2002 FIRA Robot World Congress*, Seoul, Korea, May 2002

DOERSCHUK, P., NGUYEN, V., LI, A. *Neural network control of a three-link leg*, in Proceedings of the International Conference on Tools with Artificial Intelligence, 1995, pp. 278-281 (4)

FUJIMOTO, Y., KAWAMURA, A. *Simulation of an autonomous biped walking robot including environmental force interaction*, IEEE Robotics and Automation Magazine, June 1998, pp. 33-42 (10)

GODDARD, R., ZHENG, Y., HEMAMI, H. *Control of the heel-off to toe-off motion of a dynamic biped gait*, IEEE Transactions on Systems, Man, and Cybernetics, vol. 22, no. 1, 1992, pp. 92-102 (11)

HARADA, H. *Andy-2 Visualization Video*, http://robotics.ee.uwa.edu.au/eyebot/mpg/walk-2leg/, 2006

JUNGPAKDEE, K., *Design and construction of a minimal biped walking mechanism*, B.E. Honours Thesis, The Univ. of Western Australia, Dept. of Mechanical Eng., supervised by T. Bräunl and K. Miller, 2002

KAJITA, S., YAMAURA, T., KOBAYASHI, A. *Dynamic walking control of a biped robot along a potential energy conserving orbit*, IEEE Transactions on Robotics and Automation, Aug. 1992, pp. 431-438 (8)

KUN, A., MILLER III, W. *Adaptive dynamic balance of a biped using neural networks*, in Proceedings of the 1996 IEEE International Conference on Robotics and Automation, Apr. 1996, pp. 240-245 (6)

LAM, P., BALTES, J. *Development of Walking Gaits for a Small Humanoid Robot*, Proceedings 2002 FIRA Robot World Congress, Seoul, Korea, May 2002, pp. 694-697 (4)

MILLER III, W. *Real-time neural network control of a biped walking robot*, IEEE Control Systems, Feb. 1994, pp. 41-48 (8)

MONTGOMERY, G. *Robo Crop - Inside our AI Labs*, Australian Personal Computer, Issue 274, Oct. 2001, pp. 80-92 (13)

NICHOLLS, E. *Bipedal Dynamic Walking in Robotics*, B.E. Honours Thesis, The Univ. of Western Australia, Electrical and Computer Eng., supervised by T. Bräunl, 1998

PARK, J.H., KIM, K.D. *Bipedal Robot Walking Using Gravity-Compensated Inverted Pendulum Mode and Computed Torque Control*, IEEE International Conference on Robotics and Automation, 1998, pp. 3528-3533 (6)

RÜCKERT, U., SITTE, J., WITKOWSKI, U. (Eds.) *Autonomous Minirobots for Research and Edutainment – AMiRE2001*, Proceedings of the 5th International Heinz Nixdorf Symposium, HNI-Verlagsschriftenreihe, no. 97, Univ. Paderborn, Oct. 2001

SUTHERLAND, A., BRÄUNL, T. *Learning to Balance an Unknown System*, Proceedings of the IEEE-RAS International Conference on Humanoid Robots, Humanoids 2001, Waseda University, Tokyo, Nov. 2001, pp. 385-391 (7)

TAKANISHI, A., ISHIDA, M., YAMAZAKI, Y., KATO, I. *The realization of dynamic walking by the biped walking robot WL-10RD*, in *ICAR'85*, 1985, pp. 459-466 (8)

UNKELBACH, A. *Analysis of sensor data for balancing and walking of a biped robot*, Project Thesis, Univ. Kaiserslautern / The Univ. of Western Australia, supervised by T. Bräunl and D. Henrich, 2002

WICKE, M. *Bipedal Walking*, Project Thesis, Univ. Kaiserslautern / The Univ. of Western Australia, supervised by T. Bräunl, M. Kasper, and E. von Puttkamer, 2001

ZIEGLER, J., WOLFF, K., NORDIN, P., BANZHAF, W. *Constructing a Small Humanoid Walking Robot as a Platform for the Genetic Evolution of Walking*, Proceedings of the 5th International Heinz Nixdorf Symposium, Autonomous Minirobots for Research and Edutainment, AMiRE 2001, HNI-Verlagsschriftenreihe, no. 97, Univ. Paderborn, Oct. 2001, pp. 51-59 (9)

ZIMMERMANN, J., *Balancing of a Biped Robot using Force Feedback*, Diploma Thesis, FH Koblenz / The Univ. of Western Australia, supervised by T. Bräunl, 2004

Autonomous Planes

11

B uilding an autonomous model airplane is a considerably more difficult undertaking than the previously described autonomous driving or walking robots. Model planes or helicopters require a significantly higher level of safety, not only because the model plane with its expensive equipment might be lost, but more importantly to prevent endangering people on the ground.

A number of autonomous planes or UAVs (Unmanned Aerial Vehicles) have been built in the past for surveillance tasks, for example Aerosonde [Aerosonde 2006]. These projects usually have multi-million-dollar budgets, which cannot be compared to the smaller-scale projects shown here. Two projects with similar scale and scope to the one presented here are "MicroPilot" [MicroPilot 2006], a commercial hobbyist system for model planes, and "FireMite" [Hennessey 2002], an autonomous model plane designed for competing in the International Aerial Robotics Competition [AUVS 2006].

11.1 Application

Low-budget autopilot
Our goal was to modify a remote controlled model airplane for autonomous flying to a given sequence of waypoints (*autopilot*).

- The plane takes off under remote control.
- Once in the air, the plane is switched to autopilot and flies to a previously recorded sequence of waypoints using GPS (global positioning system) data.
- The plane is switched back to remote control and landed.

So the most difficult tasks of take-off and landing are handled by a pilot using the remote control. The plane requires an embedded controller to interface to the GPS and additional sensors and to generate output driving the servos.

There are basically two design options for constructing an autopilot system for such a project (see Figure 11.1):

151

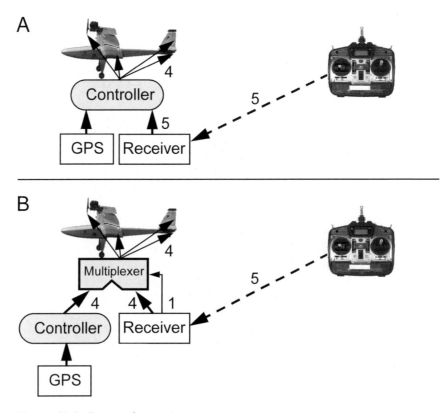

Figure 11.1: System design options

A. The embedded controller drives the plane's servos at all times. It receives sensor input as well as input from the ground transmitter.

B. A central (and remote controlled) multiplexer switches between ground transmitter control and autopilot control of the plane's servos.

Design option A is the simpler and more direct solution. The controller reads data from its sensors including the GPS and the plane's receiver. Ground control can switch between autopilot and manual operation by a separate channel. The controller is at all times connected to the plane's servos and generates their PWM control signals. However, when in manual mode, the controller reads the receiver's servo output and regenerates identical signals.

Design option B requires a four-way multiplexer as an additional hardware component. (Design A has a similar multiplexer implemented in software.) The multiplexer connects either the controller's four servo outputs or the receiver's four servo outputs to the plane's servos. A special receiver channel is used for toggling the multiplexer state under remote control.

Although design A is the superior solution in principle, it requires that the controller operates with highest reliability. Any fault in either controller hardware or controller software, for example the "hanging" of an application program, will lead to the immediate loss of all control surfaces and therefore the loss of the plane. For this reason we opted to implement design B. Although it requires a custom-built multiplexer as additional hardware, this is a rather simple electro-magnetic device that can be directly operated via remote control and is not subject to possible software faults.

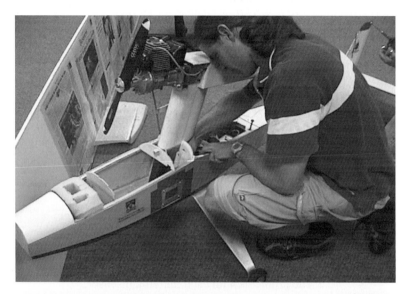

Figure 11.2: Autonomous model plane during construction and in flight

Figure 11.2 shows photos of the construction and during flight of our first autonomous plane. This plane had the EyeCon controller and the multiplexer unit mounted on opposite sides of the fuselage.

11.2 Control System and Sensors

Black box

An EyeCon system is used as on-board flight controller. Before take-off, GPS waypoints for the desired flight path are downloaded to the controller. After the landing, flight data from all on-board sensors is uploaded, similar to the operation of a "black box" data recorder on a real plane.

The EyeCon's timing processor outputs generate PWM signals that can directly drive servos. In this application, they are one set of inputs for the multiplexer, while the second set of inputs comes from the plane's receiver. Two serial ports are used on the EyeCon, one for upload/download of data and programs, and one for continuous GPS data input.

Although the GPS is the main sensor for autonomous flight, it is insufficient because it delivers a very slow update of 0.5Hz .. 1.0Hz and it cannot determine the plane's orientation. We are therefore experimenting with a number of additional sensors (see Chapter 2 for details of these sensors):

- Digital compass
  Although the GPS gives directional data, its update rates are insufficient when flying in a curve around a waypoint.

- Piezo gyroscope and inclinometer
  Gyroscopes give the rate of change, while inclinometers return the absolute orientation. The combined use of both sensor types helps reduce the problems with each individual sensor.

- Altimeter and air-speed sensor
  Both altimeter and air-speed sensor have been built by using air pressure sensors. These sensors need to be calibrated for different heights and temperatures. The construction of an air-speed sensor requires the combination of two sensors measuring the air pressure at the wing tip with a so-called "Pitot tube", and comparing the result with a third air pressure sensor inside the fuselage, which can double as a height sensor.

Figure 11.3 shows the "EyeBox", which contains most equipment required for autonomous flight, EyeCon controller, multiplexer unit, and rechargeable battery, but none of the sensors. The box itself is an important component, since it is made out of lead-alloy and is required to shield the plane's receiver from any radiation from either the controller or the multiplexer. Since the standard radio control carrier frequency of 35MHz is in the same range as the EyeCon's operating speed, shielding is essential.

Another consequence of the decision for design B is that the plane has to remain within remote control range. If the plane was to leave this range, unpredictable switching between the multiplexer inputs would occur, switching control of the plane back and forth between the correct autopilot settings and noise signals. A similar problem would exist for design A as well; however, the controller could use plausibility checks to distinguish noise from proper remote control signals. By effectively determining transmitter strength, the controller could fly the plane even outside the remote transmitter's range.

Figure 11.3: EyeBox and multiplexer unit

11.3 Flight Program

There are two principal techniques for designing a flight program and user interface of the flight system, depending on the capabilities of the GPS unit used and the desired capability and flexibility of the flight system:

A. Small and lightweight embedded GPS module
(for example Rojone MicroGenius 3 [Rojone 2002])
Using a small embedded GPS module has clear advantages in model planes. However, all waypoint entries have to be performed directly to the EyeCon and the flight path has to be computed on the EyeCon controller.

B. Standard handheld GPS with screen and buttons
(for example Magellan GPS 315 [Magellan 1999])
Standard handheld GPS systems are available at about the same cost as a GPS module, but with built-in LCD screen and input buttons. However, they are much heavier, require additional batteries, and suffer a higher risk of damage in a rough landing. Most handheld GPS systems support recording of waypoints and generation of routes, so the complete flight path can be generated on the handheld GPS without using the embedded controller in the plane. The GPS system also needs to support

the NMEA 0183 (Nautical Marine Electronics Association) data message format V2.1 GSA, a certain ASCII data format that is output via the GPS's RS232 interface. This format contains not only the current GPS position, but also the required steering angle for a previously entered route of GPS waypoints (originally designed for a boat's autohelm). This way, the on-board controller only has to set the plane's servos accordingly; all navigation is done by the GPS system.

Program 11.1 shows sample NMEA output. After the start-up sequence, we get regular code strings for position and time, but we only decode the lines starting with $GPGGA. In the beginning, the GPS has not yet logged on to a sufficient number of satellites, so it still reports the geographical position as 0 N and 0 E. The quality indicator in the sixth position (following "E") is 0, so the coordinates are invalid. In the second part of Program 11.1, the $GPRMC string has quality indicator 1 and the proper coordinates of Western Australia.

Program 11.1: NMEA sample output

```
$TOW: 0
$WK:  1151
$POS: 6378137  0         0
$CLK: 96000
$CHNL:12
$Baud rate: 4800   System clock: 12.277MHz
$HW Type: S2AR
$GPGGA,235948.000,0000.0000,N,00000.0000,E,0,00,50.0,0.0,M,,,,0000*3A
$GPGSA,A,1,,,,,,,,,,,,,50.0,50.0,50.0*05
$GPRMC,235948.000,V,0000.0000,N,00000.0000,E,,,260102,,*12
$GPGGA,235948.999,0000.0000,N,00000.0000,E,0,00,50.0,0.0,M,,,,0000*33
$GPGSA,A,1,,,,,,,,,,,,,50.0,50.0,50.0*05
$GPRMC,235948.999,V,0000.0000,N,00000.0000,E,,,260102,,*1B
$GPGGA,235949.999,0000.0000,N,00000.0000,E,0,00,50.0,0.0,M,,,,0000*32
$GPGSA,A,1,,,,,,,,,,,,,50.0,50.0,50.0*05
...
$GPRMC,071540.282,A,3152.6047,S,11554.2536,E,0.49,201.69,171202,,*11
$GPGGA,071541.282,3152.6044,S,11554.2536,E,1,04,5.5,3.7,M,,,,0000*19
$GPGSA,A,2,20,01,25,13,,,,,,,,,6.0,5.5,2.5*34
$GPRMC,071541.282,A,3152.6044,S,11554.2536,E,0.53,196.76,171202,,*1B
$GPGGA,071542.282,3152.6046,S,11554.2535,E,1,04,5.5,3.2,M,,,,0000*1E
$GPGSA,A,2,20,01,25,13,,,,,,,,,6.0,5.5,2.5*34
$GPRMC,071542.282,A,3152.6046,S,11554.2535,E,0.37,197.32,171202,,*1A
$GPGGA,071543.282,3152.6050,S,11554.2534,E,1,04,5.5,3.3,M,,,,0000*18
$GPGSA,A,2,20,01,25,13,,,,,,,,,6.0,5.5,2.5*34
$GPGSV,3,1,10,01,67,190,42,20,62,128,42,13,45,270,41,04,38,228,*7B
$GPGSV,3,2,10,11,38,008,,29,34,135,,27,18,339,,25,13,138,37*7F
$GPGSV,3,3,10,22,10,095,,07,07,254,*76
```

In our current flight system we are using approach A, to be more flexible in flight path generation. For the first implementation, we are only switching the rudder between autopilot and remote control, not all of the plane's surfaces.

Motor and elevator stay on remote control for safety reasons, while the ailerons are automatically operated by a gyroscope to eliminate any roll. Turns under autopilot therefore have to be completely flown using the rudder, which requires a much larger radius than turns using ailerons and elevator. The remaining control surfaces of the plane can be added step by step to the autopilot system.

The flight controller has to perform a number of tasks, which need to be accessible through its user interface:

Pre-flight

- Initialize and test all sensors,
  calibrate sensors.
- Initialize and test all servos,
  enable setting of zero positions of servos,
  enable setting of maximum angles of servos.
- Perform waypoint download – *(only for technique A).*

In-flight *(continuous loop)*

- Generate desired heading – *(only for technique A).*
- Set plane servos according to desired heading.
- Record flight data from sensors.

Post-flight

- Perform flight data upload.

These tasks and settings can be activated by navigating through several flight system menus, as shown in Figure 11.4 [Hines 2001]. They can be displayed and operated through button presses either directly on the EyeCon or remotely via a serial link cable on a PDA (Personal Digital Assistant, for example Compaq IPAQ).

The link between the EyeCon and the PDA has been developed to be able to remote-control (via cable) the flight controller pre-flight and post-flight, especially to download waypoints before take-off and upload flight data after landing. All pre-start diagnostics, for example correct operation of all sensors or the satellite log-on of the GPS, are transmitted from the EyeCon to the handheld PDA screen.

After completion of the flight, all sensor data from the flight together with time stamps are uploaded from the EyeCon to the PDA and can be graphically displayed. Figure 11.5 [Purdie 2002] shows an example of an uploaded flight path ([x, y] coordinates from the GPS sensor); however, all other sensor data is being logged as well for post-flight analysis.

A desirable extension to this setup is the inclusion of wireless data transmission from the plane to the ground (see also Chapter 6). This would allow us to receive instantaneous data from the plane's sensors and the controller's status as opposed to doing a post-flight analysis. However, because of interfer-

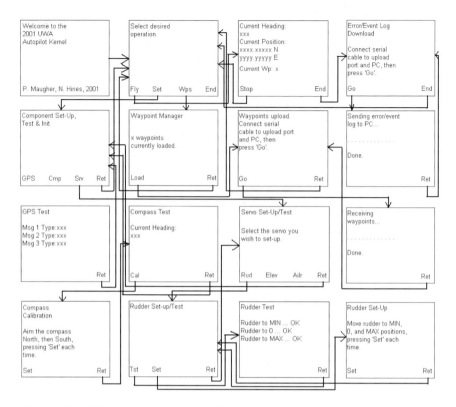

Figure 11.4: Flight system user interface

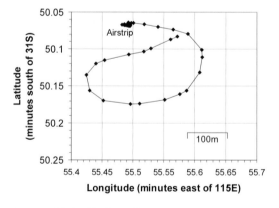

Figure 11.5: Flight path

ence problems with the other autopilot components, wireless data transmission has been left until a later stage of the project.

11.4 References

AEROSONDE, *Global Robotic Observation System Aerosonde*, `http://www.aerosonde.com`, 2006

AUVS, *International Aerial Robotics Competition*, Association for Unmanned Vehicle Systems, `http://avdil.gtri.gatech.edu/AUVS/IARCLaunchPoint.html`, 2006

HENNESSEY, G. *The FireMite Project*, `http://www.craighennessey.com/firemite/`, May 2002

HINES, N. *Autonomous Plane Project 2001 – Compass, GPS & Logging Subsystems*, B.E. Honours Thesis, The Univ. of Western Australia, Electrical and Computer Eng., supervised by T. Bräunl and C. Croft, 2001

MAGELLAN, *GPS 315/320 User Manual*, Magellan Satellite Access Technology, San Dimas CA, 1999

MICROPILOT, *MicroPilot UAV Autopilots*, `http://www.micropilot.com`, 2006

PURDIE, J. *Autonomous Flight Control for Radio Controlled Aircraft*, B.E. Honours Thesis, The Univ. of Western Australia, Electrical and Computer Eng., supervised by T. Bräunl and C. Croft, 2002

ROJONE, *MicroGenius 3 User Manual*, Rojone Pty. Ltd. CD-ROM, Sydney Australia, 2002

AUTONOMOUS VESSELS AND UNDERWATER VEHICLES

12

T he design of an autonomous vessel or underwater vehicle requires one additional skill compared to the robot designs discussed previously: watertightness. This is a challenge especially for autonomous underwater vehicles (AUVs), as they have to cope with increasing water pressure when diving and they require watertight connections to actuators and sensors outside the AUV's hull. In this chapter, we will concentrate on AUVs, since autonomous vessels or boats can be seen as AUVs without the diving functionality.

The area of AUVs looks very promising to advance commercially, given the current boom of the resource industry combined with the immense cost of either manned or remotely operated underwater missions.

12.1 Application

Unlike many other areas of mobile robots, AUVs have an immediate application area conducting various sub-sea surveillance and manipulation tasks for the resource industry. In the following, we want to concentrate on intelligent control and not on general engineering tasks such as constructing AUVs that can go to great depths, as there are industrial ROV (remotely operated vehicle) solutions available that have solved these problems.

While most autonomous mobile robot applications can also use wireless communication to a host station, this is a lot harder for an AUV. Once submerged, none of the standard communication methods work; Bluetooth or WLAN only operate up to a water depth of about 50cm. The only wireless communication method available is sonar with a very low data rate, but unfortunately these systems have been designed for the open ocean and can usually not cope with signal reflections as they occur when using them in a pool. So unless some wire-bound communication method is used, AUV applications have to be truly autonomous.

The Association for Unmanned Vehicles International (AUVSI) organizes annual competitions for autonomous aerial vehicles and for autonomous underwater vehicles [AUVSI 2006]. Unfortunately, the tasks are very demanding, so it is difficult for new research groups to enter. Therefore, we decided to develop a set of simplified tasks, which could be used for a regional or entry-level AUV competition (Figure 12.1).

We further developed the AUV simulation system SubSim (see Section 13.6), which allows to design AUVs and implement control programs for the individual tasks without having to build a physical AUV. This simulation system could serve as the platform for a simulation track of an AUV competition.

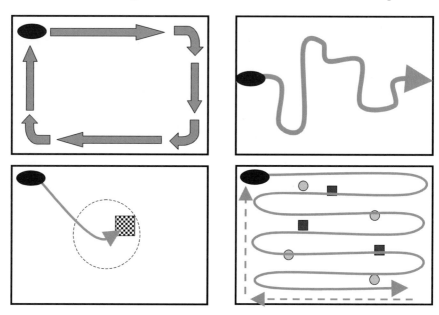

Figure 12.1: AUV competition tasks

The four suggested tasks to be completed in an olympic size swimming pool are:

1. Wall Following
   The AUV is placed close to a corner of the pool and has to follow the pool wall without touching it. The AUV should perform one lap around the pool, return to the starting position, then stop.

2. Pipeline Following
   A plastic pipe is placed along the bottom of the pool, starting on one side of the pool and terminating on the opposite side. The pipe is made out of straight pieces and 90 degree angles.
   The AUV is placed over the start of the pipe on one side of the pool and has to follow the pipe on the ground until the opposite wall has been reached.

3. Target Finding
   The AUV has to locate a target plate with a distinctive texture that is placed at a random position within a 3m diameter from the center of the pool.

4. Object Mapping
   A number of simple objects (balls or boxes of distinctive color) are placed at the bottom of the pool, distributed over the whole pool area. The AUV has to survey the whole pool area, e.g. by diving along a sweeping pattern, and record all objects found at the bottom of the pool. Finally, the AUV has to return to its start corner and upload the coordinates of all objects found.

12.2 Dynamic Model

The dynamic model of an AUV describes the AUV's motions as a result of its shape, mass distribution, forces/torques exerted by the AUV's motors, and external forces/torques (e.g. ocean currents). Since we are operating at relatively low speeds, we can disregarding the Coriolis force and present a simplified dynamic model [Gonzalez 2004]:

$$M \cdot \dot{v} + D(v) \cdot v + G = \tau$$

with:

M	mass and inertia matrix
v	linear and angular velocity vector
D	hydrodynamic damping matrix
G	gravitational and buoyancy vector
τ	force and torque vector (AUV motors and eternal forces/torques)

D can be further simplified as a diagonal matrix with zero entries for y (AUV can only move forward/backward along x, and dive/surface along z, but not move sideways), and zero entries for rotations about x and y (AUV can actively rotate only about z, while its self-righting movement, see Section 12.3, greatly eliminates rotations about x and y).

G is non-zero only in its z component, which is the sum of the AUV's gravity and buoyancy vectors.

τ is the product of the force vector combining all of an AUV's motors, with a pose matrix that defines each motor's position and orientation based on the AUV's local coordinate system.

12.3 AUV Design Mako

The Mako (Figure 12.2) was designed from scratch as a dual PVC hull containing all electronics and batteries, linked by an aluminum frame and propelled by 4 trolling motors, 2 of which are for active diving. The advantages of this design over competing proposals are [Bräunl et al. 2004], [Gonzalez 2004]:

Figure 12.2: Autonomous submarine Mako

- Ease in machining and construction due to its simple structure
- Relative ease in ensuring watertight integrity because of the lack of rotating mechanical devices such as bow planes and rudders
- Substantial internal space owing to the existence of two hulls
- High modularity due to the relative ease with which components can be attached to the skeletal frame
- Cost-effectiveness because of the availability and use of common materials and components
- Relative ease in software control implementation when compared to using a ballast tank and single thruster system
- Ease in submerging with two vertical thrusters
- Static stability due to the separation of the centers of mass and buoyancy, and dynamic stability due to the alignment of thrusters

Simplicity and modularity were key goals in both the mechanical and electrical system designs. With the vehicle not intended for use below 5m depth, pressure did not pose a major problem. The Mako AUV measures 1.34 m long, 64.5 cm wide and 46 cm tall.

The vehicle comprises two watertight PVC hulls mounted to a supporting aluminum skeletal frame. Two thrusters are mounted on the port and starboard sides of the vehicle for longitudinal movement, while two others are mounted vertically on the bow and stern for depth control. The Mako's vertical thruster diving system is not power conservative, however, when a comparison is made with ballast systems that involve complex mechanical devices, the advantages such as precision and simplicity that comes with using these two thrusters far outweighs those of a ballast system.

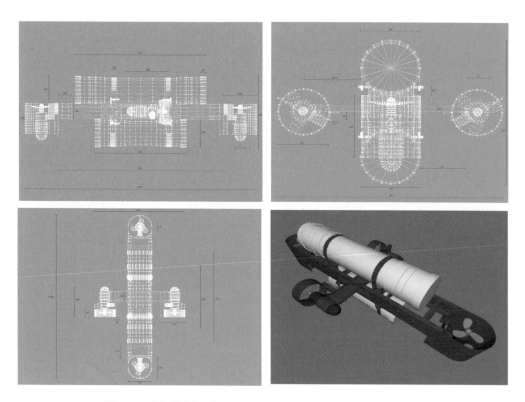

Figure 12.3: Mako design

Propulsion is provided by four modified 12V, 7A trolling motors that allow horizontal and vertical movement of the vehicle. These motors were chosen for their small size and the fact that they are intended for underwater use; a feature that minimized construction complexity substantially and provided watertight integrity.

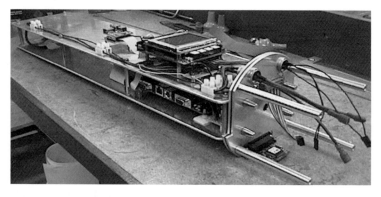

Figure 12.4: Electronics and controller setup inside Mako's top hull

The starboard and port motors provide both forward and reverse movement while the stern and bow motors provide depth control in both downward and upward directions. Roll is passively controlled by the vehicle's innate righting moment (Figure 12.5). The top hull contains mostly air besides light electronics equipment, the bottom hull contains heavy batteries. Therefore mainly a buoyancy force pulls the top cylinder up and gravity pulls the bottom cylinder down. If for whatever reason, the AUV rolls as in Figure 12.5, right, these two forces ensure that the AUV will right itself.

Overall, this provides the vehicle with 4DOF that can be actively controlled. These 4DOF provide an ample range of motion suited to accomplishing a wide range of tasks.

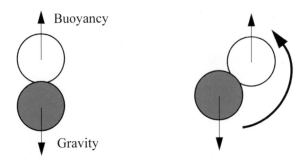

Figure 12.5: Self-righting moment of AUV

Controllers   The control system of the Mako is separated into two controllers; an EyeBot microcontroller and a mini-PC. The EyeBot's purpose is controlling the AUV's movement through its four thrusters and its sensors. It can run a completely autonomous mission without the secondary controller. The mini PC is a Cyrix 233MHz processor, 32Mb of RAM and a 5GB hard drive, running Linux. Its sole function is to provide processing power for the computationally intensive vision system.

Motor controllers designed and built specifically for the thrusters provide both speed and direction control. Each motor controller interfaces with the EyeBot controller via two servo ports. Due to the high current used by the thrusters, each motor controller produces a large amount of heat. To keep the temperature inside the hull from rising too high and damaging electronic components, a heat sink attached to the motor controller circuit on the outer hull was devised. Hence, the water continuously cools the heat sink and allows the temperature inside the hull to remain at an acceptable level.

Sensors   The sonar/navigation system utilizes an array of Navman Depth2100 echo sounders, operating at 200 kHz. One of these sensors is facing down and thereby providing an effective depth sensor (assuming the pool depth is known), while the other three sensors are used as distance sensors pointing forward, left, and right. An auxiliary control board, based on a PIC controller, has been designed to multiplex the four sonars and connect to the EyeBot [Alfirevich 2005].

Figure 12.6: Mako in operation

A low-cost Vector 2X digital magnetic compass module provides for yaw or heading control. A simple dual water detector circuit connected to analogue-to-digital converter (ADC) channels on the EyeBot controller is used to detect a possible hull breach. Two probes run along the bottom of each hull, which allows for the location (upper or lower hull) of the leak to be known. The EyeBot periodically monitors whether or not the hull integrity of the vehicle has been compromised, and if so immediately surfaces the vehicle. Another ADC input of the EyeBot is used for a power monitor that will ensure that the system voltage remains at an acceptable level. Figure 12.6 shows the Mako in operation.

12.4 AUV Design USAL

The USAL AUV uses a commercial ROV as a basis, which was heavily modified and extended (Figure 12.7). All original electronics were taken out and replaced by an EyeBot controller (Figure 12.8). The hull was split and extended by a trolling motor for active diving, which allows the AUV to hover, while the original ROV had to use active rudder control during a forward motion for diving, [Gerl 2006], [Drtil 2006]. Figure 12.8 shows USAL's complete electronics subsystem.

For simplicity and cost reasons, we decided to trial infrared PSD sensors (see Section 2.6) for the USAL instead of the echo sounders used on the Mako. Since the front part of the hull was made out of clear perspex, we were able to place the PSD sensors inside the AUV hull, so we did not have to worry about waterproofing sensors and cabling. Figure 12.9 shows the results of measurements conducted in [Drtil 2006], using this sensor setup in air (through the hull), and in different grades of water quality. Assuming good water quality, as can be expected in a swimming pool, the sensor setup returns reliable results up to a distance of about 1.1 m, which is sufficient for using it as a collision avoidance sensor, but too short for using it as a navigation aid in a large pool.

Figure 12.7: Autonomous submarine USAL

Figure 12.8: USAL controller and sensor subsystem

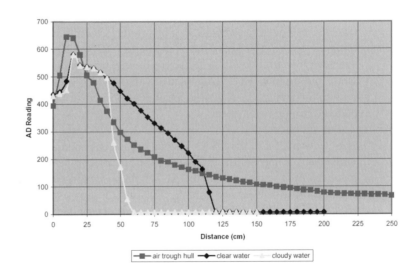

Figure 12.9: Underwater PSD measurement

The USAL system overview is shown in Figure 12.10. Numerous sensors are connected to the EyeBot on-board controller. These include a digital camera, four analog PSD infrared distance sensors, a digital compass, a three-axes solid state accelerometer and a depth pressure sensor. The Bluetooth wireless communication system can only be used when the AUV has surfaced or is diving close to the surface. The energy control subsystem contains voltage regulators and level converters, additional voltage and leakage sensors, as well as motor drivers for the stern main driving motor, the rudder servo, the diving trolling motor, and the bow thruster pump.

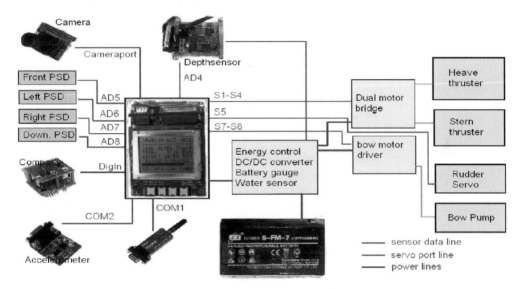

Figure 12.10: USAL system overview

Figure 12.11 shows the arrangement of the three thrusters and the stern rudder, together with a typical turning movement of the USAL.

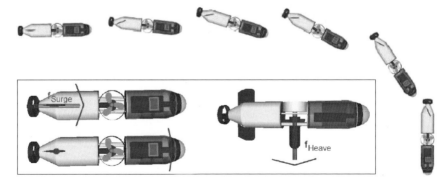

Figure 12.11: USAL thrusters and typical rudder turning maneuver

12.5 References

ALFIREVICH, E. *Depth and Position Sensing for an Autonomous Underwater Vehicle*, B.E. Honours Thesis, The Univ. of Western Australia, Electrical and Computer Eng., supervised by T. Bräunl, 2005

AUVSI, *AUVSI and ONR's 9th International Autonomous Underwater Vehicle Competition*, Association for Unmanned Vehicle Systems International, http://www.auvsi.org/competitions/water.cfm, 2006

BRÄUNL, T., BOEING, A., GONZALES, L., KOESTLER, A., NGUYEN, M., PETITT, J. *The Autonomous Underwater Vehicle Initiative – Project Mako*, 2004 IEEE Conference on Robotics, Automation, and Mechatronics (IEEE-RAM), Dec. 2004, Singapore, pp. 446-451 (6)

DRTIL, M. *Electronics and Sensor Design of an Autonomous Underwater Vehicle*, Diploma Thesis, The Univ. of Western Australia and FH Koblenz, Electrical and Computer Eng., supervised by T. Bräunl, 2006

GERL, B. *Development of an Autonomous Underwater Vehicle in an Interdisciplinary Context*, Diploma Thesis, The Univ. of Western Australia and Technical Univ. München, Electrical and Computer Eng., supervised by T. Bräunl, 2006

GONZALEZ, L. *Design, Modelling and Control of an Autonomous Underwater Vehicle*, B.E. Honours Thesis, The Univ. of Western Australia, Electrical and Computer Eng., supervised by T. Bräunl, 2004

13

SIMULATION SYSTEMS

$\mathbf{S}$ imulation is an essential part of robotics, whether for stationary manipulators, mobile robots, or for complete manufacturing processes in factory automation. We have developed a number of robot simulation systems over the years, two of which will be presented in this Chapter.

EyeSim is a simulation system for multiple interacting mobile robots. Environment modeling is 2½D, while 3D sceneries are generated with synthetic images for each robot's camera view of the scene. Robot application programs are compiled into dynamic link libraries that are loaded by the simulator. The system integrates the robot console with text and color graphics, sensors, and actuators at the level of a velocity/position control (vω) driving interface.

SubSim is a simulation system for autonomous underwater vehicles (AUVs) and contains a complete 3D physics simulation library for rigid body systems and water dynamics. It conducts simulation at a much lower level than the driving simulator EyeSim. SubSim requires a significantly higher computational cost, but is able to simulate any user-defined AUV. This means that motor number and positions, as well as sensor number and positions can be freely chosen in an XML description file, and the physics simulation engine will calculate the correct resulting AUV motion.

Both simulation systems are freely available over the Internet and can be downloaded from:

> http://robotics.ee.uwa.edu.au/eyebot/ftp/   (EyeSim)
> http://robotics.ee.uwa.edu.au/auv/ftp/   (SubSim)

13.1 Mobile Robot Simulation

Simulators can be on different levels

Quite a number of mobile robot simulators have been developed in recent years, many of them being available as freeware or shareware. The level of simulation differs considerably between simulators. Some of them operate at a very high level and let the user specify robot behaviors or plans. Others operate at a very low level and can be used to examine the exact path trajectory driven by a robot.

We developed the first mobile robot simulation system employing synthetic vision as a feedback sensor in 1996, published in [Bräunl, Stolz 1997]. This system was limited to a single robot with on-board vision system and required the use of special rendering hardware.

Both simulation systems presented below have the capability of using generated synthetic images as feedback for robot control programs. This is a significant achievement, since it allows the simulation of high-level robot application programs including image processing. While EyeSim is a multi-robot simulation system, SubSim is currently limited to a single AUV.

Most other mobile robot simulation systems, for example *Saphira* [Konolige 2001] or *3D7* [Trieb 1996], are limited to simpler sensors such as sonar, infrared, or laser sensors, and cannot deal with vision systems. [Matsumoto et al. 1999] implemented a vision simulation system very similar to our earlier system [Bräunl, Stolz 1997]. The *Webots* simulation environment [Wang, Tan, Prahlad 2000] models among others the *Khepera* mobile robot, but has only limited vision capabilities. An interesting outlook for vision simulation in motion planning for *virtual humans* is given by [Kuffner, Latombe 1999].

13.2 EyeSim Simulation System

All EyeBot programs run on EyeSim
The goal of the EyeSim simulator was to develop a tool which allows the construction of simulated robot environments and the testing of mobile robot programs before they are used on real robots. A simulation environment like this is especially useful for the programming techniques of neural networks and genetic algorithms. These have to gather large amounts of training data, which is sometimes difficult to obtain from actual vehicle runs. Potentially harmful collisions, for example due to untrained neural networks or errors, are not a concern in the simulation environment. Also, the simulation can be executed in a "perfect" environment with no sensor or actuator errors – or with certain error levels for individual sensors. This allows us to thoroughly test a robot application program's robustness in a near real-word scenario [Bräunl, Koestler, Waggershauser 2005], [Koestler, Bräunl 2004].

The simulation library has the same interface as the RoBIOS library (see Appendix B.5). This allows the creation of robot application programs for either real robots or simulation, simply by re-compiling. No changes at all are required in the application program when moving between the real robot and simulation or vice versa.

The technique we used for implementing this simulation differs from most existing robot simulation systems, which run the simulation as a separate program or process that communicates with the application by some message passing scheme. Instead, we implemented the whole simulator as a collection of system library functions which replace the sensor/actuator functions called from a robot application program. The simulator has an independent main-program, while application programs are compiled into dynamic link-libraries and are linked at run-time.

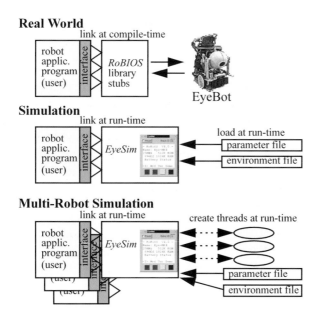

Figure 13.1: System concept

As shown in Figure 13.1 (top), a robot application program is compiled and linked to the RoBIOS library, in order to be executed on a real robot system. The application program makes system calls to individual RoBIOS library functions, for example for activating driving motors or reading sensor input. In the simulation scenario, shown in Figure 13.1 (middle), the compiled application library is now linked to the EyeSim simulator instead. The library part of the simulator implements exactly the same functions, but now activates the screen displays of robot console and robot driving environment.

In case of a multi-robot simulation (Figure 13.1 bottom), the compilation and linking process is identical to the single-robot simulation. However, at run-time, individual threads are created to simulate each of the robots concurrently. Each thread executes the robot application program individually on local data, but all threads share the common environment through which they interact.

The EyeSim user interface is split into two parts, a robot console per simulated robot and a common driving environment (Figure 13.2). The robot console models the EyeBot controller, which comprises a display and input buttons as a robot user interface. It allows direct interaction with the robot by pressing buttons and printing status messages, data values, or graphics on the screen. In the simulation environment, each robot has an individual console, while they all share the driving environment. This is displayed as a 3D view of the robot driving area. The simulator has been implemented using the *OpenGL* version *Coin3D* [Coin3D 2006], which allows panning, rotating, and zooming in 3D with familiar controls. Robot models may be supplied as *Milkshape* MS3D files [Milkshape 2006]. This allows the use of arbitrary 3D models of a

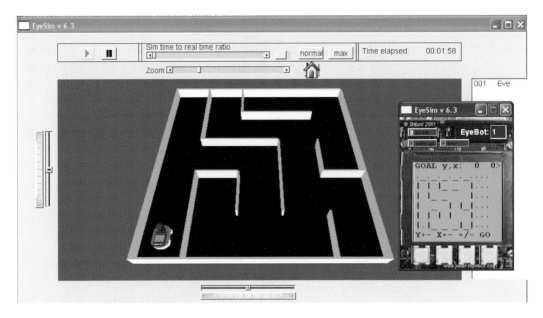

Figure 13.2: EyeSim interface

robot or even different models for different robots in the same simulation. However, these graphics object descriptions are for display purposes only. The geometric simulation model of a robot is always a cylinder.

Simulation execution can be halted via a "Pause" button and robots can be relocated or turned by a menu function. All parameter settings, including error levels, can be set via menu selections. All active robots are listed at the right of the robot environment window. Robots are assigned unique id-numbers and can run different programs.

Sensor–actuator modeling

Each robot of the EyeBot family is typically equipped with a number of actuators:

- DC driving motors with differential steering
- Camera panning motor
- Object manipulation motor ("kicker")

and sensors:

- Shaft encoders
- Tactile bumpers
- Infrared proximity sensors
- Infrared distance measurement sensors
- Digital grayscale or color camera

The real robot's RoBIOS operating system provides library routines for all of these sensor types. For the EyeSim simulator, we selected a high-level subset to work with.

Driving interfaces   Identical to the RoBIOS operating system, two driving interfaces at different abstraction levels have been implemented:

- High-level linear and rotational velocity interface (vω)
  This interface provides simple driving commands for user programs without the need for specifying individual motor speeds. In addition, this simplifies the simulation process and speeds up execution.

- Low-level motor interface
  EyeSim does not include a full physics engine, so it is not possible to place motors at arbitrary positions on a robot. However, the following three main drive mechanisms have been implemented and can be specified in the robot's parameter file. Direct motor commands can then be given during the simulation.

    - Differential drive
    - Ackermann drive
    - Mecanum wheel drive

All robots' positions and orientations are updated by a periodic process according to their last driving commands and respective current velocities.

Simulation of tactile sensors   Tactile sensors are simulated by computing intersections between the robot (modeled as a cylinder) and any obstacle in the environment (modeled as line segments) or another robot. Bump sensors can be configured as a certain sector range around the robot. That way several bumpers can be used to determine the contact position. The VWStalled function of the driving interface can also be used to detect a collision, causing the robot's wheels to stall.

Simulation of infrared sensors   The robot uses two different kinds of infrared sensors. One is a binary sensor (Proxy) which is activated if the distance to an obstacle is below a certain threshold, the other is a position sensitive device (PSD), which returns the distance value to the nearest obstacle. Sensors can be freely positioned and orientated around a robot as specified in the robot parameter file. This allows testing and comparing the performance of different sensor placements. For the simulation process, the distance between each infrared sensor at its current position and orientation toward the nearest obstacle is determined.

Synthetic camera images   The simulation system generates artificially rendered camera images from each robot's point of view. All robot and environment data is re-modeled in the object-oriented format required by the OpenInventor library and passed to the image generator of the Coin3D library [Coin3D 2006]. This allows testing, debugging, and optimizing programs for groups of robots including vision. EyeSim allows the generation of individual scene views for each of the simulated robots, while each robot receives its own camera image (Figure 13.3) and can use it for subsequent image processing tasks as in [Leclercq, Bräunl, 2001].

Error models   The simulator includes different error models, which allow to either run a simulation in a "perfect world" or to set an error rate in actuators, sensors, or communication for a more realistic simulation. This can be used for testing a robot application program's robustness against sensor noise. The error models

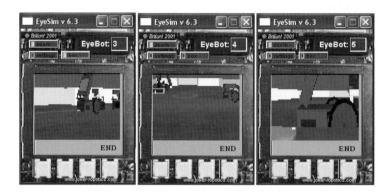

Figure 13.3: Generated camera images

use a standard Gaussian distribution for fault values added to sensor readings or actuator positions. From the user interface, error percentages can be selected for each sensor/actuator. Error values are added to distance measurement sensors (infrared, encoders) or the robot's [x, y] coordinates and orientation.

Simulated communication errors include also the partial or complete loss or corruption of a robot-to-robot data transmission.

For the generated camera images, some standard image noise methods have been implemented, which correspond to typical image transmission errors and dropouts (Figure 13.4):

- Salt-and-pepper noise
  a percentage of random black and white pixels is inserted.

- 100s&1000s noise
  a percentage of random colored pixels is inserted.

- Gaussian noise
  a percentage of pixels are changed by a zero mean random process.

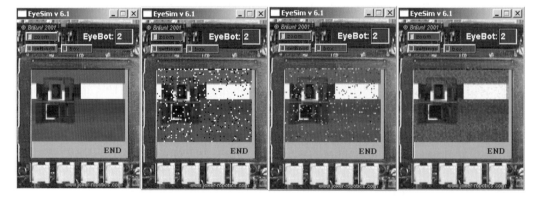

Figure 13.4: Camera image, salt-and-pepper, 100s&1000s, Gaussian noise

13.3 Multiple Robot Simulation

A multi-robot simulation can be initiated by specifying several robots in the parameter file. Concurrency occurs at three levels:

- Each robot may contain several threads for local concurrent processing.
- Several robots interact concurrently in the same simulation environment.
- System updates of all robots' positions and velocities are executed asynchronously in parallel with simulation display and user input.

Individual threads for each robot are created at run-time. *Posix* threads and semaphores are used for synchronization of the robots, with each robot receiving a unique id-number upon creation of its thread.

In a real robot scenario, each robot interacts with all other robots in two ways. Firstly, by moving around and thereby changing the environment it is part of. Secondly, by using its radio communication device to send messages to other robots. Both of these methods are also simulated in the EyeSim system.

Since the environment is defined in 2D, each line represents an obstacle of unspecified height. Each of the robot's distance sensors measures the free space to the nearest obstacle in its orientation and then returns an appropriate signal value. This does not necessarily mean that a sensor will return the physically correct distance value. For example, the infrared sensors only work within a certain range. If a distance is above or below this range, the sensor will return out-of-bounds values and the same behavior has been modeled for the simulation.

Whenever several robots interact in an environment, their respective threads are executed concurrently. All robot position changes (through driving) are made by calling a library function and will so update the common environment. All subsequent sensor readings are also library calls, which now take into account the updated robot positions (for example a robot is detected as an obstacle by another robot's sensor). Collisions between two robots or a robot and an obstacle are detected and reported on the console, while the robots involved are stopped.

Avoid global variables Since we used the more efficient thread model to implement multiple robots as opposed to separate processes, this restricts the use of global (and static) variables. Global variables can be used when simulating a single robot; however, they can cause problems with multiple robots, i.e. only a single copy exists and will be accessible to threads from different robots. Therefore, whenever possible, global and static variables should be avoided.

13.4 EyeSim Application

The sample application in Program 13.1 lets a robot drive straight until it comes too close to an obstacle. It will then stop, back up, and turn by a random angle, before it continues driving in a straight line again. Each robot is equipped with three PSD sensors (infrared distance sensors). All three sensors plus the stall function of each robot are being monitored. Figure 13.5 shows simulation experiments with six robots driving concurrently.

Program 13.1: Random drive

```
1    #include "eyebot.h"
2    #include <stdlib.h>
3    #include <math.h>
4    #define SAFETY 300
5
6    int main ()
7    { PSDHandle front, left, right;
8      VWHandle   vw;
9      float      dir;
10
11     LCDPrintf("Random Drive\n\n");
12     LCDMenu("", "", "", "END");
13     vw=VWInit(VW_DRIVE,1);
14     VWStartControl(vw, 7.0,0.3,10.0,0.1);
15     front = PSDInit(PSD_FRONT);
16     left  = PSDInit(PSD_LEFT);
17     right = PSDInit(PSD_RIGHT);
18     PSDStart(front | left | right , TRUE);
19
20     while(KEYRead() != KEY4)
21     { if ( PSDGet(left) >SAFETY && PSDGet(front)>SAFETY
22        &&  PSDGet(right)>SAFETY && !VWStalled(vw) )
23                   VWDriveStraight(vw, 0.5, 0.3);
24       else
25       { LCDPutString("back up, ");
26         VWDriveStraight(vw,-0.04,0.3);
27         VWDriveWait(vw);
28         LCDPutString("turn\n");           /* random angle */
29         dir = M_PI * (drand48() - 0.5); /* -90 .. +90   */
30         VWDriveTurn(vw, dir, 0.6);
31         VWDriveWait(vw);
32       }
33       OSWait(10);
34     }
35     VWRelease(vw);
36     return 0;
37  }
```

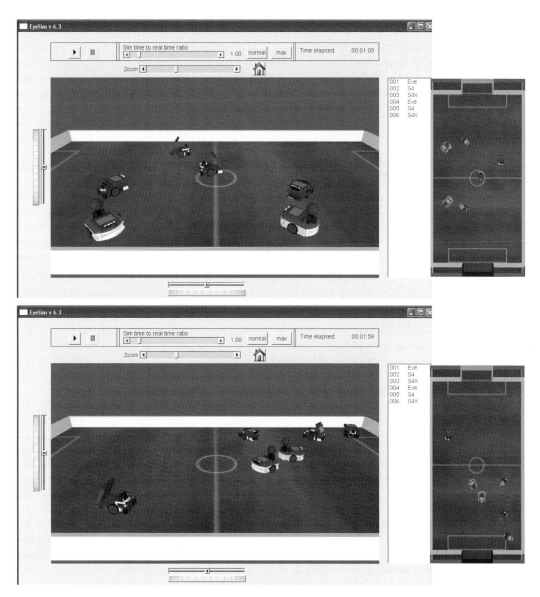

Figure 13.5: Random drive of six robots

13.5 EyeSim Environment and Parameter Files

All environments are modeled by 2D line segments and can be loaded from text files. Possible formats are either the *world format* used in the *Saphira* robot operating system [Konolige 2001] or the *maze format* developed by Bräunl following the widely used "Micro Mouse Contest" notation [Bräunl 1999].

World format   The environment in world format is described by a text file. It specifies walls as straight line segments by their start and end points with dimensions in millimeters. An implicit stack allows the specification of a substructure in local coordinates, without having to translate and rotate line segments. Comments are allowed following a semicolon until the end of a line.

The world format starts by specifying the total world size in mm, for example:

```
width   4680
height 3240
```

Wall segments are specified as 2D lines $[x_1,y_1, x_2,y_2]$, so four integers are required for each line, for example:

```
;rectangle
0 0 0 1440
0 0 2880 0
0 1440 2880 1440
2880 0 2880 1440
```

Through an implicit stack, local *poses* (position and orientation) can be set. This allows an easier description of an object in object coordinates, which may be offset and rotated in world coordinates. To do so, the definition of an object (a collection of line segments) is enclosed within a push and pop statement, which may be nested. Push requires the pose parameters [x, y, phi], while pop does not have any parameters. For example:

```
;two lines translated to [100,100] and rotated by 45 deg.
push 100 100 45
0 0 200   0
0 0 200 200
pop
```

The starting position and orientation of a robot may be specified by its pose [x, y, φ], for example:

```
position 180 1260 -90
```

Maze format   The maze format is a very simple input format for environments with orthogonal walls only, such as the Micro Mouse competitions. We wanted the simulator to be able to read typical *natural* graphics ASCII maze representations, which are available from the web, like the one below.

Each wall is specified by single characters within a line. A "|" (at odd positions in a line, 1, 3, 5, ..) denotes a wall segment in the y-direction, a "_" (at even positions in a line, 2, 4, 6, ..) is a wall segment in the x-direction. So, each line contains in fact the horizontal walls of its coordinate and the vertical wall segments of the line above it.

```
 _____      ___
|    _____   |   |   |
|  | _____  | |  _|   |
|  | |_____   | | | | | |
|  | _    _|__|__|   _|
|  |_|_____  |
|  |____      |   _    | |
|   _   |  |__|  | |  _|
| | | |     |  _____   |
|S|_____|_____|_|
```

The example below defines a rectangle with two dividing walls:

```
 __ __ _
|      _|
|_|_ _|
```

The following shows the same example in a slightly different notation, which avoids gaps in horizontal lines (in the ASCII representation) and therefore looks nicer:

```
 _____
|     _|
|_|___|
```

Extra characters may be added to a maze to indicate starting positions of one or multiple robots. Upper-case characters assume a wall below the character, while lower-case letters do not. The letters U (or S), D, L, R may be used in the maze to indicate a robot's start position and orientation: up (equal to start), down, left, or right. In the last line of the maze file, the size of a wall segment can be specified in mm (default value 360mm) as a single integer number.

A ball can be inserted by using the symbol "o", a box can be inserted with the symbol "x". The robots can then interact with the ball or box by pushing or kicking it (see Figure 13.6).

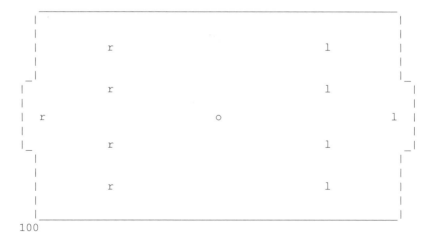

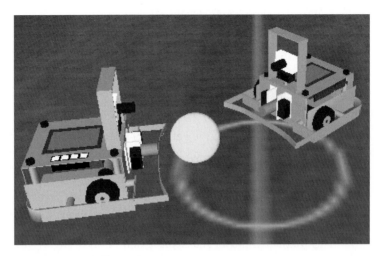

Figure 13.6: Ball simulation

A number of parameter files are used for the EyeSim simulator, which determine simulation parameters, physical robot description, and robot sensor layout, as well as the simulation environment and graphical representation:

- `myfile.sim`
  Main simulation description file, contains links to environment and robot application binary.

- `myfile.c (or .cpp) and myfile.dll`
  Robot application source file and compiled binary as dynamic link library (DLL).

The following parameter files can be supplied by the application programmer, but do not have to be. A number of environment, as well as robot description and graphics files are available as a library:

- `myenvironment.maz` or `myenvironment.wld`
  Environment file in maze or world format (see Section 13.5).

- `myrobot.robi`
  Robot description file, physical dimensions, location of sensors, etc.

- `myrobot.ms3d`
  Milkshape graphics description file for 3D robot shape (graphics representation only).

SIM parameter file Program 13.2 shows an example for a ".`sim`" file. It specifies which environment file (here: "`maze1.maz`") and which robot description file (here: `S4.robi`") are being used.

The robot's starting position and orientation may be specified in the "`robi`" line as optional parameters. This is required for environments that do not specify a robot starting position. E.g.:

```
robi  S4.robi  DriveDemo.dll  400 400 90
```

Program 13.2: EyeSim parameter file ".sim"

```
1   # world description file (either maze or world)
2   maze  maze1.maz
3
4   # robot description file
5   robi  S4.robi  DriveDemo.dll
```

ROBI
parameter file
There is a clear distinction between robot and simulation parameters, which is expressed by using different parameter files. This split allows the use of different robots with different physical dimensions and equipped with different sensors in the same simulation.

Program 13.3: Robot parameter file ".robi" for S4 soccer robot

```
1   # the name of the robi
2   name S4
3
4   # robot diameter in mm
5   diameter 186
6
7   # max linear velocity in mm/s
8   speed    600
9
10  # max rotational velocity in deg/s
11  turn     300
12
13  # file name of the graphics model used for this robi
14  model    S4.ms3d
15
16  # psd sensor definition: (id-number from "hdt_sem.h")
17  # "psd", name, id, relative position to robi center(x,y,z)
18  # in mm, angle in x-y plane in deg
19  psd PSD_FRONT  -200   60  20  30      0
20  psd PSD_LEFT   -205   56  45  30     90
21  psd PSD_RIGHT  -210   56 -45  30    -90
22
23  # color camera sensor definition:
24  # "camera", relative position to robi center (x,y,z),
25  # pan-tilt-angle (pan, tilt), max image resolution
26  camera  62 0 60    0  -5    80  60
27
28  # wheel diameter [mm], max. rotational velocity [deg/s],
29  # encoder ticks/rev., wheel-base distance [mm]
30  wheel   54  3600  1100   90
31
32  # motors and encoders for low level drive routines
33  # Diff.-drive: left motor, l. enc, right motor, r. enc
34  drive  DIFFERENTIAL_DRIVE  MOTOR_LEFT QUAD_LEFT
35                             MOTOR_RIGHT QUAD_RIGHT
```

Each robot type is described by two files: the ".robi" parameter file, which contains all parameters of a robot relevant to the simulation, and the default Milkshape ".ms3d" graphics file, which contains the robot visualization as a colored 3D model (see next section). With this distinction, we can have a number of physically identical robots with different shapes or color representation in the simulation visualization. Program 13.3 shows an example of a typical ".robi" file, which contains:

- Robot type name
- Physical size
- Maximum speed and rotational speed
- Default visualization file (may be changed in ".sim" file)
- PSD sensor placement
- Digital camera placement and camera resolution in pixels
- Wheel velocity and dimension
- Drive system to enable low-level (motor- or wheel-level) driving, supported drive systems are DIFFERENTIAL_DRIVE, ACKERMANN_DRIVE, and OMNI_DRIVE

With the help of the robot parameter file, we can run the same simulation with different robot sensor settings. For example, we can change the sensor mounting positions in the parameter file and find the optimal solution for a given problem by repeated simulation runs.

13.6 SubSim Simulation System

SubSim is a simulation system for Autonomous Underwater Vehicles (AUVs) and therefore requires a full 3D physics simulation engine. The simulation software is designed to address a broad variety of users with different needs, such as the structure of the user interface, levels of abstraction, and the complexity of physics and sensor models. As a result, the most important design goal for the software is to produce a simulation tool that is as extensible and flexible as possible. The entire system was designed with a plug-in based architecture. Entire components, such as the end-user API, the user interface and the physics simulation library can be exchanged to accommodate the users' needs. This allows the user to easily extend the simulation system by adding custom plug-ins written in any language supporting dynamic libraries, such as standard C or C++.

The simulation system provides a software developer kit (SDK) that contains the framework for plug-in development, and tools for designing and visualizing the submarine. The software packages used to create the simulator include:

- wxWidgets [wxWidgets 2006] (formerly wxWindows)
  A mature and comprehensive open source cross platform C++ GUI framework. This package was selected as it greatly simplifies the task

of cross platform interface development. It also offers straightforward plug-in management and threading libraries.

- TinyXML [tinyxml 2006]
  This XML parser was chosen because it is simple to use and small enough to distribute with the simulation.

- Newton Game Dynamics Engine [Newton 2006]
  The physics simulation library is exchangeable and can be selected by the user. However, the Newton system, a fast and deterministic physics solver, is SubSim's default physics engine.

Physics simulation The underlying low-level physics simulation library is responsible for calculating the position, orientation, forces, torques and velocities of all bodies and joints in the simulation. Since the low-level physics simulation library performs most of the physics calculations, the higher-level physics abstraction layer (PAL) is only required to simulate motors and sensors. The PAL allows custom plug-ins to be incorporated to the existing library, allowing custom sensor and motor models to replace, or supplement the existing implementations.

Application programmer interface The simulation system implements two separate application programmer interfaces (APIs). The low-level API is the internal API, which is exposed to developers so that they can encapsulate the functionality of their own controller API. The high-level API is the RoBIOS API (see Appendix B.5), a user friendly API that mirrors the functionality present on the EyeBot controller used in both the Mako and USAL submarines.

The internal API consists of only five functions:

```
SSID InitDevice(char *device_name);
SSERROR QueryDevice (SSID device, void *data);
SSERROR SetData(SSID device, void *data);
SSERROR GetData(SSID device, void *data);
SSERROR GetTime(SSTIME time);
```

The function `InitDevice` initializes the device given by its name and stores it in the internal registry. It returns a unique handle that can be used to further reference the device (e.g. sensors, motors). `QueryDevice` stores the state of the device in the provided data structure and returns an error if the execution failed. `GetTime` returns a time stamp holding the execution time of the submarine's program in ms. In case of failure an error code is returned.

The functions that are actually manipulating the sensors and actuators and therefore affect the interaction of the submarine with its environment are either the `GetData` or `SetData` function. While the first one retrieves the data (e.g. sensor readings) the latter one changes the internal state of a device by passing control and/or information data to the device. Both functions return appropriate error codes if the operation fails.

185

13.7 Actuator and Sensor Models

Propulsion model The motor model (propulsion model) implemented in the simulation is based on the standard armature controlled DC motor model [Dorf, Bishop 2001]. The transfer function for the motor in terms of an input voltage (V) and output rotational speed (θ) is:

$$\frac{\theta}{V} = \frac{K}{(Js + b)(Ls + R) + K^2}$$

Where:
J is the moment of inertia of the rotor,
s is the complex Laplace parameter,
b is the damping ratio of the mechanical system,
L is the rotor electrical inductance,
R is the terminal electrical resistance,
K is the electro-motive force constant.

Thruster model The default thruster model implemented is based on the lumped parameter dynamic thruster model developed by [Yoerger, Cook, Slotine 1991]. The thrust produced is governed by:

$$\text{Thrust} = C_t \cdot \Omega \cdot |\Omega|$$

Where:
Ω is the propeller angular velocity,
C_t is the proportionality constant.

Control surfaces Simulation of control surfaces (e.g. rudder) is required for AUV types such as USAL. The model used to determine the lift from diametrically opposite fins [Ridley, Fontan, Corke 2003] is given by:

$$L_{fin} = \frac{1}{2}\rho C_{L_{\delta f}} S_{fin} \delta_e v_e^2$$

Where:
L_{fin} is the lift force,
ρ is the density,
$C_{L_{\delta f}}$ is the rate of change of lift coefficient with respect to fin effective angle of attack,
S_{fin} is the fin platform area,
δ_e is the effective fin angle,
v_e is the effective fin velocity

SubSim also provides a much simpler model for the propulsion system in the form of an interpolated look-up table. This allows a user to experimentally collect input values and measure the resulting thrust force, applying these forces directly to the submarine model.

Sensor models   The PAL already simulates a number of sensors. Each sensor can be coupled with an error model to allow the user to simulate a sensor that returns data similar to the accuracy of the physical equipment they are trying to simulate. Many of the position and orientation sensors can be directly modeled from the data available from the lower level physics library. Every sensor is attached to a body that represents a physical component of an AUV.

The simulated inclinometer sensor calculates its orientation from the orientation of the body that it is attached to, relative to the inclinometers own initial orientation. Similarly, the simulated gyroscope calculates its orientation from the attached body's angular velocity and its own axis of rotation. The velocimeter calculates the velocity in a given direction from its orientation axis and the velocity information from the attached body.

Contact sensors are simulated by querying the collision detection routines of the low-level physics library for the positions where collisions occurred. If the collisions queried occur within the domain of the contact sensors, then these collisions are recorded.

Distance measuring sensors, such as echo-sounders and Position Sensitive Devices (PSDs) are simulated by traditional ray casting techniques, provided the low level physics library supports the necessary data structures.

A realistic synthetic camera image is being generated by the simulation system as well. With this, user application programs can use image processing for navigating the simulated AUV. Camera user interface and implementation are similar to the EyeSim mobile robot simulation system.

Environments   Detailed modeling of the environment is necessary to recreate the complex tasks facing the simulated AUV. Dynamic conditions force the AUV to continually adjust its behavior. E.g. introducing (ocean) currents causes the submarine to permanently adapt its position, poor lighting and visibility decreases image quality and eventually adds noise to PSD and vision sensors. The terrain is an essential part of the environment as it defines the universe the simulation takes part in as well as physical obstacles the AUV may encounter.

Error models   Like all the other components of the simulation system, error models are provided as plug-in extensions. All models either apply characteristic, random, or statistically chosen noise to sensor readings or the actuators' control signals. We can distinguish two different types of errors: Global errors and local errors. Global errors, such as voltage gain, affect all connected devices. Local errors only affect a certain device at a certain time. In general, local errors can be data dropouts, malfunctions or device specific errors that occur when the device constraints are violated. For example, the camera can be affected by a number of errors such as detector, Gaussian, and salt-and-pepper noise. Voltage gains (either constant or time dependent) can interfere with motor controls as well as sensor readings.

Also to be considered are any peculiarities of the simulated medium, e.g. refraction due to glass/water transitions and condensation due to temperature differences on optical instruments inside the hull.

13.8 SubSim Application

The example in Program 13.4 is written using the high-level RoBIOS API (see Appendix B.5). It implements a simple wall following task for an AUV, swimming on the water surface. Only a single PSD sensor is used (PSD_LEFT) for wall following using a bang-bang controller (see Section 4.1). No obstacle detection is done in front of the AUV.

The Mako AUV first sets both forward motors to medium speed. In an endless loop, it then continuously evaluates its PSD sensor to the left and sets left/right motor speeds accordingly, in order to avoid a wall collision.

Program 13.4: Sample AUV control program

```
 1   #include <eyebot.h>
 2
 3   int main(int argc, char* argv[])
 4   { PSDHandle psd;
 5     int distance;
 6     MotorHandle left_motor;
 7     MotorHandle right_motor;
 8     psd = PSDInit(PSD_LEFT);
 9     PSDStart(psd, 1);
10     left_motor = MOTORInit(MOTOR_LEFT);
11     right_motor= MOTORInit(MOTOR_RIGHT);
12     MOTORDrive(right_motor, 50);   /* medium speed */
13     MOTORDrive(left_motor,  50);
14     while(1)   /* endless loop */
15     { distance = PSDGet(psd);   /* distance to left */
16       if (distance < 100) MOTORDrive(left_motor, 90);
17       else if (distance>200) MOTORDrive(right_motor, 90);
18         else { MOTORDrive(right_motor, 50);
19                MOTORDrive(left_motor,  50);
20              }
21     }
22   }
```

The graphical user interface (GUI) is best demonstrated by screen shots of some simulation activity. Figure 13.7 shows Mako doing a pipeline inspection in ocean terrain, using vision feedback for detecting the pipeline. The controls of the main simulation window allow the user to rotate, pan, and zoom the scene, while it is also possible to link the user's view to the submarine itself. The console window shows the EyeBot controller with the familiar buttons and LCD, where the application program's output in text and graphics are displayed.

Figure 13.8 shows USAL hovering at the pool surface with sensor visualization switched on. The camera viewing direction and opening angle is shown as the viewing frustrum at the front end of the submarine. The PSD distance sensors are visualized by rays emitted from the submarine up to the next obstacle or pool wall (see also downward rays in pipeline example Figure 13.7).

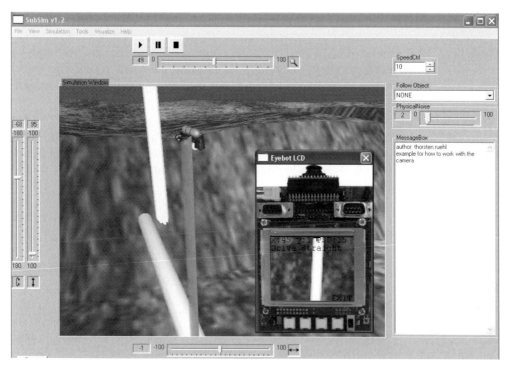

Figure 13.7: Mako pipeline following

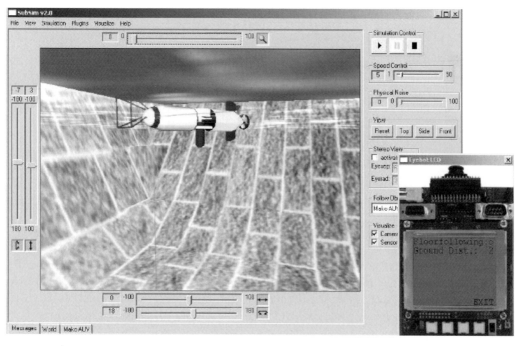

Figure 13.8: USAL pool mission

13.9 SubSim Environment and Parameter Files

XML (Extensible Markup Language) [Quin 2006] has been chosen as the basis for all parameter files in SubSim. These include parameter files for the overall simulation setup (.sub), the AUV and any passive objects in the scenery (.xml), and the environment/terrain itself (.xml).

Simulation file SUB

The general simulation parameter file (.sub) is shown in Program 13.5. It specifies the environment to be used (inclusion of a world file), the submarine to be used for the simulation (here: link to Mako.xml), any passive objects in the simulation (here: buoy.xml), and a number of general simulator settings (here: physics, view, and visualize).

The file extension ".sub" is being entered in the Windows registry, so a double-click on this parameter file will automatically start SubSim and the associated application program.

Program 13.5: Overall simulation file (.sub)

```
 1   <Simulation>
 2       <Docu text="FloorFollowing Example"/>
 3       <World file="terrain.xml" />
 4
 5       <WorldObjects>
 6           <Submarine file="./mako/mako.xml"
 7                      hdtfile="./mako/mako.hdt">
 8               <Client file="./floorfollowing.dll" />
 9           </Submarine>
10           <WorldObject file="./buoy/buoy.xml" />
11       </WorldObjects>
12
13       <SimulatorSettings>
14           <Physics noise="0.002" speed="40"/>
15           <View rotx="0" roty="0" strafex="0" strafey="0"
16                 zoom="40" followobject="Mako AUV"/>
17           <Visualize psd="dynamic"/>
18       </SimulatorSettings>
19   </Simulation>
```

Object file XML

The object xml file format (see Program 13.6) is being used for active objects, i.e. the AUV that is being controlled by a program, as well as inactive objects, e.g. floating buoys, submerged pipelines, or passive vessels.

The graphics section defines the AUV's or object's graphics appearance by linking to an *ms3d* graphics model, while the physics section deals with all simulation aspects of the object. Within the physics part, the primitives section specifies the object's position, orientation, dimensions, and mass. The subsequent sections on sensors and actuators apply only to (active) AUVs. Here, relevant details about each of the AUV's sensors and actuators are defined.

Program 13.6: AUV object file for the Mako

```
1   <?xml version="1.0"?>
2   <Submarine name="Mako AUV">
3    <Origin x="10" y="0" z="0"/>
4    <Graphics>
5      <Origin x="0" y="0" z="0"/>
6      <Scale x="0.18" y="0.18" z="0.18" />
7      <Modelfile="mako.ms3d" />
8    </Graphics>
9    <Noise>
10     <WhiteNoise strength="5.0" connects="psd_down" />
11     <WhiteNoise strength="0.5" connects="psd_front"/>
12   </Noise>
13
14   <Physics>
15     <Primitives>
16       <Box name="Mako AUV">
17         <Position x="0" y="0" z="0" />
18         <Dimensions width="1.8" height="0.5"
19                     depth="0.5" />
20         <Mass mass= "0.5"> </Mass>
21       </Box>
22     </Primitives>
23
24     <Sensors>
25         <Velocimeter name="vel0">
26           <Axis x="1" y="0" z="0"></Axis>
27           <Connection connects="Mako AUV">
28           </Connection>
29         </Velocimeter>
30   ...
31     </Sensors>
32
33     <Actuators>
34       <ImpulseActuator name="fakeprop1">
35         <Position x="0" y="0" z="0.25"></Position>
36         <Axis x="1" y="0" z="0"></Axis>
37         <Connection connects="Mako AUV"></Connection>
38       </ImpulseActuator>
39   ...
40       <Propeller name="prop0">
41         <Position x="0" y="0" z="0.25"></Position>
42         <Axis x="1" y="0" z="0"></Axis>
43         <Connection connects="Mako AUV"></Connection>
44         <Alpha lumped="0.05"></Alpha>
45       </Propeller>
46   ...
47     </Actuators>
48   </Physics>
49   </Submarine>
```

World file
XML

The world xml file format (see Program 13.7) allows the specification of typical underwater scenarios, e.g. a swimming pool or a general subsea terrain with arbitrary depth profile.

The sections on physics and water set relevant simulation parameters. The terrain section sets the world's dimensions and links to both a height map and a texture file for visualization. The visibility section affects both the graphics representation of the world, and the synthetic image that AUVs can see through their simulated on-board cameras. The optional section `WorldObjects` allows to specify passive objects that should always be present in this world setting (here a buoy). Individual objects can also be specified in the ".sub" simulation parameter file.

Program 13.7: World file for a swimming pool

```
 1   <?xml version="1.0"?>
 2   <World>
 3     <Environment>
 4       <Physics>
 5         <Engine engine="Newton" />
 6         <Gravity x="0" y="-9.81" z="0" />
 7       </Physics>
 8
 9       <Water density="1030.0"linear_viscosity="0.00120"
10                     angular_viscosity="0.00120">
11         <Dimensions width="24" length="49" />
12         <Texture file="water.bmp" />
13       </Water>
14
15       <Terrain>
16         <Origin x="0" y="-3" z="0" />
17         <Dimensions width="25" length="50" depth="4" />
18         <Heightmap file="pool.bmp" />
19         <Texture file="stone.bmp" />
20       </Terrain>
21
22       <Visibility>
23         <Fog density="0.0" depth="100" />
24       </Visibility>
25     </Environment>
26
27     <WorldObjects>
28       <WorldObject file="buoy/buoy.xml" />
29     </WorldObjects>
30   </World>
```

13.10 References

BRÄUNL, T. *Research Relevance of Mobile Robot Competitions*, IEEE Robotics and Automation Magazine, vol. 6, no. 4, Dec. 1999, pp. 32-37 (6)

BRÄUNL, T., STOLZ, H. *Mobile Robot Simulation with Sonar Sensors and Cameras*, Simulation, vol. 69, no. 5, Nov. 1997, pp. 277-282 (6)

BRÄUNL, T. KOESTLER, A. WAGGERSHAUSER, A. *Mobile Robots between Simulation & Reality*, Servo Magazine, vol. 3, no. 1, Jan. 2005, pp. 43-50 (8)

COIN3D, *The Coin Source*, http://www.Coin3D.org, 2006

DORF, R. BISHOP, R. *Modern Control Systems*, Prentice-Hall, Englewood Cliffs NJ, Ch. 4, 2001, pp 174-223 (50)

KOESTLER, A., BRÄUNL, T., *Mobile Robot Simulation with Realistic Error Models*, International Conference on Autonomous Robots and Agents, ICARA 2004, Dec. 2004, Palmerston North, New Zealand, pp. 46-51 (6)

KONOLIGE, K. *Saphira Version 6.2 Manual*, [originally: Internal Report, SRI, Stanford CA, 1998], http://www.ai.sri.com/~konolige/saphira/, 2001

KUFFNER, J., LATOMBE, J.-C. *Fast Synthetic Vision, Memory, and Learning Models for Virtual Humans*, Proceedings of Computer Animation, IEEE, 1999, pp. 118-127 (10)

LECLERCQ, P., BRÄUNL, T. *A Color Segmentation Algorithm for Real-Time Object Localization on Small Embedded Systems*, in R. Klette, S. Peleg, G. Sommer (Eds.), Robot Vision 2001, Lecture Notes in Computer Science, no. 1998, Springer-Verlag, Berlin Heidelberg, 2001, pp. 69-76 (8)

MATSUMOTO, Y., MIYAZAKI, T., INABA, M., INOUE, H. *View Simulation System: A Mobile Robot Simulator using VR Technology*, Proceedings of the International Conference on Intelligent Robots and Systems, IEEE/RSJ, 1999, pp. 936-941 (6)

MILKSHAPE, *Milkshape 3D*, http://www.swissquake.ch/chumbalum-soft, 2006

NEWTON, *Newton Game Dynamics*, http://www.physicsengine.com, 2006

QUIN, L., *Extensible Markup Language (XML)*, W3C Architecture Domain, http://www.w3.org/XML/, 2006

RIDLEY, P., FONTAN, J., CORKE, P. *Submarine Dynamic Modeling*, Australasian Conference on Robotics and Automation, CD-ROM Proceedings, 2003, pp. (6)

TINYXML, *tinyxml*, http://tinyxml.sourceforge.net, 2006

TRIEB, R. *Simulation as a tool for design and optimization of autonomous mobile robots* (in German), Ph.D. Thesis, Univ. Kaiserslautern, 1996

WANG, L., TAN, K., PRAHLAD, V. *Developing Khepera Robot Applications in a Webots Environment*, 2000 International Symposium on Micromechatronics and Human Science, IEEE, 2000, pp. 71-76 (6)

WXWIDGETS, *wxWidgets, the open source, cross-platform native UI framework*, http://www.wxwidgets.org, 2006

YOERGER, D., COOKE, J, SLOTINE, J. *The Influence of Thruster Dynamics on Underwater Vehicle Behaviours and their Incorporation into Control System Design*, IEEE Journal on Oceanic Engineering, vol. 15, no. 3, 1991, pp. 167-178 (12)

PART III:
MOBILE ROBOT APPLICATIONS

LOCALIZATION AND NAVIGATION

14

L ocalization and navigation are the two most important tasks for mobile robots. We want to know where we are, and we need to be able to make a plan for how to reach a goal destination. Of course these two problems are not isolated from each other, but rather closely linked. If a robot does not know its exact position at the start of a planned trajectory, it will encounter problems in reaching the destination.

After a variety of algorithmic approaches were proposed in the past for localization, navigation, and mapping, probabilistic methods that minimize uncertainty are now applied to the whole problem complex at once (e.g. SLAM, simultaneous localization and mapping).

14.1 Localization

One of the central problems for driving robots is localization. For many application scenarios, we need to know a robot's position and orientation at all times. For example, a cleaning robot needs to make sure it covers the whole floor area without repeating lanes or getting lost, or an office delivery robot needs to be able to navigate a building floor and needs to know its position and orientation relative to its starting point. This is a non-trivial problem in the absence of global sensors.

The localization problem can be solved by using a global positioning system. In an outdoor setting this could be the satellite-based GPS. In an indoor setting, a global sensor network with infrared, sonar, laser, or radio beacons could be employed. These will give us directly the desired robot coordinates as shown in Figure 14.1.

Let us assume a driving environment that has a number of synchronized beacons that are sending out sonar signals at the same regular time intervals, but at different (distinguishable) frequencies. By receiving signals from two or

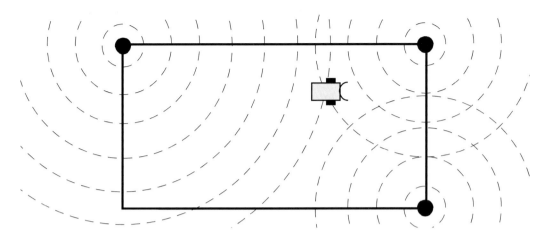

Figure 14.1: Global positioning system

three different beacons, the robot can determine its local position from the time difference of the signals' arrival times.

Using two beacons can narrow down the robot position to two possibilities, since two circles have two intersection points. For example, if the two signals arrive at exactly the same time, the robot is located in the middle between the two transmitters. If, say, the left beacon's signal arrives before the right one, then the robot is closer to the left beacon by a distance proportional to the time difference. Using local position coherence, this may already be sufficient for global positioning. However, to be able to determine a 2D position without local sensors, three beacons are required.

Only the robot's position can be determined by this method, not its orientation. The orientation has to be deducted from the change in position (difference between two subsequent positions), which is exactly the method employed for satellite-based GPS, or from an additional compass sensor.

Using global sensors is in many cases not possible because of restrictions in the robot environment, or not desired because it limits the autonomy of a mobile robot (see the discussion about overhead or global vision systems for robot soccer in Chapter 18). On the other hand, in some cases it is possible to convert a system with global sensors as in Figure 14.1 to one with local sensors. For example, if the sonar sensors can be mounted on the robot and the beacons are converted to reflective markers, then we have an autonomous robot with local sensors.

Homing beacons   Another idea is to use light emitting homing beacons instead of sonar beacons, i.e. the equivalent of a lighthouse. With two light beacons with different colors, the robot can determine its position at the intersection of the lines from the beacons at the measured angle. The advantage of this method is that the robot can determine its position *and* orientation. However, in order to do so, the robot has either to perform a 360° rotation, or to possess an omni-directional vision system that allows it to determine the angle of a recognized light beacon.

For example, after doing a 360° rotation in Figure 14.2, the robot knows it sees a green beacon at an angle of 45° and a red beacon at an angle of 165° in its local coordinate system.

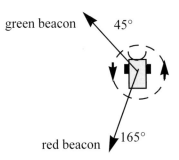

Figure 14.2: Beacon measurements

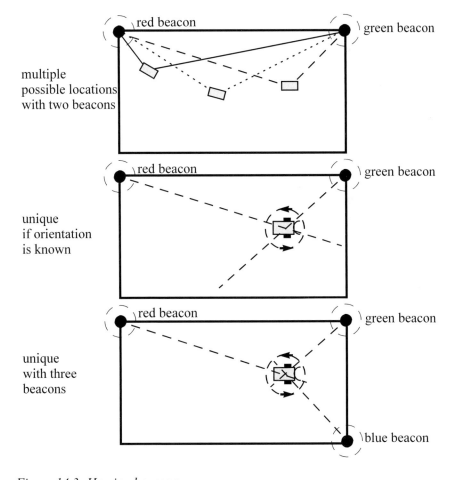

Figure 14.3: Homing beacons

We still need to fit these two vectors in the robot's environment with known beacon positions (see Figure 14.3). Since we do not know the robot's distance from either of the beacons, all we know is the angle difference under which the robot sees the beacons (here: 165°– 45° = 120°).

As can be seen in Figure 14.3, top, knowing only two beacon angles is not sufficient for localization. If the robot in addition knows its global orientation, for example by using an on-board compass, localization is possible (Figure 14.3, middle). When using three light beacons, localization is also possible without additional orientation knowledge (Figure 14.3, bottom).

Dead reckoning   In many cases, driving robots have to rely on their wheel encoders alone for short-term localization, and can update their position and orientation from time to time, for example when reaching a certain waypoint. So-called "dead reckoning" is the standard localization method under these circumstances. Dead reckoning is a nautical term from the 1700s when ships did not have modern navigation equipment and had to rely on vector-adding their course segments to establish their current position.

Dead reckoning can be described as *local polar coordinates*, or more practically as *turtle graphics geometry*. As can be seen in Figure 14.4, it is required to know the robot's starting position and orientation. For all subsequent driving actions (for example straight sections or rotations on the spot or curves), the robot's current position is updated as per the feedback provided from the wheel encoders.

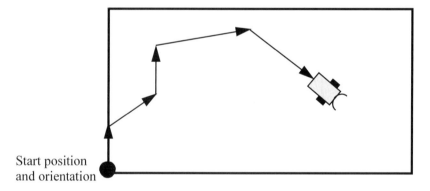

Start position
and orientation

Figure 14.4: Dead reckoning

Obviously this method has severe limitations when applied for a longer time. All inaccuracies due to sensor error or wheel slippage will add up over time. Especially bad are errors in orientation, because they have the largest effect on position accuracy.

This is why an on-board compass is very valuable in the absence of global sensors. It makes use of the earth's magnetic field to determine a robot's absolute orientation. Even simple digital compass modules work indoors and outdoors and are accurate to about 1° (see Section 2.7).

14.2 Probabilistic Localization

All robot motions and sensor measurements are affected by a certain degree of noise. The aim of probabilistic localization is to provide the best possible estimate of the robot's current configuration based on all previous data and their associated distribution functions. The final estimate will be a probability distribution because of the inherent uncertainty [Choset et al. 2005].

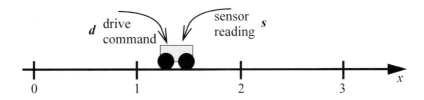

Figure 14.5: Uncertainty in actual position

Example   Assume a robot is driving in a straight line along the x axis, starting at the true position $x=0$. The robot executes driving commands with distance d, where d is an integer, and it receives sensor data from its on-board global (absolute) positioning system s (e.g. a GPS receiver), where s is also an integer. The values for d and $\Delta s = s - s'$ (current position measurement minus position measurement before executing driving command) may differ from the true position $\Delta x = x - x'$.

The robot's driving accuracy from an arbitrary starting position has to be established by extensive experimental measurements and can then be expressed by a PMF (probability mass function), e.g.:

$$p(\Delta x = d-1) = 0.2; \quad p(\Delta x = d) = 0.6; \quad p(\Delta x = d+1) = 0.2$$

Note that in this example, the robot's true position can only deviate by plus or minus one unit (e.g. cm); all position data are discrete.

In a similar way, the accuracy of the robot's position sensor has to be established by measurements, before it can be expressed as a PMF. In our example, there will again only be a possible deviation from the true position by plus or minus one unit:

$$p(x = s-1) = 0.1; \quad p(x = s) = 0.8; \quad p(x = s+1) = 0.1$$

Assuming the robot has executed a driving command with $d=2$ and after completion of this command, its local sensor reports its position as $s=2$. The probabilities for its actual position x are as follows, with n as normalization factor:

$$
\begin{aligned}
p(x=1) &= n \cdot p(s=2 \mid x=1) \cdot p(x=1 \mid d=2, x'=0) \cdot p(x'=0) \\
&= n \cdot 0.1 \cdot 0.2 \cdot 1 = 0.02n \\
p(x=2) &= n \cdot p(s=2 \mid x=2) \cdot p(x=2 \mid d=2, x'=0) \cdot p(x'=0) \\
&= n \cdot 0.8 \cdot 0.6 \cdot 1 = 0.48n \\
p(x=3) &= n \cdot p(s=2 \mid x=3) \cdot p(x=3 \mid d=2, x'=0) \cdot p(x'=0) \\
&= n \cdot 0.1 \cdot 0.2 \cdot 1 = 0.02n
\end{aligned}
$$

Positions 1, 2 and 3 are the only ones the robot can be at after a driving command with distance 2, since our PMF has probability 0 for all deviations greater than plus or minus one. Therefore, the three probabilities must add up to one, and we can use this fact to determine the normalization factor n:

$0.02n + 0.48n + 0.02n = 1$
$\rightarrow n = 1.92$

Robot's belief Now, we can calculate the probabilities for the three positions, which reflect the robot's *belief*:

$p(x=1) = 0.04;$
$p(x=2) = 0.92;$
$p(x=3) = 0.04$

So the robot is most likely to be in position 2, but it remembers all probabilities at this stage.

Continuing with the example, let us assume the robot executes a second driving command, this time with $d=1$, but after execution its sensor still reports $s=2$. The robot will now recalculate its position belief according to the conditional probabilities, with x denoting the robot's true position after driving and x' before driving:

$p(x=1) = n \cdot p(s=2 \mid x=1) \cdot$
$\qquad [\; p(x=1 \mid d=1, x'=1) \cdot p(x'=1)$
$\qquad +p(x=1 \mid d=1, x'=2) \cdot p(x'=2)$
$\qquad +p(x=1 \mid d=1, x'=3) \cdot p(x'=3)\;]$
$\qquad = n \cdot 0.1 \cdot (0.2 \cdot 0.04 + \mathbf{0} \cdot 0.92 + \mathbf{0} \cdot 0.04)$
$\qquad = 0.0008n$

$p(x=2) = n \cdot p(s=2 \mid x=2) \cdot$
$\qquad [\; p(x=2 \mid d=1, x'=1) \cdot p(x'=1)$
$\qquad +p(x=2 \mid d=1, x'=2) \cdot p(x'=2)$
$\qquad +p(x=2 \mid d=1, x'=3) \cdot p(x'=3)\;]$
$\qquad = n \cdot 0.8 \cdot (0.6 \cdot 0.04 + 0.2 \cdot 0.92 + \mathbf{0} \cdot 0.04)$
$\qquad = 0.1664n$

$p(x=3) = n \cdot p(s=2 \mid x=3) \cdot$
$\qquad [\; p(x=3 \mid d=1, x'=1) \cdot p(x'=1)$
$\qquad +p(x=3 \mid d=1, x'=2) \cdot p(x'=2)$
$\qquad +p(x=3 \mid d=1, x'=3) \cdot p(x'=3)\;]$
$\qquad = n \cdot 0.1 \cdot (0.2 \cdot 0.04 + 0.6 \cdot 0.92 + 0.2 \cdot 0.04)$
$\qquad = 0.0568n$

Note that only states $x = 1, 2$ and 3 were computed since the robot's true position can only differ from the sensor reading by one. Next, the probabilities are normalized to 1.

$\qquad 0.0008n + 0.1664n + 0.0568n = 1$
$\rightarrow \qquad n = 4.46$

$$\rightarrow \quad \begin{aligned} p(x=1) &= 0.0036 \\ p(x=2) &= 0.743 \\ p(x=3) &= 0.254 \end{aligned}$$

These final probabilities are reasonable because the robot's sensor is more accurate than its driving, hence $p(x=2) > p(x=3)$. Also, there is a very small chance the robot is in position 1, and indeed this is represented in its belief.

The biggest problem with this approach is that the configuration space must be discrete. That is, the robot's position can only be represented discretely. A simple technique to overcome this is to set the discrete representation to the minimum resolution of the driving commands and sensors, e.g. if we may not expect driving or sensors to be more accurate than 1cm, we can then express all distances in 1cm increments. This will, however, result in a large number of measurements and a large number of discrete distances with individual probabilities.

Particle filters   A technique called *particle filters* can be used to address this problem and will allow the use of non-discrete configuration spaces. The key idea in particle filters is to represent the robot's belief as a set of N particles, collectively known as M. Each particle consists of a robot configuration x and a weight $w \in [0, 1]$.

After driving, the robot updates the j-th particle's configuration x_j by first sampling the PDF (probability density function) of $p(x_j \mid d, x_j')$; typically a Gaussian distribution. After that, the robot assigns a new weight $w_j = p(s \mid x_j)$ for the j-th particle. Then, weight normalization occurs such that the sum of all weights is one. Finally, resampling occurs such that only the most likely particles remain. A standard resampling algorithm [Choset et al. 2005] is shown below:

```
M = { }
R = rand(0, 1/N)
c = w[0]
i = 0
for j=0 to N-1 do
        u = R + j/N
        while u > c do
                i = i + 1
                c = c + w[i]
        end while
        M = M + { (x[i], 1/N) }   /* add particle to set */
end for
```

Example   Like in the previous example the robots starts at $x=0$, but this time the PDF for driving is a uniform distribution specified by:

$$p(\Delta x = d + b) = \begin{cases} 1 & \text{for } b \in [-0.5, 0.5] \\ 0 & \text{otherwise} \end{cases}$$

The sensor PDF is specified by:

$$p(x=s+b) = \begin{cases} 16b + 4 & \text{for } b \in [-0.25, 0] \\ -16b + 4 & \text{for } b \in [0, 0.25] \\ 0 & \text{otherwise} \end{cases}$$

The PDF for x'=0 and d=2 is shown in Figure 14.6, left, the PDF for s=2 is shown in Figure 14.6, right.

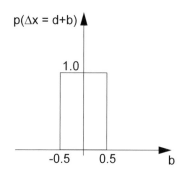
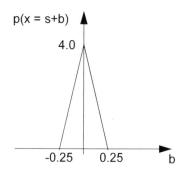

Figure 14.6: Probability density functions

Assuming the initial configuration $x=0$ is known with absolute certainty and our system consists of 4 particles (this is a very small number; in practice around 10,000 particles are used). Then the initial set is given by:

$$M = \{(0, 0.25), (0, 0.25), (0, 0.25), (0, 0.25)\}$$

Now, the robot is given a driving command d=2 and after completion, its sensors report the position as s=2. The robot first updates the configuration of each particle by sampling the PDF in Figure 14.6, left, four times. One possible result of sampling is: 1.6, 1.8, 2.2 and 2.1. Hence, M is updated to:

$$M = \{(1.6, 0.25), (1.8, 0.25), (2.2, 0.25), (2.1, 0.25)\}$$

Now, the weights for the particles are updated according to the PDF shown in Figure 14.6, right. This results in:

$p(x=1.6) = 0, \quad p(x=1.8) = 0.8, \quad p(x=2.2) = 0.8, \quad p(x=2.1) = 2.4.$

Therefore, M is updated to:

$M = \{(1.6, 0), (1.8, 0.8), (2.2, 0.8), (2.1, 2.4)\}$

After that, the weights are normalized to add up to one. This gives:

$M = \{(1.6, 0), (1.8, 0.2), (2.2, 0.2), (2.1, 0.6)\}$

Finally, the resampling procedure is applied with R=0.1 . The new M will then be:

$$M = \{(1.8, 0.25), (2.2, 0.25), (2.1, 0.25), (2.1, 0.25)\}$$

Note that the particle value 2.1 occurs twice because it is the most likely, while 1.6 drops out. If we need to know the robot's position estimate P at any time, we can simply calculate the weighted sum of all particles. In the example this comes to:

$$P = 1.8 \cdot 0.25 + 2.2 \cdot 0.25 + 2.1 \cdot 0.25 + 2.1 \cdot 0.25 = 2.05$$

14.3 Coordinate Systems

Local and global coordinate systems   We have seen how a robot can drive a certain distance or turn about a certain angle in its *local coordinate system*. For many applications, however, it is important to first establish a map (in an unknown environment) or to plan a path (in a known environment). These path points are usually specified in *global* or *world coordinates*.

Transforming local to global coordinates   Translating *local* robot coordinates to *global* world coordinates is a 2D transformation that requires a translation and a rotation, in order to match the two coordinate systems (Figure 14.7).

Assume the robot has the global position $[r_x, r_y]$ and has global orientation φ. It senses an object at local coordinates $[o_{x'}, o_{y'}]$. Then the global coordinates $[o_x, o_y]$ can be calculated as follows:

$$[o_x, o_y] = \text{Trans}(r_x, r_y) \cdot \text{Rot}(\varphi) \cdot [o_{x'}, o_{y'}]$$

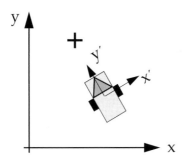

Figure 14.7: Global and local coordinate systems

For example, the marked position in Figure 14.7 has local coordinates [0, 3]. The robot's position is [5, 3] and its orientation is 30°. The global object position is therefore:

$$\begin{aligned}
[o_x, o_y] &= \text{Trans}(5, 3) \cdot \text{Rot}(30°) \cdot [0, 3] \\
&= \text{Trans}(5, 3) \cdot [-1.5, 2.6] \\
&= [3.5, 5.6]
\end{aligned}$$

Homogeneous coordinates   Coordinate transformations such as this can be greatly simplified by using "homogeneous coordinates". Arbitrary long 3D transformation sequences can be summarized in a single 4×4 matrix [Craig 1989]. In the 2D case above, a 3×3 matrix is sufficient:

$$\begin{bmatrix} o_x \\ o_y \\ 1 \end{bmatrix} = \begin{bmatrix} 1 & 0 & 5 \\ 0 & 1 & 3 \\ 0 & 0 & 1 \end{bmatrix} \cdot \begin{bmatrix} \cos\alpha & -\sin\alpha & 0 \\ \sin\alpha & \cos\alpha & 0 \\ 0 & 0 & 1 \end{bmatrix} \cdot \begin{bmatrix} 0 \\ 3 \\ 1 \end{bmatrix}$$

$$\begin{bmatrix} o_x \\ o_y \\ 1 \end{bmatrix} = \begin{bmatrix} \cos\alpha & -\sin\alpha & 5 \\ \sin\alpha & \cos\alpha & 3 \\ 0 & 0 & 1 \end{bmatrix} \cdot \begin{bmatrix} 0 \\ 3 \\ 1 \end{bmatrix}$$

for $\alpha = 30°$ this comes to:

$$\begin{bmatrix} o_x \\ o_y \\ 1 \end{bmatrix} = \begin{bmatrix} 0.87 & -0.5 & 5 \\ 0.5 & 0.87 & 3 \\ 0 & 0 & 1 \end{bmatrix} \cdot \begin{bmatrix} 0 \\ 3 \\ 1 \end{bmatrix} = \begin{bmatrix} 3.5 \\ 5.6 \\ 1 \end{bmatrix}$$

Navigation algorithms
Navigation, however, is much more than just driving to a certain specified location – it all depends on the particular task to be solved. For example, are the destination points known or do they have to be searched, are the dimensions of the driving environment known, are all objects in the environment known, are objects moving or stationary, and so on?

There are a number of well-known navigation algorithms, which we will briefly touch on in the following. However, some of them are of a more theoretical nature and do not closely match the real problems encountered in practical navigation scenarios. For example, some of the shortest path algorithms require a set of node positions and full information about their distances. But in many practical applications there are no *natural* nodes (e.g. large empty driving spaces) or their location or existence is unknown, as for partially or completely unexplored environments.

See [Arkin 1998] for more details and Chapters 15 and 16 for related topics.

14.4 Dijkstra's Algorithm

Reference [Dijkstra 1959]

Description Algorithm for computing *all* shortest paths from a given starting node in a fully connected graph. Time complexity for naive implementation is $O(e + v^2)$, and can be reduced to $O(e + v \cdot \log v)$, for e edges and v nodes.
Distances between neighboring nodes are given as edge(n,m).

Required Relative distance information between all nodes; distances must not be negative.

Dijkstra's Algorithm

Algorithm Start "ready set" with start node. In loop select node with shortest distance in every step, then compute distances to all of its neighbors and store path predecessors. Add current node to "ready set"; loop finishes when all nodes are included.

1. Init
Set start distance to 0, dist[s]=0,
others to infinite: dist[i]=∞ (for $i{\neq}s$),
Set Ready = { } .

2. Loop until all nodes are in Ready
Select node n with shortest known distance that is not in Ready set
Ready = Ready + {n} .

FOR each neighbor node m of n
  IF dist[n]+edge(n,m) < dist[m]   /* shorter path found */
    THEN { dist[m] = dist[n]+edge(n,m);
                  pre[m] = n;
              }

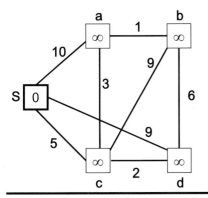

From s to:	S	a	b	c	d
Distance	0	∞	∞	∞	∞
Predecessor	-	-	-	-	-

Step 0: Init list, no predecessors
Ready = {}

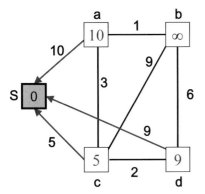

From s to:	S	a	b	c	d
Distance	0	10	∞	5	9
Predecessor	-	S	-	S	S

Step 1: Closest node is s, add to Ready

Update distances and pred. to all neighbors of s
Ready = {S}

Figure 14.8: Dijkstra's algorithm step 0 and 1

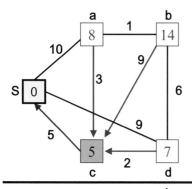

From s to:	**S**	a	b	**c**	d
Distance	0	~~10~~ 8	14	5	~~9~~ 7
Predecessor	-	~~s~~ c	c	s	~~s~~ c

Step 2: Next closest node is *c*, add to Ready
Update distances and pred. for *a* and *d*
Ready = {*S, c*}

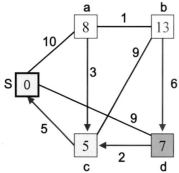

From s to:	**S**	a	b	**c**	**d**
Distance	0	8	~~14~~ 13	5	7
Predecessor	-	c	~~c~~ d	s	c

Step 3: Next closest node is *d*, add to Ready
Update distance and pred. for *b*
Ready = {*s, c, d*}

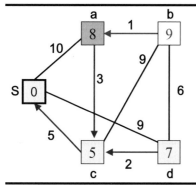

From s to:	**S**	**a**	b	**c**	**d**
Distance	0	8	~~13~~ 9	5	7
Predecessor	-	c	~~d~~ a	s	c

Step 4: Next closest node is *a*, add to Ready
Update distance and pred. for *b*
Ready = {*S, a, c, d*}

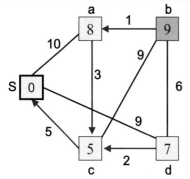

From s to:	**S**	**a**	**b**	**c**	**d**
Distance	0	8	9	5	7
Predecessor	-	c	a	s	c

Step 5: Closest node is *b*, add to Ready
check all neighbors of *s*
Ready = {*S, a, b, c, d*} ***complete!***

Figure 14.9: Dijkstra's algorithm steps 2-5

Example Consider the nodes and distances in Figure 14.8. On the left hand side is the distance graph, on the right-hand side is the table with the shortest distances found so far and the immediate path predecessors that lead to this distance.

In the beginning (initialization step), we only know that start node *S* is reachable with distance 0 (by definition). The distances to all other nodes are infinite and we do not have a path predecessor recorded yet. Proceeding from step 0 to step 1, we have to select the node with the shortest distance from all nodes that are not yet included in the Ready set. Since Ready is still empty, we have to look at all nodes. Clearly *S* has the shortest distance (0), while all other nodes still have an infinite distance.

For step 1, Figure 14.8 bottom, *S* is now included into the Ready set and the distances and path predecessors (equal to *S*) for all its neighbors are being updated. Since S is neighbor to nodes *a*, *c*, and *d*, the distances for these three nodes are being updated and their path predecessor is being set to *S*.

When moving to step 2, we have to select the node with the shortest path among *a*, *b*, *c*, *d*, as *S* is already in the Ready set. Among these, node *c* has the shortest path (5). The table is updated for all neighbors of *c*, which are *S*, *a*, *b*, and *d*. As shown in Figure 14.9, new shorter distances are found for *a*, *b*, and *d*, each entering *c* as their immediate path predecessor.

In the following steps 3 through 5, the algorithm's loop is repeated, until finally, all nodes are included in the Ready set and the algorithm terminates. The table now contains the shortest path from the start node to each of the other nodes, as well as the path predecessor for each node, allowing us to reconstruct the shortest path.

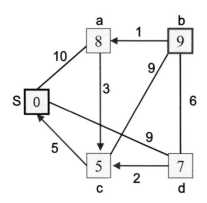

From s to:	S	a	b	c	d
Distance	0	8	9	5	7
Predecessor	-	c	a	S	c

Example: Find shortest path $S \to b$

dist[b] = 9
pre[b] = a
  pre[a] = c
    pre[c] = S

Shortest path: $S \to c \to a \to b$, length is 9

Figure 14.10: Determine shortest path

Figure 14.10 shows how to construct the shortest path from each node's predecessor. For finding the shortest path between *S* and *b*, we already know the shortest distance (9), but we have to reconstruct the shortest path backwards from *b*, by following the predecessors:

pre[b]=a, pre[a]=c, pre[c]=S

Therefore, the shortest path is: $S \to c \to a \to b$

14.5 A* Algorithm

Reference [Hart, Nilsson, Raphael 1968]

Description Pronounced "A-Star"; heuristic algorithm for computing the shortest path from *one* given start node to *one* given goal node. Average time complexity is $O(k \cdot \log_k v)$ for v nodes with branching factor k, but can be quadratic in worst case.

Required Relative distance information between all nodes plus **lower bound** of distance to goal from each node (e.g. air-line or linear distance).

Algorithm Maintain sorted list of paths to goal, in every step expand only the currently shortest path by adding adjacent node with shortest distance (including estimate of remaining distance to goal).

Example Consider the nodes and local distances in Figure 14.11. Each node has also a lower bound distance to the goal (e.g. using the Euclidean distance from a global positioning system).

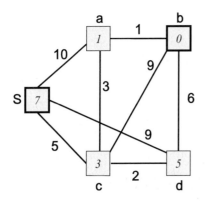

Node values are lower bound distances to goal b (e.g. linear distances)

Arc values are distances between neighboring nodes

Figure 14.11: A example*

For the first step, there are three choices:

- {S, a} with min. length $10 + 1 = 11$
- {S, c} with min. length $5 + 3 = $ **8**
- {S, d} with min. length $9 + 5 = 14$

Using a "best-first" algorithm, we explore the shortest estimated path first: {S, c}. Now the next expansion from partial path {S, c} are:

- {S, c, a} with min. length $5 + 3 + 1 = $ **9**
- {S, c, b} with min. length $5 + 9 + 0 = 14$
- {S, c, d} with min. length $5 + 2 + 5 = 12$

As it turns out, the currently shortest partial path is $\{S, c, a\}$, which we will now expand further:

- $\{S, c, a, b\}$ with min. length $5 + 3 + 1 + 0 = \mathbf{9}$

There is only a single possible expansion, which reaches the goal node b and is the shortest path found so far, so the algorithm terminates. The shortest path and the corresponding distance have been found.

Note This algorithm may look complex since there seems to be the need to store incomplete paths and their lengths at various places. However, using a recursive best-first search implementation can solve this problem in an elegant way without the need for explicit path storing.

The quality of the lower bound goal distance from each node greatly influences the timing complexity of the algorithm. The closer the given lower bound is to the true distance, the shorter the execution time.

14.6 Potential Field Method

References [Arbib, House 1987], [Koren, Borenstein 1991], [Borenstein, Everett, Feng 1998]

Description Global map generation algorithm with virtual forces.

Required Start and goal position, positions of all obstacles and walls.

Algorithm Generate a map with virtual attracting and repelling forces. Start point, obstacles, and walls are repelling, goal is attracting; force strength is inverse to object distance; robot simply follows force field.

Example Figure 14.12 shows an example with repelling forces from obstacles and walls, plus a superimposed general field direction from start to goal.

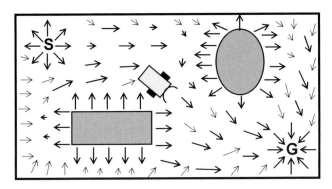

Figure 14.12: Potential field

Figure 14.13 exemplifies the potential field generation steps in the form of 3D surface plots. A ball placed on the start point of this surface would roll

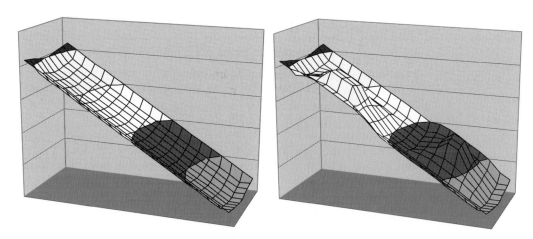

Figure 14.13: Potential fields as 3D surfaces

toward the goal point – this demonstrates the derived driving path of a robot. The 3D surface on the left only represents the force vector field between start and goal as a potential (height) difference, as well as repelling walls. The 3D surface on the right has the repelling forces for two obstacles added.

Problem The robot can get stuck in local minima. In this case the robot has reached a spot with zero force (or a level potential), where repelling and attracting forces cancel each other out. So the robot will stop and never reach the goal.

14.7 Wandering Standpoint Algorithm

Reference [Puttkamer 2000]

Description Local path planning algorithm.

Required Local distance sensor.

Algorithm Try to reach goal from start in direct line. When encountering an obstacle, measure avoidance angle for turning left and for turning right, turn to smaller angle. Continue with boundary-following around the object, until goal direction is clear again.

Example Figure 14.14 shows the subsequent robot positions from *Start* through 1..6 to *Goal*. The goal is not directly reachable from the start point. Therefore, the robot switches to boundary-following mode until, at point 1, it can drive again unobstructed toward the goal. At point 2, another obstacle has been reached, so the robot once again switches to boundary-following mode. Finally at point 6, the goal is directly reachable in a straight line without further obstacles.

Realistically, the actual robot path will only approximate the waypoints but not exactly drive through them.

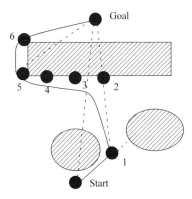

Figure 14.14: Wandering standpoint

Problem The algorithm can lead to an endless loop for extreme obstacle placements. In this case the robot keeps driving, but never reaches the goal.

14.8 DistBug Algorithm

Reference [Kamon, Rivlin 1997]

Description Local planning algorithm that guarantees convergence and will find path if one exists.

Required Own position (odometry), goal position, and distance sensor data.

Algorithm Drive straight towards the goal when possible, otherwise do boundary-following around an obstacle. If this brings the robot back to the same previous collision point with the obstacle, then the goal is unreachable.
Below is our version of an algorithmic translation of the original paper.

Constant: STEP min. distance of two leave points, e.g. 1cm

Variables: P current robot position (x, y)
G goal position (x, y)
Hit location where current obstacle was first hit
Min_dist minimal distance to goal during boundary following

1. Main program
    Loop
        "drive towards goal" /* non-blocking, proc. continues while driv. */
        if P=G **then** {**"success"**; **terminate**;}
        if "obstacle collision" { Hit = P; **call** follow; }
    End loop

2. Subroutine follow

Min_dist = ∞; /* init */
Turn left; /* to align with wall */
Loop

"drive following obstacle boundary"; /* non-block., cont. proc. */
D = dist(P, G) /* air-line distance from current position to goal */
F = free(P, G) /* space in direction of goal, e.g. PSD measurement */
if D < Min_dist **then** Min_dist = D;

if F ≥ D **or** D–F ≤ Min_dist – STEP **then** return;
/* goal is directly reachable **or** a point closer to goal is reachable */
if P = Hit **then** { **"goal unreachable"**; **terminate**; }
End loop

Problem Although this algorithm has nice theoretical properties, it is not very usable in practice, as the positioning accuracy and sensor distance required for the success of the algorithm are usually not achievable. Most variations of the Dist-Bug algorithm suffer from a lack of robustness against noise in sensor readings and robot driving/positioning.

Examples Figure 14.15 shows two standard DistBug examples, here simulated with the EyeSim system. In the example on the left hand side, the robot starts in the main program loop, driving forward towards the goal, until it hits the U-shaped obstacle. A hit point is recorded and subroutine *follow* is called. After a left turn, the robot follows the boundary around the left leg, at first getting further away from the goal, then getting closer and closer. Eventually, the free space in goal direction will be greater or equal to the remaining distance to the goal (this happens at the leave point). Then the boundary follow subroutine returns to the main program, and the robot will for the second time drive directly towards the goal. This time the goal is reached and the algorithm terminates.

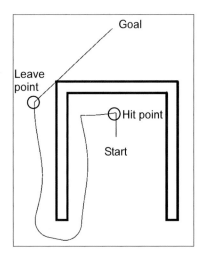
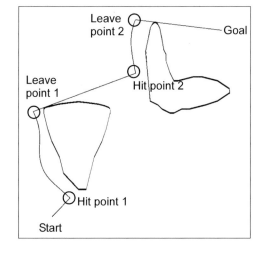

Figure 14.15: Distbug examples

Figure 14.15, right, shows another example. The robot will stop boundary following at the first leave point, because its sensors have detected that it can reach a point closer to the goal than before. After reaching the second hit point, boundary following is called a second time, until at the second leave point the robot can drive directly to the goal.

Figure 14.16 shows two more examples that further demonstrate the Dist-Bug algorithm. In Figure 14.16, left, the goal is inside the E-shaped obstacle and cannot be reached. The robot first drives straight towards the goal, hits the obstacle and records the hit point, then starts boundary following. After completion of a full circle around the obstacle, the robot returns to the hit point, which is its termination condition for an unreachable goal.

Figure 14.16, right, shows a more complex example. After the hit point has been reached, the robot surrounds almost the whole obstacle until it finds the entry to the maze-like structure. It continues boundary following until the goal is directly reachable from the leave point.

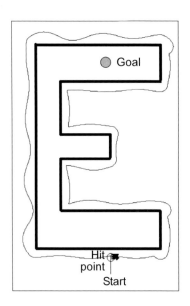
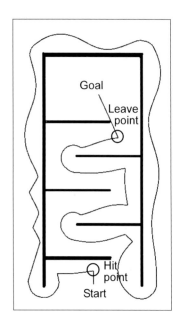

Figure 14.16: Complex Distbug examples

14.9 References

ARBIB, M., HOUSE, D. *Depth and Detours: An Essay on Visually Guided Behavior*, in M. Arbib, A. Hanson (Eds.), Vision, Brain and Cooperative Computation, MIT Press, Cambridge MA, 1987, pp. 129-163 (35)

ARKIN, R. *Behavior-Based Robotics*, MIT Press, Cambridge MA, 1998

BORENSTEIN, J., EVERETT, H., FENG, L. *Navigating Mobile Robots: Sensors and Techniques*, AK Peters, Wellesley MA, 1998

CHOSET H., LYNCH, K., HUTCHINSON, S., KANTOR, G., BURGARD, W., KAVRAKI, L., THRUN, S. *Principles of Robot Motion: Theory, Algorithms, and Implementations*, MIT Press, Cambridge MA, 2005

CRAIG, J. *Introduction to Robotics – Mechanics and Control*, 2nd Ed., Addison-Wesley, Reading MA, 1989

DIJKSTRA, E. *A note on two problems in connexion with graphs*, Numerische Mathematik, Springer-Verlag, Heidelberg, vol. 1, pp. 269-271 (3), 1959

HART, P., NILSSON, N., RAPHAEL, B. *A Formal Basis for the Heuristic Determination of Minimum Cost Paths*, IEEE Transactions on Systems Science and Cybernetics, vol. SSC-4, no. 2, 1968, pp. 100-107 (8)

KAMON, I., RIVLIN, E. *Sensory-Based Motion Planning with Global Proofs*, IEEE Transactions on Robotics and Automation, vol. 13, no. 6, Dec. 1997, pp. 814-822 (9)

KOREN, Y., BORENSTEIN, J. *Potential Field Methods and Their Inherent Limitations for Mobile Robot Navigation*, Proceedings of the IEEE Conference on Robotics and Automation, Sacramento CA, April 1991, pp. 1398-1404 (7)

PUTTKAMER, E. VON. *Autonome Mobile Roboter*, Lecture notes, Univ. Kaiserslautern, Fachbereich Informatik, 2000

MAZE EXPLORATION

Mobile robot competitions have been around for over 20 years, with the Micro Mouse Contest being the first of its kind in 1977. These competitions have inspired generations of students, researchers, and laypersons alike, while consuming vast amounts of research funding and personal time and effort. Competitions provide a goal together with an objective performance measure, while extensive media coverage allows participants to present their work to a wider forum.

As the robots in a competition evolved over the years, becoming faster and smarter, so did the competitions themselves. Today, interest has shifted from the "mostly solved" maze contest toward robot soccer (see Chapter 18).

15.1 Micro Mouse Contest

Start: 1977 in New York

"The stage was set. A crowd of spectators, mainly engineers, were there. So were reporters from the Wall Street Journal, the New York Times, other publications, and television. All waited in expectancy as Spectrum's Mystery Mouse Maze was unveiled. Then the color television cameras of CBS and NBC began to roll; the moment would be recreated that evening for viewers of the Walter Cronkite and John Chancellor-David Brinkley news shows" [Allan 1979].

This report from the first "Amazing Micro-Mouse Maze Contest" demonstrates the enormous media interest in the first mobile robot competition in New York in 1977. The academic response was overwhelming. Over 6,000 entries followed the announcement of Don Christiansen [Christiansen 1977], who originally suggested the contest.

The task is for a robot mouse to drive from the start to the goal in the fastest time. Rules have changed somewhat over time, in order to allow exploration of the whole maze and then to compute the shortest path, while also counting exploration time at a reduced factor.

The first mice constructed were rather simple – some of them did not even contain a microprocessor as controller, but were simple "wall huggers" which

Figure 15.1: Maze from Micro Mouse Contest in London 1986

would find the goal by always following the left (or the right) wall. A few of these scored even higher than some of the *intelligent* mice, which were mechanically slower.

John Billingsley [Billingsley 1982] made the Micro Mouse Contest popular in Europe and called for the first rule change: starting in a corner, the goal should be in the center and not in another corner, to eliminate wall huggers. From then on, more intelligent behavior was required to solve a maze (Figure 15.1). Virtually all robotics labs at that time were building micromice in one form or another – a real micromouse craze was going around the world. All of a sudden, people had a goal and could share ideas with a large number of colleagues who were working on exactly the same problem.

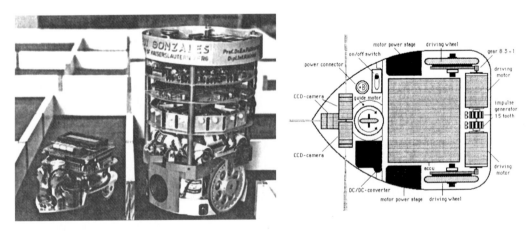

Figure 15.2: Two generations of micromice, Univ. Kaiserslautern

Micromouse technology evolved quite a bit over time, as did the running time. A typical sensor arrangement was to use three sensors to detect any walls in front, to the left, and to the right of the mouse. Early mice used simple micro-switches as touch sensors, while later on sonar, infrared, or even optical sensors [Hinkel 1987] became popular (Figure 15.2).

While the mouse's size is restricted by the maze's wall distance, smaller and especially lighter mice have the advantage of higher acceleration/deceleration and therefore higher speed. Even smaller mice became able to drive in a straight diagonal line instead of going through a sequence of left/right turns, which exist in most mazes.

Figure 15.3: Micromouse, Univ. of Queensland

One of today's fastest mice comes from the University of Queensland, Australia (see Figure 15.3 – the Micro Mouse Contest has survived until today!), using three extended arms with several infrared sensors each for reliable wall distance measurement. By and large, it looks as if the micromouse problem has been solved, with the only possible improvement being on the mechanics side, but not in electronics, sensors, or software [Bräunl 1999].

15.2 Maze Exploration Algorithms

For maze exploration, we will develop two algorithms: a simple iterative procedure that follows the left wall of the maze ("wall hugger"), and an only slightly more complex recursive procedure to explore the full maze.

15.2.1 Wall-Following

Our first naive approach for the exploration part of the problem is to always follow the left wall. For example, if a robot comes to an intersection with several open sides, it follows the leftmost path. Program 15.1 shows the implementation of this function `explore_left`. The start square is assumed to be at position [0,0], the four directions north, west, south, and east are encoded as integers 0, 1, 2, 3.

The procedure `explore_left` is very simple and comprises only a few lines of code. It contains a single while-loop that terminates when the goal square is

Program 15.1: Explore-Left

```
1   void explore_left(int goal_x, int goal_y)
2   { int x=0, y=0, dir=0; /* start position */
3     int front_open, left_open, right_open;
4
5     while (!(x==goal_x && y==goal_y)) /* goal not reached */
6     { front_open = PSDGet(psd_front) > THRES;
7       left_open  = PSDGet(psd_left)  > THRES;
8       right_open = PSDGet(psd_right) > THRES;
9
10      if (left_open) turn(+1, &dir); /* turn left */
11        else if (front_open);            /* drive straight*/
12          else if (right_open) turn(-1, &dir);/* turn right */
13            else turn(+2, &dir);         /* dead end - back up */
14      go_one(&x,&y,dir);        /* go one step in any case */
15    }
16  }
```

reached (x and y coordinates match). In each iteration, it is determined by
reading the robot's infrared sensors whether a wall exists on the front, left-, or
right-hand side (boolean variables front_open, left_open, right_open).
The robot then selects the "leftmost" direction for its further journey. That is, if
possible it will always drive left, if not it will try driving straight, and only if
the other two directions are blocked, will it try to drive right. If none of the
three directions are free, the robot will turn on the spot and go back one square,
since it has obviously arrived at a dead-end.

Program 15.2: Driving support functions

```
1   void turn(int change, int *dir)
2   {   VWDriveTurn(vw, change*PI/2.0, ASPEED);
3       VWDriveWait(vw);
4       *dir = (*dir+change +4) % 4;
5   }
```

```
1   void go_one(int *x, int *y, int dir)
2   { switch (dir)
3     { case 0: (*y)++; break;
4       case 1: (*x)--; break;
5       case 2: (*y)--; break;
6       case 3: (*x)++; break;
7     }
8     VWDriveStraight(vw, DIST, SPEED);
9     VWDriveWait(vw);
10  }
```

The support functions for turning multiples of 90° and driving one square
are quite simple and shown in Program 15.2. Function turn turns the robot by
the desired angle (±90° or 180°), and then updates the direction parameter dir.

Function `go_one` updates the robot's position in x and y, depending on the current direction `dir`. It then drives the robot a fixed distance forward.

This simple and elegant algorithm works very well for most mazes. However, there are mazes where this algorithm does not work As can be seen in Figure 15.4, a maze can be constructed with the goal in the middle, so a wall-following robot will never reach it. The recursive algorithm shown in the following section, however, will be able to cope with arbitrary mazes.

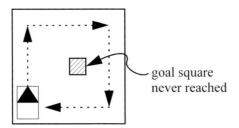

Figure 15.4: Problem for wall-following

15.2.2 Recursive Exploration

The algorithm for full maze exploration guarantees that each reachable square in the maze will be visited, independent of the maze construction. This, of course, requires us to generate an internal representation of the maze and to maintain a bit-field for marking whether a particular square has already been visited. Our new algorithm is structured in several stages for exploration and navigation:

1. Explore the whole maze:
    Starting at the start square, visit all reachable squares in the maze, then return to the start square (this is accomplished by a recursive algorithm).

2. Compute the shortest distance from the start square to any other square by using a "flood fill" algorithm.

3. Allow the user to enter the coordinates of a desired destination square:
    Then determine the shortest driving path by reversing the path in the flood fill array from the destination to the start square.

The difference between the wall-following algorithm and this recursive exploration of all paths is sketched in Figure 15.5. While the wall-following algorithm only takes a single path, the recursive algorithm explores all possible paths subsequently. Of course this requires some bookkeeping of squares already visited to avoid an infinite loop.

Program 15.3 shows an excerpt from the central recursive function `explore`. Similar to before, we determine whether there are walls in front and to the left and right of the current square. However, we also mark the current square as visited (data structure `mark`) and enter any walls found into our inter-

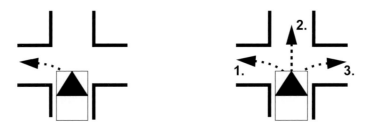

Figure 15.5: Left wall-following versus recursive exploration

Program 15.3: Explore

```
 1  void explore()
 2  { int front_open, left_open, right_open;
 3    int old_dir;
 4
 5    mark[rob_y][rob_x] = 1;    /* set mark */
 6    PSDGet(psd_left), PSDGet(psd_right));
 7    front_open = PSDGet(psd_front) > THRES;
 8    left_open  = PSDGet(psd_left)  > THRES;
 9    right_open = PSDGet(psd_right) > THRES;
10    maze_entry(rob_x,rob_y,rob_dir,         front_open);
11    maze_entry(rob_x,rob_y,(rob_dir+1)%4, left_open);
12    maze_entry(rob_x,rob_y,(rob_dir+3)%4, right_open);
13    old_dir = rob_dir;
14
15    if (front_open  && unmarked(rob_y,rob_x,old_dir))
16      { go_to(old_dir);    /* go 1 forward */
17        explore();         /* recursive call */
18        go_to(old_dir+2); /* go 1 back */
19      }
20  if (left_open && unmarked(rob_y,rob_x,old_dir+1))
21      { go_to(old_dir+1); /* go 1 left */
22        explore();         /* recursive call */
23        go_to(old_dir-1); /* go 1 right */
24      }
25  if (right_open && unmarked(rob_y,rob_x,old_dir-1))
26      { go_to(old_dir-1); /* go 1 right */
27        explore();         /* recursive call */
28        go_to(old_dir+1); /* go 1 left */
29      }
30  }
```

nal representation using auxiliary function `maze_entry`. Next, we have a maximum of three recursive calls, depending on whether the direction front, left, or right is open (no wall) and the next square in this direction has not been visited before. If this is the case, the robot will drive into the next square and the procedure `explore` will be called recursively. Upon termination of this call, the robot will return to the previous square. Overall, this will result in the robot exploring the whole maze and returning to the start square upon completion of the algorithm.

A possible extension of this algorithm is to check in every iteration if all surrounding walls of a new, previously unvisited square are already known (for example if the surrounding squares have been visited). In that case, it is not required for the robot to actually visit this square. The trip can be saved and the internal database can be updated.

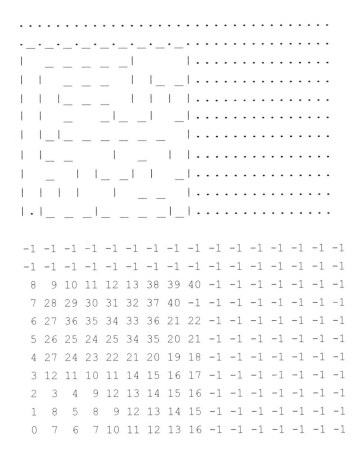

```
-1  -1  -1  -1  -1  -1  -1  -1  -1  -1  -1  -1  -1  -1  -1  -1
-1  -1  -1  -1  -1  -1  -1  -1  -1  -1  -1  -1  -1  -1  -1  -1
 8   9  10  11  12  13  38  39  40  -1  -1  -1  -1  -1  -1  -1
 7  28  29  30  31  32  37  40  -1  -1  -1  -1  -1  -1  -1  -1
 6  27  36  35  34  33  36  21  22  -1  -1  -1  -1  -1  -1  -1
 5  26  25  24  25  34  35  20  21  -1  -1  -1  -1  -1  -1  -1
 4  27  24  23  22  21  20  19  18  -1  -1  -1  -1  -1  -1  -1
 3  12  11  10  11  14  15  16  17  -1  -1  -1  -1  -1  -1  -1
 2   3   4   9  12  13  14  15  16  -1  -1  -1  -1  -1  -1  -1
 1   8   5   8   9  12  13  14  15  -1  -1  -1  -1  -1  -1  -1
 0   7   6   7  10  11  12  13  16  -1  -1  -1  -1  -1  -1  -1
```

Figure 15.6: Maze algorithm output

We have now completed the first step of the algorithm, sketched in the beginning of this section. The result can be seen in the top of Figure 15.6. We now know for each square whether it can be reached from the start square or not, and we know all walls for each reachable square.

Flood fill algorithm   In the second step, we want to find the minimum distance (in squares) of each maze square from the start square. Figure 15.6, bottom, shows the shortest distances for each point in the maze from the start point. A value of –1 indicates a position that cannot be reached (for example outside the maze boundaries). We are using a flood fill algorithm to accomplish this (see Program 15.4).

Program 15.4: Flood fill

```
1      for (i=0; i<MAZESIZE; i++) for (j=0; j<MAZESIZE; j++)
2      {   map [i][j] = -1;   /* init */
3          nmap[i][j] = -1;
4      }
5      map [0][0] = 0;
6      nmap[0][0] = 0;
7      change = 1;
8
9      while (change)
10     { change = 0;
11       for (i=0; i<MAZESIZE; i++) for (j=0; j<MAZESIZE; j++)
12       { if (map[i][j] == -1)
13         { if (i>0)
14             if (!wall[i][j][0]   && map[i-1][j] != -1)
15               { nmap[i][j] = map[i-1][j] + 1; change = 1; }
16           if (i<MAZESIZE-1)
17             if (!wall[i+1][j][0] && map[i+1][j] != -1)
18               { nmap[i][j] = map[i+1][j] + 1; change = 1; }
19           if (j>0)
20             if (!wall[i][j][1]   && map[i][j-1] != -1)
21               { nmap[i][j] = map[i][j-1] + 1; change = 1; }
22           if (j<MAZESIZE-1)
23             if (!wall[i][j+1][1] && map[i][j+1] != -1)
24               { nmap[i][j] = map[i][j+1] + 1; change = 1; }
25         }
26       }
27       for (i=0; i<MAZESIZE; i++) for (j=0; j<MAZESIZE; j++)
28         map[i][j] = nmap[i][j];   /* copy back */
29     }
```

The program uses a 2D integer array `map` for recording the distances plus a copy, `nmap`, which is used and copied back after each iteration. In the beginning, each square (array element) is marked as unreachable (−1), except for the start square [0,0], which can be reached in zero steps. Then, we run a while-loop as long as at least one map entry changes. Since each "change" reduces the number of unknown squares (value −1) by at least one, the upper bound of loop iterations is the total number of squares ($MAZESIZE^2$). In each iteration we use two nested `for`-loops to examine all unknown squares. If there is a path (no wall to north, south, east, or west) to a known square (value ≠ −1), the new distance is computed and entered in distance array `nmap`. Additional `if`-selections are required to take care of the maze borders. Figure 15.7 shows the stepwise generation of the distance map.

In the third and final step of our algorithm, we can now determine the shortest path to any maze square from the start square. We already have all wall information and the shortest distances for each square (see Figure 15.6). If the

```
-1 -1 -1 -1 -1 -1        -1 -1 -1 -1 -1 -1        -1 -1 -1 -1 -1 -1
-1 -1 -1 -1 -1 -1        -1 -1 -1 -1 -1 -1        -1 -1 -1 -1 -1 -1
-1 -1 -1 -1 -1 -1        -1 -1 -1 -1 -1 -1        -1 -1 -1 -1 -1 -1
-1 -1 -1 -1 -1 -1        -1 -1 -1 -1 -1 -1         2 -1 -1 -1 -1 -1
-1 -1 -1 -1 -1 -1         1 -1 -1 -1 -1 -1         1 -1 -1 -1 -1 -1
 0 -1 -1 -1 -1 -1         0 -1 -1 -1 -1 -1         0 -1 -1 -1 -1 -1

-1 -1 -1 -1 -1 -1        -1 -1 -1 -1 -1 -1         5 -1 -1 -1 -1 -1
-1 -1 -1 -1 -1 -1         4 -1 -1 -1 -1 -1         4 -1 -1 -1 -1 -1
 3 -1 -1 -1 -1 -1         3 -1 -1 -1 -1 -1         3 -1 -1 -1 -1 -1
 2  3 -1 -1 -1 -1         2  3  4 -1 -1 -1         2  3  4 -1 -1 -1
 1 -1 -1 -1 -1 -1         1 -1 -1 -1 -1 -1         1 -1  5 -1 -1 -1
 0 -1 -1 -1 -1 -1         0 -1 -1 -1 -1 -1         0 -1 -1 -1 -1 -1
```

Figure 15.7: Distance map development (excerpt)

user wants the robot to drive to, say, maze square [1,2] (row 1, column 2, assuming the start square is at [0,0]), then we know already from our distance map (see Figure 15.7, bottom right) that this square can be reached in five steps. In order to find the shortest driving path, we can now simply trace back the path from the desired goal square [1,2] to the start square [0,0]. In each step we select a connected neighbor square from the current square (no wall between them) that has a distance of one less than the current square. That is, if the current square has distance d, the new selected neighbor square has to have a distance of d–1 (Figure 15.8).

```
X-᷂ Eyebot Console 0    •□x    X-᷂ Eyebot Console 0    •□x    X-᷂ Eyebot Console 0    •□x    X-᷂ Eyebot Console 0    •□x
GOAL y,x:   0   0             Map distances                 Map distances                 Path..done
                                                             -1 -1 -1 -1
._._._._._...                 .............._...            4 -1 -1 -1                    ._._._._._....
| | |_ _   |...               xxxxx.........                3 12 11 10                    | | |_ _  G|...
| |_ _     |...               xxxxx.........                2  3  4  9                    | |_ _ * *|...
|    _ | |_|...               xxxxx.........                1  8  5  8                    |* * *|*|_|...
| | | |    |...               xxxxx.........                0  7  6  7                    |*| |*|* |...
|.|_ _ _|_|...                xxxxx.........                                              |S|_ * *|_|...
Y+- X+- +/- GO                Map Mrk Maz DRV               Map Mrk Maz DRV               Map Mrk Maz DRV
```

Figure 15.8: Screen dumps: exploration, visited cells, distances, shortest path

Program 15.5 shows the program to find the path, Figure 15.9 demonstrates this for the example [1,2]. Since we already know the length of the path (entry in `map`), a simple for-loop is sufficient to construct the shortest path. In each iteration, all four sides are checked whether there is a path and the neighbor square has a distance one less than the current square. This must be the case for at least one side, or there would be an error in our data structure. There could be more than one side for which this is true. In this case, multiple shortest paths exist.

Program 15.5: Shortest path

```
1   void build_path(int i, int j)
2   { int k;
3     for (k = map[i,j]-1; k>=0; k--)
4     {
5       if (i>0 && !wall[i][j][0] && map[i-1][j] == k)
6       { i--;
7         path[k] = 0;  /* north */
8       }
9     else
10      if (i<MAZESIZE-1  &&!wall[i+1][j][0] &&map[i+1][j]==k)
11      { i++;
12        path[k] = 2;  /* south */
13      }
14    else
15      if (j>0  && !wall[i][j][1] && map[i][j-1]==k)
16      { j--;
17        path[k] = 3;  /* east */
18      }
19    else
20      if (j<MAZESIZE-1  &&!wall[i][j+1][1] &&map[i][j+1]==k)
21      { j++;
22        path[k] = 1;  /* west */
23      }
24    else
25      { LCDPutString("ERROR\a");
26      }
27    }
28  }
```

15.3 Simulated versus Real Maze Program

Simulations are never enough: the real world contains real problems! We first implemented and tested the maze exploration problem in the EyeSim simulator before running the same program unchanged on a real robot [Koestler, Bräunl 2005]. Using the simulator initially for the higher levels of program design and debugging is very helpful, because it allows us to concentrate on the logic issues and frees us from all real-world robot problems. Once the logic of the exploration and path planning algorithm has been verified, one can concentrate on the lower-level problems like fault-tolerant wall detection and driving accuracy by adding sensor/actuator noise to error levels typically encountered by real robots. *This basically transforms a Computer Science problem into a Computer Engineering problem.*

Now we have to deal with false and inaccurate sensor readings and dislocations in robot positioning. We can still use the simulator to make the necessary

```
5 -1 -1 -1 -1 -1        5 -1 -1 -1 -1 -1        5 -1 -1 -1 -1 -1
4 -1 -1 -1 -1 -1        4 -1 -1 -1 -1 -1        4 -1 -1 -1 -1 -1
3 -1 -1 -1 -1 -1        3 -1 -1 -1 -1 -1        3 -1 -1 -1 -1 -1
2  3  4 -1 -1 -1        2  3  4 -1 -1 -1        2  3  4 -1 -1 -1
1 -1  5 -1 -1 -1        1 -1  5 -1 -1 -1        1 -1  5 -1 -1 -1
0 -1 -1 -1 -1 -1        0 -1 -1 -1 -1 -1        0 -1 -1 -1 -1 -1
```
Path: {}   **Path: {S}**   **Path: {E,S}**

```
5 -1 -1 -1 -1 -1        5 -1 -1 -1 -1 -1        5 -1 -1 -1 -1 -1
4 -1 -1 -1 -1 -1        4 -1 -1 -1 -1 -1        4 -1 -1 -1 -1 -1
3 -1 -1 -1 -1 -1        3 -1 -1 -1 -1 -1        3 -1 -1 -1 -1 -1
2  3  4 -1 -1 -1        2  3  4 -1 -1 -1        2  3  4 -1 -1 -1
1 -1  5 -1 -1 -1        1 -1  5 -1 -1 -1        1 -1  5 -1 -1 -1
0 -1 -1 -1 -1 -1        0 -1 -1 -1 -1 -1        0 -1 -1 -1 -1 -1
```
Path: {E,E,S}   **Path: {N,E,E,S}**   **Path: {N,N,E,E,S}**

Figure 15.9: Shortest path for position [y,x] = [1,2]

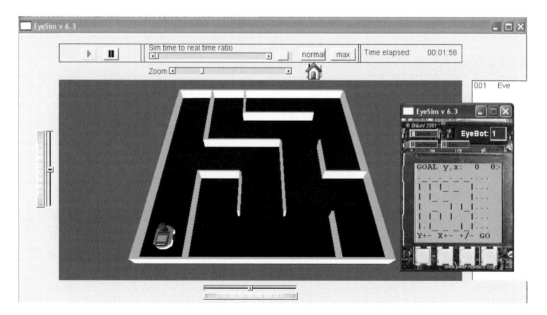

Figure 15.10: Simulated maze solving

changes to improve the application's robustness and fault tolerance, before we eventually try it on a real robot in a maze environment.

What needs to be added to the previously shown maze program can be described by the term *fault tolerance*. We must not assume that a robot has turned 90° degrees after we give it the corresponding command. It may in fact

227

have turned only 89° or 92°. The same holds for driving a certain distance or for sensor readings. The logic of the program does not need to be changed, only the driving routines.

The best way of making our maze program fault tolerant is to have the robot continuously monitor its environment while driving (Figure 15.10). Instead of issuing a command to drive straight for a certain distance and wait until it is finished, the robot should constantly measure the distance to the left and right wall while driving and continuously correct its steering while driving forward. This will correct driving errors as well as possible turning errors from a previous change of direction. If there is no wall to the left or the right or both, this information can be used to adjust the robot's position inside the square, i.e. avoiding driving too far or too short (Figure 15.11). For the same reason, the robot needs to constantly monitor its distance to the front, in case there is a wall.

Figure 15.11: Adaptive driving using three sensors

15.4 References

ALLAN, R. *The amazing micromice: see how they won*, IEEE Spectrum, Sept. 1979, vol. 16, no. 9, pp. 62-65 (4)

BILLINGSLEY, J. *Micromouse - Mighty mice battle in Europe*, Practical Computing, Dec. 1982, pp. 130-131 (2)

BRÄUNL, T. *Research Relevance of Mobile Robot Competitions*, IEEE Robotics and Automation Magazine, vol. 6, no. 4, Dec. 1999, pp. 32-37 (6)

CHRISTIANSEN, C. *Announcing the amazing micromouse maze contest*, IEEE Spectrum, vol. 14, no. 5, May 1977, p. 27 (1)

HINKEL, R. *Low-Level Vision in an Autonomous Mobile Robot*, EUROMICRO 1987; 13th Symposium on Microprocessing and Microprogramming, Portsmouth England, Sept. 1987, pp. 135-139 (5)

KOESTLER, A., BRÄUNL, T., *Mobile Robot Simulation with Realistic Error Models*, International Conference on Autonomous Robots and Agents, ICARA 2004, Dec. 2004, Palmerston North, New Zealand, pp. 46-51 (6)

MAP GENERATION

16

Mapping an unknown environment is a more difficult task than the maze exploration shown in the previous chapter. This is because, in a maze, we know for sure that all wall segments have a certain length and that all angles are at 90°. In the general case, however, this is not the case. So a robot having the task to explore and map an arbitrary unknown environment has to use more sophisticated techniques and requires higher-precision sensors and actuators as for a maze.

16.1 Mapping Algorithm

Several map generation systems have been presented in the past [Chatila 1987], [Kampmann 1990], [Leonard, Durrant-White 1992], [Melin 1990], [Piaggio, Zaccaria 1997], [Serradilla, Kumpel 1990]. While some algorithms are limited to specific sensors, for example sonar sensors [Bräunl, Stolz 1997], many are of a more theoretical nature and cannot be directly applied to mobile robot control. We developed a practical map generation algorithm as a combination of configuration space and occupancy grid approaches. This algorithm is for a 2D environment with static obstacles. We assume no prior information about obstacle shape, size, or position. A version of the "DistBug" algorithm [Kamon, Rivlin 1997] is used to determine routes around obstacles. The algorithm can deal with imperfect distance sensors and localization errors [Bräunl, Tay 2001].

For implementation, we are using the robot *Eve*, which is the first EyeBot-based mobile robot. It uses two different kinds of infrared sensors: one is a binary sensor (IR-proxy) which is activated if the distance to an obstacle is below a certain threshold, the other is a position sensitive device (IR-PSD) which returns a distance value to the nearest obstacle. In principle, sensors can be freely positioned and oriented around the robot. This allows testing and comparison of the performance of different sensor positions. The sensor configuration used here is shown in Figure 16.1.

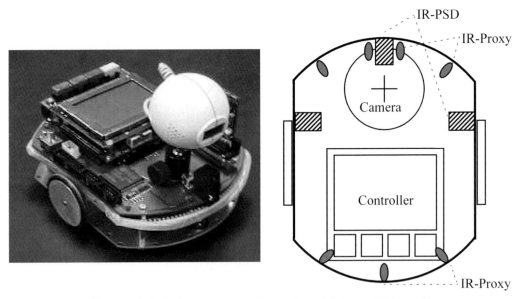

Figure 16.1: Robot sensor configuration with three PSDs and seven proxies

Accuracy and speed are the two major criteria for map generation. Although the quad-tree representation seems to fit both criteria, it was unsuitable for our setup with limited accuracy sensors. Instead, we used the approach of visibility graphs [Sheu, Xue 1993] with configuration space representation. Since the typical environments we are using have only few obstacles and lots of free space, the configuration space representation is more efficient than the free space representation. However, we will not use the configuration space approach in the form published by Sheu and Xue. We modified the approach to record exact obstacle outlines in the environment, and do not add safety areas around them, which results in a more accurate mapping.

The second modification is to employ a grid structure for the environment, which simplifies landmark localization and adds a level of fault tolerance. The use of a dual map representation eliminates many of the problems of each individual approach, since they complement each other. The configuration space representation results in a well-defined map composed of line segments, whereas the grid representation offers a very efficient array structure that the robot can use for easy navigation.

The basic tasks for map generation are to explore the environment and list all obstacles encountered. In our implementation, the robot starts exploring its environment. If it locates any obstacles, it drives toward the nearest one, performs a boundary-following algorithm around it, and defines a map entry for this obstacle. This process continues until all obstacles in the local environment are explored. The robot then proceeds to an unexplored region of the physical environment and repeats this process. Exploration is finished when the internal map of the physical environment is fully defined.

The task planning structure is specified by the structogram in Figure 16.2.

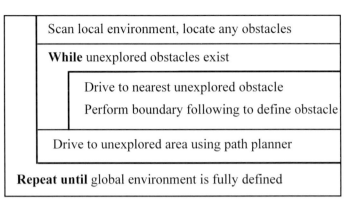

Figure 16.2: Mapping algorithm

16.2 Data Representation

The physical environment is represented in two different map systems, the configuration space representation and the occupancy grid.

Configuration space was introduced by Lozano-Perez [Lozano-Perez 1982] and modified by Fujimura [Fujimura 1991] and Sheu and Xue [Sheu, Xue 1993]. Its data representation is a list of vertex–edge (V,E) pairs to define obstacle outlines. Occupancy grids [Honderd, Jongkind, Aalst 1986] divide the 2D space into square areas, whose number depends on the selected resolution. Each square can be either free or occupied (being part of an obstacle).

Since the configuration space representation allows a high degree of accuracy in the representation of environments, we use this data structure for storing the results of our mapping algorithm. In this representation, data is only entered when the robot's sensors can be operated at optimum accuracy, which in our case is when it is close to an object. Since the robot is closest to an object during execution of the boundary-following algorithm, only then is data entered into configuration space. The configuration space is used for the following task in our algorithm:

- Record obstacle locations.

The second map representation is the occupancy grid, which is continually updated while the robot moves in its environment and takes distance measurements using its infrared sensors. All grid cells along a straight line in the measurement direction up to the measured distance (to an obstacle) or the sensor limit can be set to state "free". However, because of sensor noise and position error, we use the state "preliminary free" or "preliminary occupied" instead, when a grid cell is explored for the first time. This value can at a later time be either confirmed or changed when the robot revisits the same area at a closer range. Since the occupancy grid is not used for recording the generated map (this is done by using the configuration space), we can use a rather small, coarse, and efficient grid structure. Therefore, each grid cell may represent a rather large area in our environment. The occupancy grid fulfils the following tasks in our algorithm:

- Navigating to unexplored areas in the physical environment.
- Keeping track of the robot's position.
- Recording grid positions as free or occupied (preliminary or final).
- Determining whether the map generation has been completed.

Figure 16.3 shows a downloaded screen shot of the robot's on-board LCD after it finished exploring a maze environment and generated a configuration space map.

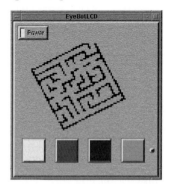

Figure 16.3: EyeBot screen shot

16.3 Boundary-Following Algorithm

When a robot encounters an obstacle, a boundary-following algorithm is activated, to ensure that all paths around the obstacle are checked. This will locate all possible paths to navigate around the obstacle.

In our approach, the robot follows the edge of an obstacle (or a wall) until it returns close to its starting position, thereby fully enclosing the obstacle. Keeping track of the robot's exact position and orientation is crucial, because this may not only result in an imprecise map, but also lead to mapping the same obstacle twice or failing to return to an initial position. This task is non-trivial, since we work without a global positioning system and allow imperfect sensors.

Care is also taken to keep a minimum distance between robot and obstacle or wall. This should ensure the highest possible sensor accuracy, while avoiding a collision with the obstacle. If the robot were to collide, it would lose its position and orientation accuracy and may also end up in a location where its maneuverability is restricted.

Obstacles have to be stationary, but no assumptions are made about their size and shape. For example, we do not assume that obstacle edges are straight lines or that edges meet in rectangular corners. Due to this fact, the boundary-following algorithm must take into account any variations in the shape and angle of corners and the curvature of obstacle surfaces.

Path planning is required to allow the robot to reach a destination such as an unexplored area in the physical environment. Many approaches are available to calculate paths from existing maps. However, as in this case the robot is just in the process of generating a map, this limits the paths considered to areas that it has explored earlier. Thus, in the early stages, the generated map will be incomplete, possibly resulting in the robot taking sub-optimal paths.

Our approach relies on a path planning implementation that uses minimal map data like the DistBug algorithm [Kamon, Rivlin 1997] to perform path planning. The DistBug algorithm uses the direction toward a target to chart its course, rather than requiring a complete map. This allows the path planning algorithm to traverse through unexplored areas of the physical environment to reach its target. The algorithm is further improved by allowing the robot to choose its boundary-following direction freely when encountering an obstacle.

16.4 Algorithm Execution

A number of states are defined for each cell in the occupancy grid. In the beginning, each cell is set to the initial state "unknown". Whenever a free space or an obstacle is encountered, this information is entered in the grid data structure, so cells can either be "free" or contain an "obstacle". In addition we introduce the states "preliminary free" and "preliminary obstacle" to deal with sensor error and noise. These temporal states can later be either confirmed or changed when the robot passes within close proximity through that cell. The algorithm stops when the map generation is completed; that is, when no more preliminary states are in the grid data structure (after an initial environment scan).

Grid cell states

- Unknown ○
- Preliminary free ▫
- Free ◻
- Preliminary obstacle ▪
- Obstacle ■

Our algorithm uses a rather coarse occupancy grid structure for keeping track of the space the robot has already visited, together with the much higher-resolution configuration space to record all obstacles encountered. Figure 16.4 shows snapshots of pairs of grid and space during several steps of a map generation process.

In step A, the robot starts with a completely unknown occupancy grid and an empty corresponding configuration space (i.e. no obstacles). The first step is to do a 360° scan around the robot. For this, the robot performs a rotation on the spot. The angle the robot has to turn depends on the number and location of its range sensors. In our case the robot will rotate to +90° and –90°. Grid A shows a snapshot after the initial range scan. When a range sensor returns a

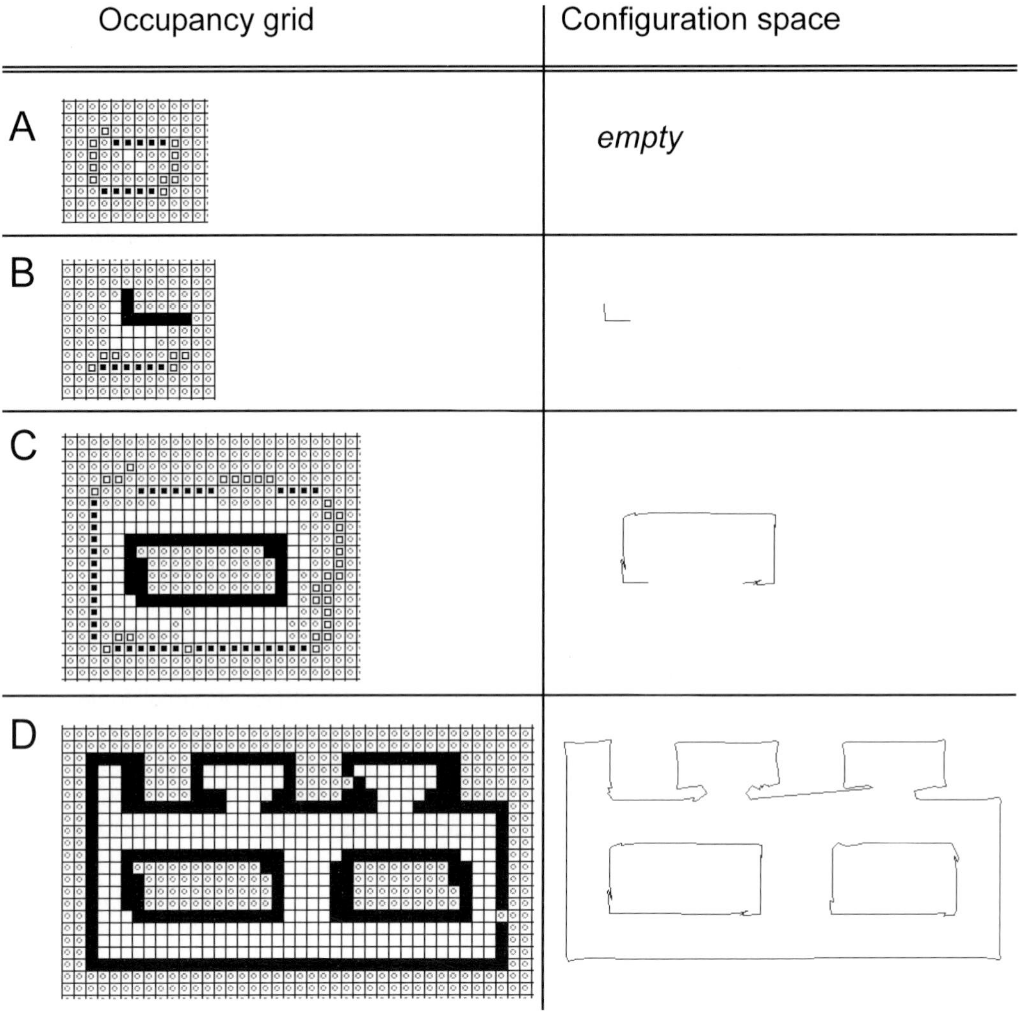

Occupancy grid	Configuration space
A	*empty*
B	
C	
D	

Figure 16.4: Stepwise environment exploration with corresponding occupancy grids and configuration spaces

value less than the maximum measurement range, then a preliminary obstacle is entered in the cell at the measured distance. All cells between this obstacle and the current robot position are marked as preliminary empty. The same is entered for all cells in line of a measurement that does not locate an obstacle; all other cells remain "unknown". Only final obstacle states are entered into the configuration space, therefore space A is still empty.

In step B, the robot drives to the closest obstacle (here the wall toward the top of the page) in order to examine it closer. The robot performs a wall-following behavior around the obstacle and, while doing so, updates both grid and space. Now at close range, preliminary obstacle states have been changed to final obstacle states and their precise location has been entered into configuration space B.

In step C, the robot has completely surrounded one object by performing the wall-following algorithm. The robot is now close again to its starting position. Since there are no preliminary cells left around this rectangular obstacle, the algorithm terminates the obstacle-following behavior and looks for the nearest preliminary obstacle.

In step D, the whole environment has been explored by the robot, and all preliminary states have been eliminated by a subsequent obstacle-following routine around the rectangular obstacle on the right-hand side. One can see the match between the final occupancy grid and the final configuration space.

16.5 Simulation Experiments

Experiments were conducted first using the EyeSim simulator (see Chapter 13), then later on the physical robot itself. Simulators are a valuable tool to test and debug the mapping algorithm in a controlled environment under various constraints, which are hard to maintain in a physical environment. In particular, we are able to employ several error models and error magnitudes for the robot sensors. However, simulations can never be a substitute for an experiment in a physical environment [Bernhardt, Albright 1993], [Donald 1989].

Collision detection and avoidance routines are of greater importance in the physical environment, since the robot relies on dead reckoning using wheel encoders for determining its exact position and orientation. A collision may cause wheel slippage, and therefore invalidate the robot's position and orientation data.

The first test of the mapping algorithm was performed using the EyeSim simulator. For this experiment, we used the world model shown in Figure 16.5 (a). Although this is a rather simple environment, it possesses all the required characteristic features to test our algorithm. The outer boundary surrounds two smaller rooms, while some other corners require sharp 180° turns during the boundary-following routine.

Figures 16.5 (b-d) show various maps that were generated for this environment. Varying error magnitudes were introduced to test our implementation of the map generation routine. From the results it can be seen that the configuration space representation gradually worsens with increasing error magnitude. Especially corners in the environment become less accurately defined. Nevertheless, the mapping system maintained the correct shape and structure even though errors were introduced.

We achieve a high level of fault tolerance by employing the following techniques:

- Controlling the robot to drive as close to an obstacle as possible.
- Limiting the error deviation of the infrared sensors.
- Selectively specifying which points are included in the map.

By limiting the points entered into the configuration space representation to critical points such as corners, we do not record any deviations along a straight edge of an obstacle. This makes the mapping algorithm insensitive to small disturbances along straight lines. Furthermore, by maintaining a close distance to the obstacles while performing boundary-following, any sudden changes in sensor readings from the infrared sensors can be detected as errors and eliminated.

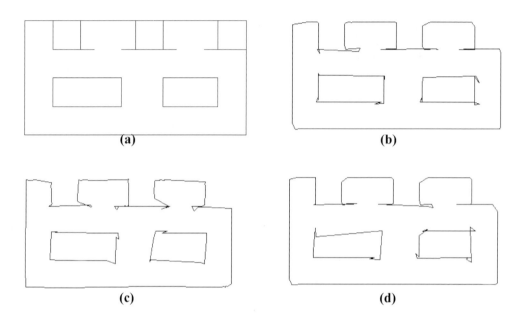

Figure 16.5: Simulation experiment:
(a) simulated environment (b) generated map with zero error model
(c) with 20% PSD error (d) with 10% PSD error and 5% positioning error

16.6 Robot Experiments

Using a physical maze with removable walls, we set up several environments to test the performance of our map generation algorithm. The following figures show a photograph of the environment together with a measured map and the generated map after exploration through the robot.

Figure 16.6 represents a simple environment, which we used to test our map generation implementation. A comparison between the measured and generated map shows a good resemblance between the two maps.

Figure 16.7 displays a more complicated map, requiring more frequent turns by the robot. Again, both maps show the same shape, although the angles are not as closely mapped as in the previous example.

The final environment in Figure 16.8 contains non-rectangular walls. This tests the algorithm's ability to generate a map in an environment without the assumption of right angles. Again, the generated maps show distinct shape similarities.

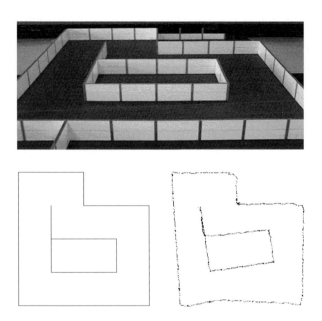

Figure 16.6: Experiment 1 photograph, measured map, and generated map

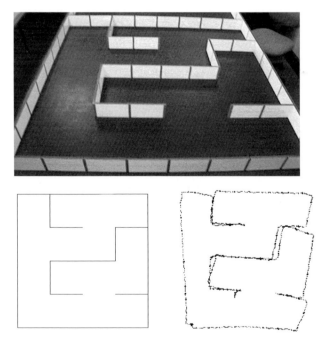

Figure 16.7: Experiment 2 photograph, measured map, and generated map

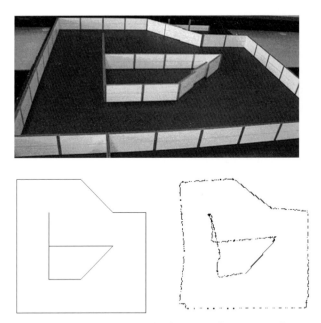

Figure 16.8: Experiment 3 photograph, measured map, and generated map

The main source of error in our map generation algorithm can be traced back to the infrared sensors. These sensors only give reliable measurements within the range of 50 to 300 millimeters. However, individual readings can deviate by as much as 10 millimeters. Consequently, these errors must be considered and included in our simulation environment.

The second source of error is the robot positioning using dead reckoning. With every move of the robot, small inaccuracies due to friction and wheel slippage lead to errors in the perceived robot position and orientation. Although these errors are initially small, the accumulation of these inaccuracies can lead to a major problem in robot localization and path planning, which will directly affect the accuracy of the generated map.

For comparing the measured map with the generated map, a relation between the two maps must be found. One of the maps must be translated and rotated so that both maps share the same reference point and orientation. Next, an objective measure for map similarity has to be found. Although it might be possible to compare every single point of every single line of the maps to determine a similarity measure, we believe that such a method would be rather inefficient and not give satisfying results. Instead, we identify key points in the measured map and compare them with points in the generated map. These key points are naturally corners and vertices in each map and can be identified as the end points of line segments. Care has to be taken to eliminate points from the correct map which lie in regions which the robot cannot sense or reach, since processing them would result in an incorrectly low matching level of the generated map. Corner points are matched between the two maps using the smallest Euclidean distance, before their deviation can be measured.

16.7 Results

Table 16.1 summarizes the map accuracy generated for Figure 16.5. Errors are specified as median error per pixel and as median error relative to the robot's size (approximately 170mm) to provide a more graphic measure.

	Median Error	**Median Error Relative to Robot Size**
Figure b (no error)	21.1mm	12.4%
Figure c (20% PSD)	29.5mm	17.4%
Figure d (10% PSD, 5% pos.)	28.4mm	16.7%

Table 16.1: Results of simulation

From these results we can observe that the mapping error stays below 17% of the robot's size, which is satisfactory for our experiments. An interesting point to note is that the map error is still above 10% in an environment with perfect sensors. This is an inaccuracy of the EyeSim simulator, and mainly due to two factors:

- We assume the robot is obtaining data readings continuously. However, this is not the case as the simulator obtains readings at time-discrete intervals. This may result in the robot simulation missing exact corner positions.

- We assume the robot performs many of its functions simultaneously, such as moving and obtaining sensor readings. However, this is not the case as programs run sequentially and some time delay is inevitable between actions. This time delay is increased by the large number of computations that have to be performed by the simulator, such as its graphics output and the calculations of updated robot positions.

	Median Error	**Median Error Relative to Robot Size**
Experiment 1	39.4mm	23.2%
Experiment 2	33.7mm	19.8%
Experiment 3	46.0mm	27.1%

Table 16.2: Results of real robot

For the experiments in the physical environment (summarized in Table 16.2), higher error values were measured than obtained from the simulation.

However, this was to be expected since our simulation system cannot model all aspects of the physical environment perfectly. Nevertheless, the median error values are about 23% of the robot's size, which is a good value for our implementation, taking into consideration limited sensor accuracy and a noisy environment.

16.8 References

BERNHARDT, R., ALBRIGHT, S. (Eds.) *Robot Calibration*, Chapman & Hall, London, 1993, pp. 19-29 (11)

BRÄUNL, T., STOLZ, H. *Mobile Robot Simulation with Sonar Sensors and Cameras*, Simulation, vol. 69, no. 5, Nov. 1997, pp. 277-282 (6)

BRÄUNL, T., TAY, N. *Combining Configuration Space and Occupancy Grid for Robot Navigation*, Industrial Robot International Journal, vol. 28, no. 3, 2001, pp. 233-241 (9)

CHATILA, R. *Task and Path Planning for Mobile Robots*, in A. Wong, A. Pugh (Eds.), Machine Intelligence and Knowledge Engineering for Robotic Applications, no. 33, Springer-Verlag, Berlin, 1987, pp. 299-330 (32)

DONALD, B. (Ed.) *Error Detection and Recovery in Robotics*, Springer-Verlag, Berlin, 1989, pp. 220-233 (14)

FUJIMURA, K. *Motion Planning in Dynamic Environments*, 1st Ed., Springer-Verlag, Tokyo, 1991, pp. 1-6 (6), pp. 10-17 (8), pp. 127-142 (16)

HONDERD, G., JONGKIND, W., VAN AALST, C. *Sensor and Navigation System for a Mobile Robot*, in L. Hertzberger, F. Groen (Eds.), Intelligent Autonomous Systems, Elsevier Science Publishers, Amsterdam, 1986, pp. 258-264 (7)

KAMON, I., RIVLIN, E. *Sensory-Based Motion Planning with Global Proofs*, IEEE Transactions on Robotics and Automation, vol. 13, no. 6, 1997, pp. 814-821 (8)

KAMPMANN, P. *Multilevel Motion Planning for Mobile Robots Based on a Topologically Structured World Model*, in T. Kanade, F. Groen, L. Hertzberger (Eds.), Intelligent Autonomous Systems 2, Stichting International Congress of Intelligent Autonomous Systems, Amsterdam, 1990, pp. 241-252 (12)

LEONARD, J., DURRANT-WHITE, H. *Directed Sonar Sensing for Mobile Robot Navigation*, Rev. Ed., Kluwer Academic Publishers, Boston MA, 1992, pp. 101-111 (11), pp. 127-128 (2), p. 137 (1)

LOZANO-PÉREZ, T. *Task Planning*, in M. Brady, J. HollerBach, T. Johnson, T. Lozano-Pérez, M. Mason (Eds.), Robot Motion Planning and Control, MIT Press, Cambridge MA, 1982, pp. 473-498 (26)

References

MELIN, C. *Exploration of Unknown Environments by a Mobile Robot*, in T. Kanade, F. Groen, L. Hertzberger (Eds.), Intelligent Autonomous Systems 2, Stichting International Congress of Intelligent Autonomous Systems, Amsterdam, 1990, pp. 715-725 (11)

PIAGGIO, M., ZACCARIA, R. *Learning Navigation Situations Using Roadmaps*, IEEE Transactions on Robotics and Automation, vol. 13, no. 6, 1997, pp. 22-27 (6)

SERRADILLA, F., KUMPEL, D. *Robot Navigation in a Partially Known Factory Avoiding Unexpected Obstacles*, in T. Kanade, F. Groen, L. Hertzberger (Eds.), Intelligent Autonomous Systems 2, Stichting International Congress of Intelligent Autonomous Systems, Amsterdam, 1990, pp. 972-980 (9)

SHEU, P., XUE, Q. (Eds.) *Intelligent Robotic Planning Systems*, World Scientific Publishing, Singapore, 1993, pp. 111-127 (17), pp. 231-243 (13)

REAL-TIME IMAGE PROCESSING

E very digital consumer camera today can read images from a sensor chip and (optionally) display them in some form on a screen. However, what we want to do is implement an *embedded vision system*, so reading and maybe displaying image data is only the necessary first step. We want to extract information from an image in order to steer a robot, for example following a colored object. Since both the robot and the object may be moving, we have to be fast. Ideally, we want to achieve a frame rate of 10 fps (frames per second) for the whole perception–action cycle. Of course, given the limited processing power of an embedded controller, this restricts us in the choice of both the image resolution and the complexity of the image processing operations.

In the following, we will look at some basic image processing routines. These will later be reused for more complex robot applications programs, like robot soccer in Chapter 18.

For further reading in robot vision see [Klette, Peleg, Sommer 2001] and [Blake, Yuille 1992]. For general image processing textbooks see [Parker 1997], [Gonzales, Woods 2002], [Nalwa 1993], and [Faugeras 1993]. A good practical introduction is [Bässmann, Besslich 1995].

17.1 Camera Interface

Since camera chip development advances so rapidly, we have already had five camera chip generations interfaced to the EyeBot and have implemented the corresponding camera drivers. For a robot application program, however, this is completely transparent. The routines to access image data are:

- `CAMInit(NORMAL)`
  Initializes camera, independent of model. Older camera models supported modes different to "normal".

- `CAMRelease()`
  Disable camera.

- `CAMGetFrame (image *buffer)`
  Read a single grayscale image from the camera, save in buffer.

- `CAMGetColFrame (colimage *buffer, int convert)`
  Read a single color image from the camera. If "convert" equals 1, the image is immediately converted to 8bit grayscale.

- `int CAMSet (int para1, int para2, int para3)`
  Set camera parameters. Parameter meaning depends on camera model (see Appendix B.5.4).

- `int CAMGet (int *para1, int *para2 ,int *para3)`
  Get camera parameters. Parameter meaning depends on camera model (see Appendix B.5.4).

The first important step when using a camera is setting its focus. The Eye-Cam C2 cameras have an analog grayscale video output, which can be directly plugged into a video monitor to view the camera image. The lens has to be focussed for the desired object distance.

Sieman's star   Focussing is also possible with the EyeBot's black and white display only. For this purpose we place a focus pattern like the so-called "Sieman's star" in Figure 17.1 at the desired distance in front of the camera and then change the focus until the image appears sharp.

Figure 17.1: Focus pattern

17.2 Auto-Brightness

The auto-brightness function adapts a cameras's aperture to the continuously changing brightness conditions. If a sensor chip does support aperture settings but does not have an auto-brightness feature, then it can be implemented in software. The first idea for implementing the auto-brightness feature for a grayscale image is as follows:

1. Compute the average of all gray values in the image.

2.a If the average is below threshold no. 1: open aperture.

2.b If the average is above threshold no. 2: close aperture.

Figure 17.2: Auto-brightness using only main diagonal

So far, so good, but considering that computing the average over all pixels is quite time consuming, the routine can be improved. Assuming that to a certain degree the gray values are evenly distributed among the image, using just a cross-section of the whole image, for example the main diagonal (Figure 17.2), should be sufficient.

Program 17.1: Auto-brightness

```
1   typedef BYTE image    [imagerows][imagecolumns];
2   typedef BYTE colimage[imagerows][imagecolumns][3];
```

```
1   #define THRES_LO  70
2   #define THRES_HI 140
3
4   void autobrightness(image orig)
5   { int i,j, brightness = 100, avg =0;
6     for (i=0; i<imagerows; i++) avg += orig[i][i];
7     avg = avg/imagerows;
8
9     if (avg<THRES_LO)
10    { brightness = MIN(brightness * 1.05, 200);
11      CAMSet(brightness, 100, 100)
12    }
13    else if (avg>THRES_HI)
14    { brightness = MAX(brightness / 1.05,  50);
15      CAMSet(brightness, 100, 100)
16    }
17  }
```

Program 17.1 shows the pre-defined data types for grayscale images and color images and the implementation for auto-brightness, assuming that the number of rows is less than or equal to the number of columns in an image (in this implementation: 60 and 80). The CAMSet routine adjusts the brightness setting of the camera to the new calculated value, the two other parameters (here: offset and contrast) are left unchanged. This routine can now be called in regular intervals (for example once every second, or for every 10th image, or even for every image) to update the camera's brightness setting. Note that this program only works for the QuickCam, which allows aperture settings, but does not have auto-brightness.

17.3 Edge Detection

One of the most fundamental image processing operations is edge detection. Numerous algorithms have been introduced and are being used in industrial applications; however, for our purposes very basic operators are sufficient. We will present here the Laplace and Sobel edge detectors, two very common and simple edge operators.

The Laplace operator produces a local derivative of a grayscale image by taking four times a pixel value and subtracting its left, right, top, and bottom neighbors (Figure 17.3). This is done for every pixel in the whole image.

	−1	
−1	4	−1
	−1	

Figure 17.3: Laplace operator

The coding is shown in Program 17.2 with a single loop running over all pixels. There are no pixels beyond the outer border of an image and we need to avoid an access error by accessing array elements outside defined bounds. Therefore, the loop starts at the second row and stops at the last but one row. If required, these two rows could be set to zero. The program also limits the maximum value to white (255), so that any result value remains within the byte data type.

The Sobel operator that is often used for robotics applications is only slightly more complex [Bräunl 2001].

In Figure 17.4 we see the two filter operations the Sobel filter is made of. The Sobel-x only finds discontinuities in the x-direction (vertical lines), while Sobel-y only finds discontinuities in the y-direction (horizontal lines). Combining these two filters is done by the formulas shown in Figure 17.4, right, which give the edge strength (depending on how large the discontinuity is) as well as the edge direction (for example a dark-to-bright transition at 45° from the x-axis).

Program 17.2: Laplace edge operator

```
1    void Laplace(BYTE * imageIn, BYTE * imageOut)
2    { int i, delta;
3    for (i = width; i < (height-1)*width; i++)
4        { delta = abs(4 * imageIn[i]
5                        -imageIn[i-1]      -imageIn[i+1]
6                        -imageIn[i-width] -imageIn[i+width]);
7            if (delta > white) imageOut[i] = white;
8            else imageOut[i] = (BYTE)delta;
9        }
10   }
```

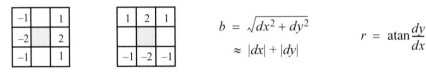

$$b = \sqrt{dx^2 + dy^2}$$
$$\approx |dx| + |dy|$$

$$r = \operatorname{atan}\frac{dy}{dx}$$

Figure 17.4: Sobel-x and Sobel-y masks, formulas for strength and angle

For now, we are only interested in the edge strength, and we also want to avoid time consuming functions such as square root and any trigonometric functions. We therefore approximate the square root of the sum of the squares by the sum of the absolute values of dx and dy.

Program 17.3: Sobel edge operator

```
1    void Sobel(BYTE *imageIn, BYTE *imageOut)
2    { int i, grad, delaX, deltaY;
3
4        memset(imageOut, 0, width); /* clear first row */
5        for (i = width; i < (height-1)*width; i++)
6        { deltaX = 2*imageIn[i+1] + imageIn[i-width+1]
7                    + imageIn[i+width+1]  - 2*imageIn[i-1]
8                    - imageIn[i-width-1]  - imageIn[i+width-1];
9
10           deltaY =   imageIn[i-width-1] + 2*imageIn[i-width]
11                    + imageIn[i-width+1]  - imageIn[i+width-1]
12                    - 2*imageIn[i+width]  - imageIn[i+width+1];
13
14           grad = (abs(deltaX) + abs(deltaY)) / 3;
15           if (grad > white) grad = white;
16           imageOut[i] = (BYTE)grad;
17       }
18       memset(imageOut + i, 0, width); /* clear last line */
19   }
```

The coding is shown in Program 17.3. Only a single loop is used to run over all pixels. Again, we neglect a one-pixel-wide borderline; pixels in the first and last row of the result image are set to zero. The program already applies a heuristic scaling (divide by three) and limits the maximum value to white (255), so the result value remains a single byte.

17.4 Motion Detection

The idea for a very basic motion detection algorithm is to subtract two subsequent images (see also Figure 17.5):

1. Compute the absolute value for grayscale difference for all pixel pairs of two subsequent images.

2. Compute the average over all pixel pairs.

3. If the average is above a threshold, then motion has been detected.

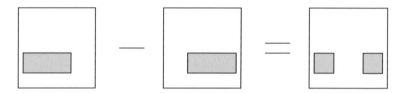

Figure 17.5: Motion detection

This method only detects the presence of motion in an image pair, but does not determine any direction or area. Program 17.4 shows the implementation of this problem with a single loop over all pixels, summing up the absolute differences of all pixel pairs. The routine returns 1 if the average difference per pixel is greater than the specified threshold, and 0 otherwise.

Program 17.4: Motion detection

```
1   int motion(image im1, image im2, int threshold)
2   { int diff=0;
3     for (i = 0; i < height*width; i++)
4       diff += abs(i1[i][j] - i2[i][j]);
5     return (diff > threshold*height*width); /* 1 if motion*/
6   }
```

This algorithm could also be extended to calculate motion separately for different areas (for example the four quarters of an image), in order to locate the rough position of the motion.

17.5 Color Space

Before looking at a more complex image processing algorithm, we take a side-step and look at different color representations or "color spaces". So far we have seen grayscale and RGB color models, as well as Bayer patterns (RGGB). There is not one superior way of representing color information, but a number of different models with individual advantages for certain applications.

17.5.1 Red Green Blue (RGB)

The RGB space can be viewed as a 3D cube with red, green, and blue being the three coordinate axes (Figure 17.6). The line joining the points (0, 0, 0) and (1, 1, 1) is the main diagonal in the cube and represents all shades of gray from black to white. It is usual to normalize the RGB values between 0 and 1 for floating point operations or to use a byte representation from 0 to 255 for integer operations. The latter is usually preferred on embedded systems, which do not possess a hardware floating point unit.

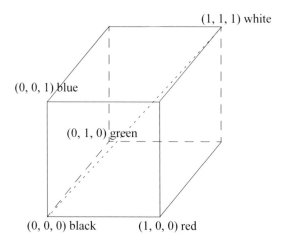

Figure 17.6: RGB color cube

In this color space, a color is determined by its red, green, and blue components in an additive synthesis. The main disadvantage of this color space is that the color hue is not independent of intensity and saturation of the color.

Luminosity L in the RGB color space is defined as the sum of all three components:

$$L = R + G + B$$

Luminosity is therefore dependent on the three components R, G, and B.

17.5.2 Hue Saturation Intensity (HSI)

The HSI color space (see Figure 17.7) is a cone where the middle axis represents luminosity, the phase angle represents the hue of the color, and the radial distance represents the saturation. The following set of equations specifies the conversion from RGB to HSI color space:

$$I = \frac{1}{3}(R+G+B)$$

$$S = 1 - \frac{3}{(R+G+B)}[\min(R,G,B)]$$

$$H = \cos^{-1}\left\{\frac{\frac{1}{2}[(R-G)+(R-B)]}{[(R-G)^2+(R-B)(G-B)]^{1/2}}\right\}$$

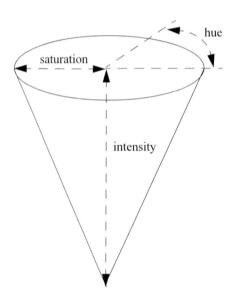

Figure 17.7: HSI color cone

The advantage of this color space is to de-correlate the intensity information from the color information. A grayscale value is represented by an intensity, zero saturation, and *arbitrary* hue value. So it can simply be differentiated between chromatic (color) and achromatic (grayscale) pixels, only by using the saturation value. On the other hand, because of the same relationship it is not sufficient to use the hue value alone to identify pixels of a certain color. The saturation has to be above a certain threshold value.

17.5.3 Normalized RGB (rgb)

Most camera image sensors deliver pixels in an RGB-like format, for example Bayer patterns (see Section 2.9.2). Converting all pixels from RGB to HSI might be too intensive a computing operation for an embedded controller. Therefore, we look at a faster alternative with similar properties.

One way to make the RGB color space more robust with regard to lighting conditions is to use the "normalized RGB" color space (denoted by "rgb") defined as:

$$r = \frac{R}{R+G+B} \qquad g = \frac{G}{R+G+B} \qquad b = \frac{B}{R+G+B}$$

This normalization of the RGB color space allows us to describe a certain color independently of the luminosity (sum of all components). This is because the luminosity in rgb is always equal to one:

$$r + g + b = 1 \ \forall \ (r, g, b)$$

17.6 Color Object Detection

If it is guaranteed for a robot environment that a certain color only exists on one particular object, then we can use color detection to find this particular object. This assumption is widely used in mobile robot competitions, for example the AAAI'96 robot competition (collect yellow tennis balls) or the RoboCup and FIRA robot soccer competitions (kick the orange golf ball into the yellow or blue goal). See [Kortenkamp, Nourbakhsh, Hinkle 1997], [Kaminka, Lima, Rojas 2002], and [Cho, Lee 2002].

The following hue-histogram algorithm for detecting colored objects was developed by Bräunl in 2002. It requires minimal computation time and is therefore very well suited for embedded vision systems. The algorithm performs the following steps:

1. Convert the RGB color image to a hue image (HSI model).
2. Create a histogram over all image columns of pixels matching the object color.
3. Find the maximum position in the column histogram.

The first step only simplifies the comparison whether two color pixels are similar. Instead of comparing the differences between three values (red, green, blue), only a single *hue* value needs to be compared (see [Hearn, Baker 1997]). In the second step we look at each image column separately and record how many pixels are similar to the desired ball color. For a 60×80 image, the histogram comprises just 80 integer values (one for each column) with values between 0 (no similar pixels in this column) and 60 (all pixels similar to the ball color).

At this level, we are not concerned about continuity of the matching pixels in a column. There may be two or more separate sections of matching pixels, which may be due to either occlusions or reflections on the same object – or there might be two different objects of the same color. A more detailed analysis of the resulting histogram could distinguish between these cases.

Program 17.5: RGB to hue conversion

```
1    int RGBtoHue(BYTE r, BYTE g, BYTE b)
2    /* return hue value for RGB color */
3    #define NO_HUE -1
4    { int hue, delta, max, min;
5
6        max   = MAX(r, MAX(g,b));
7        min   = MIN(r, MIN(g,b));
8        delta = max - min;
9        hue =0; /* init hue*/
10
11       if (2*delta <= max) hue = NO_HUE; /* gray, no color */
12       else {
13          if (r==max) hue = 42 + 42*(g-b)/delta;      /* 1*42 */
14          else if (g==max) hue = 126 +42*(b-r)/delta; /* 3*42 */
15          else if (b==max) hue = 210 +42*(r-g)/delta; /* 5*42 */
16       }
17       return hue; /* now: hue is in range [0..252] */
18   }
```

Program 17.5 shows the conversion of an RGB image to an image (hue, saturation, value), following [Hearn, Baker 1997]. We drop the saturation and value components, since we only need the hue for detecting a colored object like a ball. However, they are used to detect invalid hues (NO_HUE) in case of a too low saturation (r, g, and b having similar or identical values for grayscales), because in these cases arbitrary hue values can occur.

Input image
with sample column marked

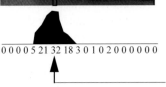

0 0 0 0 5 21 32 18 3 0 1 0 2 0 0 0 0 0 0

Histogram
with counts of matching pixels per column

Column with maximum number of matches

Figure 17.8: Color detection example

The next step is to generate a histogram over all x-positions (over all columns) of the image, as shown in Figure 17.8. We need two nested loops going over every single pixel and incrementing the histogram array in the corresponding position. The specified threshold limits the allowed deviation from the desired object color hue. Program 17.6 shows the implementation.

Program 17.6: Histogram generation

```
1    int GenHistogram(image hue_img, int obj_hue,
2                        line histogram, int thres)
3    /* generate histogram over all columns */
4    { int x,y;
5      for (x=0;x<imagecolumns;x++)
6      { histogram[x] = 0;
7        for (y=0;y<imagerows;y++)
8          if (hue_img[y][x] != NO_HUE &&
9              (abs(hue_img[y][x] - obj_hue) < thres ||
10             253 - abs(hue_img[y][x] - obj_hue) < thres)
11               histogram[x]++;
12     }
13   }
```

Finally, we need to find the maximum position in the generated histogram. This again is a very simple operation in a single loop, running over all positions of the histogram. The function returns both the maximum position and the maximum value, so the calling program can determine whether a sufficient number of matching pixels has been found. Program 17.7 shows the implementation.

Program 17.7: Object localization

```
1    void FindMax(line histogram, int *pos, int *val)
2    /* return maximum position and value of histogram */
3    int x;
4    { *pos = -1; *val = 0;  /* init */
5      for (x=0; x<imagecolumns; x++)
6        if (histogram[x] > *val)
7          { *val = histogram[x]; *pos = x; }
8    }
```

Programs 17.6 and 17.7 can be combined for a more efficient implementation with only a single loop and reduced execution time. This also eliminates the need for explicitly storing the histogram, since we are only interested in the maximum value. Program 17.8 shows the optimized version of the complete algorithm.

For demonstration purposes, the program draws a line in each image column representing the number of matching pixels, thereby optically visualizing the histogram. This method works equally well on the simulator as on the real

Program 17.8: Optimized color search

```
1   void ColSearch(colimage img, int obj_hue, int thres,
2                   int *pos, int *val)
3   /* find x position of color object, return pos and value*/
4   { int x,y, count, h, distance;
5     *pos = -1; *val = 0;   /* init */
6     for (x=0;x<imagecolumns;x++)
7     { count = 0;
8       for (y=0;y<imagerows;y++)
9       { h = RGBtoHue(img[y][x][0],img[y][x][1],
10                     img[y][x][2]);
11        if (h != NO_HUE)
12        { distance = abs((int)h-obj_hue); /* hue dist. */
13          if (distance > 126) distance = 253-distance;
14          if (distance < thres) count++;
15        }
16      }
17      if (count > *val) { *val = count; *pos = x; }
18      LCDLine(x,53, x, 53-count, 2); /* visualization only*/
19    }
20  }
```

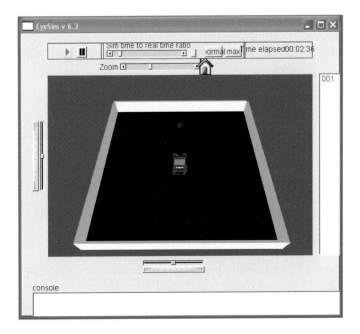
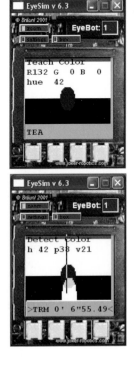

Figure 17.9: Color detection on EyeSim simulator

robot. In Figure 17.9 the environment window with a colored ball and the console window with displayed image and histogram can be seen.

Program 17.9: Color search main program

```
1    #define X 40   // ball coordinates for teaching
1    #define Y 40
2
3    int main()
4    { colimage c;
5      int hue, pos, val;
6
7      LCDPrintf("Teach Color\n");
8      LCDMenu("TEA","","","");
9      CAMInit(NORMAL);
10     while (KEYRead() != KEY1)
11     { CAMGetColFrame(&c,0);
12       LCDPutColorGraphic(&c);
13       hue = RGBtoHue(c[Y][X][0], c[Y][X][1], c[Y][X][2]);
14       LCDSetPos(1,0);
15       LCDPrintf("R%3d G%3d B%3d\n",
16                    c[Y][X][0], c[Y][X][1], c[Y][X][2]);
17       LCDPrintf("hue %3d\n", hue);
18                    OSWait(100);
19     }
20
21     LCDClear();
22     LCDPrintf("Detect Color\n");
23     LCDMenu("","","","END");
24     while (KEYRead() != KEY4)
25     { CAMGetColFrame(&c,0);
26       LCDPutColorGraphic(&c);
27       ColSearch(c, hue, 10, &pos, &val);   /* search image */
28       LCDSetPos(1,0);
29       LCDPrintf("h%3d p%2d v%2d\n", hue, pos, val);
30       LCDLine  (pos, 0, pos, 53, 2);   /* vertical line */
31     }
32     return 0;
33   }
```

The main program for the color search is shown in Program 17.9. In its first phase, the camera image is constantly displayed together with the RGB value and hue value of the middle position. The user can record the hue value of an object to be searched. In the second phase, the color search routine is called with every image displayed. This will display the color detection histogram and also locate the object's x-position.

This algorithm only determines the x-position of a colored object. It could easily be extended to do the same histogram analysis over all lines (instead of over all columns) as well and thereby produce the full [x, y] coordinates of an object. To make object detection more robust, we could further extend this algorithm by asserting that a detected object has more than a certain minimum number of similar pixels per line or per column. By returning a start and finish value for the line diagram and the column diagram, we will get $[x_1, y_1]$ as the object's start coordinates and $[x_2, y_2]$ as the object's finish coordinates. This rectangular area can be transformed into object center and object size.

17.7 Image Segmentation

Detecting a single object that differs significantly either in shape or in color from the background is relatively easy. A more ambitious application is segmenting an image into disjoint regions. One way of doing this, for example in a grayscale image, is to use connectivity and edge information (see Section 17.3, [Bräunl 2001], and [Bräunl 2006] for an interactive system). The algorithm shown here, however, uses color information for faster segmentation results [Leclercq, Bräunl 2001].

This color segmentation approach transforms all images from RGB to rgb (normalized RGB) as a pre-processing step. Then, a color class lookup table is constructed that translates each rgb value to a "color class", where different color classes ideally represent different objects. This table is a three-dimensional array with (rgb) as indices. Each entry is a reference number for a certain "color class".

17.7.1 Static Color Class Allocation

Optimized for fixed application If we know the number and characteristics of the color classes to be distinguished beforehand, we can use a static color class allocation scheme. For example, for robot soccer (see Chapter 18), we need to distinguish only three color classes: orange for the ball and yellow and blue for the two goals. In a case like this, the location of the color classes can be calculated to fill the table. For example, "blue goal" is defined for all points in the 3D color table for which blue dominates, or simply:

$$b > \text{threshold}_b$$

In a similar way, we can distinguish orange and yellow, by a combination of thresholds on the red and green component:

$$\text{colclass} = \begin{cases} \text{blueGoal} & \text{if } b > \text{thres}_b \\ \text{yellowGoal} & \text{if } r > \text{thres}_r \text{ and } g > \text{thres}_g \\ \text{orangeBall} & \text{if } r > \text{thres}_r \text{ and } g < \text{thres}_g \end{cases}$$

If (rgb) were coded as 8bit values, the table would comprise $(2^8)^3$ entries, which comes to 16MB when using 1 byte per entry. This is too much memory for a small embedded system, and also too high a resolution for this color segmentation task. Therefore, we only use the five most significant bits of each color component, which comes to a more manageable size of $(2^5)^3 = 32KB$.

In order to determine the correct threshold values, we start with an image of the blue goal. We keep changing the blue threshold until the recognized rectangle in the image matches the right projected goal dimensions. The thresholds for red and green are determined in a similar manner, trying different settings until the best distinction is found (for example the orange ball should not be classified as the yellow goal and vice versa). With all thresholds determined, the corresponding color class (for example 1 for ball, 2 or 3 for goals) is calculated and entered for each rgb position in the color table. If none of the criteria is fulfilled, then the particular rgb value belongs to none of the color classes and 0 is entered in the table. In case that more than one criterion is fulfilled, then the color classes have not been properly defined and there is an overlap between them.

17.7.2 Dynamic Color Class Allocation

General technique However, in general it is also possible to use a dynamic color class allocation, for example by *teaching* a certain color class instead of setting up fixed topological color borders. A simple way of defining a color space is by specifying a sub-cube of the full rgb cube, for example allowing a certain offset from the desired (*taught*) value r′g′b′ :

$$r \in [r'-\delta .. r'+\delta]$$
$$g \in [g'-\delta .. g'+\delta]$$
$$b \in [b'-\delta .. b'+\delta]$$

Starting with an empty color table, each new sub-cube can be entered by three nested loops, setting all sub-cube positions to the new color class identifier. Other topological entries are also possible, of course, depending on the desired application.

A new color can simply be added to previously taught colors by placing a sample object in front of the camera and averaging a small number of center pixels to determine the object hue. A median filter of about 4×4 pixels will be sufficient for this purpose.

17.7.3 Object Localization

Having completed the color class table, segmenting an image seems simple. All we have to do is look up the color class for each pixel's rgb value. This gives us a situation as sketched in Figure 17.10. Although to a human observer, coherent color areas and therefore objects are easy to detect, it is not trivial to extract this information from the 2D segmented output image.

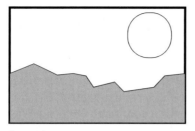

Input image   Segmented image

Figure 17.10: Segmentation example

If, as for many applications, identifying rectangular areas is sufficient, then the task becomes relatively simple. For now, we assume there is at most a single coherent object of each color class present in the image. For more objects of the same color class, the algorithm has to be extended to check for coherence. In the simple case, we only need to identify four parameters for each color class, namely top left and bottom right corner, or in coordinates:

$$[x_{tl}, y_{tl}], \; [x_{br}, y_{br}]$$

Finding these coordinates for each color class still requires a loop over all pixels of the segmented image, comparing the indices of the current pixel position with the determined extreme (top/left, bottom/right) positions of the previously visited pixels of the same color class.

17.8 Image Coordinates versus World Coordinates

Image coordinates Whenever an object is identified in an image, all we have is its *image coordinates*. Working with our standard 60×80 resolution, all we know is that our desired object is, say, at position [50, 20] (i.e. bottom left) and has a size of 5×7 pixels. Although this information might already be sufficient for some simple applications (we could already steer the robot in the direction of the object), for many applications we would like to know more precisely the object's location in *world coordinates* relative from our robot in meters in the x- and y-direction (see Figure 17.11).

World coordinates

For now, we are only interested in the object's position in the robot's local coordinate system {x′, y′}, not in the global word coordinate system {x, y}. Once we have determined the coordinates of the object in the robot coordinate system and also know the robot's (absolute) position and orientation, we can transform the object's local coordinates to global world coordinates.

As a simplification, we are looking for objects with rotational symmetry, such as a ball or a can, because they look the same (or at least similar) from any viewing angle. The second simplification is that we assume that objects are not floating in space, but are resting on the ground, for example the table the robot is driving on. Figure 17.12 demonstrates this situation with a side

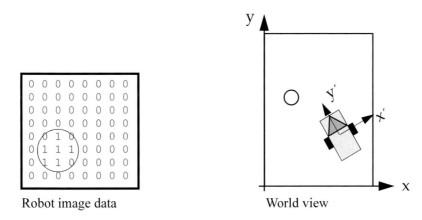

Robot image data	World view

Figure 17.11: Image and world coordinates

view and a top view from the robot's local coordinate system. What we have to determine is the relationship between the ball position in local coordinates $[x', y']$ and the ball position in image coordinates $[j, i]$:

$$y' = f(i, h, \alpha, f, d)$$
$$x' = g(j, 0, \beta, f, d)$$

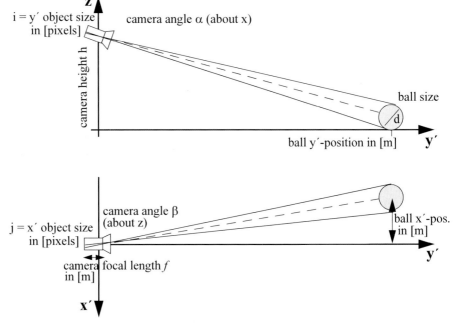

Figure 17.12: Camera position and orientation

It is obvious that f and g are the same function, taking as parameters:

- One-dimensional distance in image coordinates
  (object's length in image rows or columns in pixels)
- Camera offset
  (height in y´z´ view, 0 side offset in x´y´ view)
- Camera rotation angle
  (tilt or pan)
- Camera focal length
  (distance between lens and sensor array)
- Ball size
  (diameter d)

Provided that we know the detected object's true physical size (for example golf ball for robot soccer), we can use the intercept theorem to calculate its local displacement. With a zero camera offset and a camera angle of zero (no tilting or panning), we have the proportionality relationships:

$$\frac{y'}{f} \sim \frac{d}{i} \qquad\qquad \frac{x'}{f} \sim \frac{d}{j}$$

These can be simplified when introducing a camera-specific parameter $g = k \cdot f$ for converting between pixels and meters:

$$y' = g \cdot d / i$$
$$x' = g \cdot d / j$$

So in other words, the larger the image size in pixels, the closer the object is. The transformation is just a constant linear factor; however, due to lens distortions and other sources of noise these ideal conditions will not be observed in an experiment. It is therefore better to provide a lookup table for doing the transformation, based on a series of distance measurements.

With the camera offset, either to the side or above the driving plane, or placed at an angle, either panning about the z-axis or tilting about the x-axis, the trigonometric formulas become somewhat more complex. This can be solved either by adding the required trigonometric functions to the formulas and calculating them for every image frame, or by providing separate lookup tables from all camera viewing angles used. In Section 18.5 this method is applied to robot soccer.

17.9 References

BÄSSMANN, H., BESSLICH, P. *Ad Oculos: Digital Image Processing*, International Thompson Publishing, Washington DC, 1995

BLAKE, A., YUILLE, A. (Eds.) *Active Vision*, MIT Press, Cambridge MA, 1992

BRÄUNL, T. *Parallel Image Processing*, Springer-Verlag, Berlin Heidelberg, 2001

BRÄUNL, T. *Improv – Image Processing for Robot Vision*, http://robotics.ee.uwa.edu.au/improv, 2006

CHO, H., LEE., J.-J. (Ed.) *2002 FIRA Robot World Congress*, Proceedings, Korean Robot Soccer Association, Seoul, May 2002

FAUGERAS, O. *Three-Dimensional Computer Vision*, MIT Press, Cambridge MA, 1993

GONZALES, R., WOODS, R., *Digital Image Processing*, 2nd Ed., Prentice Hall, Upper Saddle River NJ, 2002

HEARN, D., BAKER, M. *Computer Graphics - C Version*, Prentice Hall, Upper Saddle River NJ, 1997

KAMINKA, G. LIMA, P., ROJAS, R. (Eds.) *RoboCup 2002: Robot Soccer World Cup VI*, Proccedings, Fukuoka, Japan, Springer-Verlag, Berlin Heidelberg, 2002

KLETTE, R., PELEG, S., SOMMER, G. (Eds.) *Robot Vision*, Proceedings of the International Workshop RobVis 2001, Auckland NZ, Lecture Notes in Computer Science, no. 1998, Springer-Verlag, Berlin Heidelberg, Feb. 2001

KORTENKAMP, D., NOURBAKHSH, I., HINKLE, D. *The 1996 AAAI Mobile Robot Competition and Exhibition*, AI Magazine, vol. 18, no. 1, 1997, pp. 25-32 (8)

LECLERCQ, P., BRÄUNL, T. *A Color Segmentation Algorithm for Real-Time Object Localization on Small Embedded Systems*, Robot Vision 2001, International Workshop, Auckland NZ, Lecture Notes in Computer Science, no. 1998, Springer-Verlag, Berlin Heidelberg, Feb. 2001, pp. 69-76 (8)

NALWA, V. *A Guided Tour of Computer Vision*, Addison-Wesley, Reading MA, 1993

PARKER, J. *Algorithms for Image Processing and Computer Vision*, John Wiley & Sons, New York NY, 1997

ROBOT SOCCER

18

F ootball, or soccer as it is called in some countries, is often referred to as
"the world game". No other sport is played and followed by as many
nations around the world. So it did not take long to establish the idea of
robots playing soccer against each other. As has been described earlier on the
Micro Mouse Contest, robot competitions are a great opportunity to share new
ideas and actually see good concepts at work.

Robot soccer is more than one robot generation beyond simpler competitions like solving a maze. In soccer, not only do we have a lack of environment
structure (less walls), but we now have teams of robots playing an opposing
team, involving moving targets (ball and other players), requiring planning,
tactics, and strategy – all in real time. So, obviously, this opens up a whole new
dimension of problem categories. Robot soccer will remain a great challenge
for years to come.

18.1 RoboCup and FIRA Competitions

See details at:
www.fira.net
www.robocup.org Today, there are two world organizations involved in robot soccer, FIRA and
RoboCup. FIRA [Cho, Lee 2002] organized its first robot tournament in 1996
in Korea with Jong-Hwan Kim. RoboCup [Asada 1998] followed with its first
competition in 1997 in Japan with Asada, Kuniyoshi, and Kitano [Kitano et al.
1997], [Kitano et al. 1998].

FIRA's "MiroSot" league (Micro-Robot World Cup Soccer Tournament)
has the most stringent size restrictions [FIRA 2006]. The maximum robot size
is a cube of 7.5cm side length. An overhead camera suspended over the playing field is the primary sensor. All image processing is done centrally on an
off-board workstation or PC, and all driving commands are sent to the robots
via wireless remote control. Over the years, FIRA has added a number of different leagues, most prominently the "SimuroSot" simulation league and the
"RoboSot" league for small autonomous robots (without global vision). In
2002, FIRA introduced "HuroSot", the first league for humanoid soccer playing robots. Before that all robots were wheel-driven vehicles.

RoboCup started originally with the "Small-Size League", "Middle-Size League", and "Simulation League" [RoboCup 2006]. Robots of the small-size league must fit in a cylinder of 18cm diameter and have certain height restrictions. As for MiroSot, these robots rely on an overhead camera over the playing field. Robots in the middle-size league abolished global vision after the first two years. Since these robots are considerably larger, they are mostly using commercial robot bases equipped with laptops or small PCs. This gives them at least one order of magnitude higher processing power; however, it also drives up the cost for putting together such a robot soccer team. In later years, RoboCup added the commentator league (subsequently dropped), the rescue league (not related to soccer), the "Sony 4-legged league" (which, unfortunately, only allows the robots of one company to compete), and finally in 2002 the "Humanoid League".

The following quote from RoboCup's website may in fact apply to both organizations [RoboCup 2006]:

> *"RoboCup is an international joint project to promote AI, robotics, and related fields. It is an attempt to foster AI and intelligent robotics research by providing a standard problem where a wide range of technologies can be integrated and examined. RoboCup chose to use the soccer game as a central topic of research, aiming at innovations to be applied for socially significant problems and industries. The ultimate goal of the RoboCup project is: By 2050, develop a team of fully autonomous humanoid robots that can win against the human world champion team in soccer."*

Real robots don't use global vision! We will concentrate here on robot soccer played by wheeled robots (humanoid robot soccer is still in its infancy) *without* the help of *global vision*. The RoboCup Small-Size League, but not the Middle-Size League or FIRA RoboSot, allows the use of an overhead camera suspended above the soccer field. This leads teams to use a single central workstation that does the image processing and planning for all robots. There are no occlusions: ball, robots, and goals are always perfectly visible. Driving commands are then issued via wireless links to individual "robots", which are not autonomous at all and in some respect reduced to remote control toy cars. Consequently, the "AllBots" team from Auckland, New Zealand does in fact use toy cars as a low-budget alternative [Baltes 2001a]. Obviously, *global vision soccer* is a completely different task to *local vision soccer*, which is much closer to common research areas in robotics, including vision, self-localization, and distributed planning.

The robots of our team "CIIPS Glory" carry EyeCon controllers to perform local vision on-board. Some other robot soccer teams, like "4 Stooges" from Auckland, New Zealand, use EyeCon controllers as well [Baltes 2001b].

Robot soccer teams play five-a-side soccer with rules that are freely adapted from FIFA soccer. Since there is a boundary around the playing field, the game is actually closer to ice hockey. The big challenge is not only that reliable image processing has to be performed in real time, but also that a team of five robots/actors has to be organized. In addition, there is an opposing team which will change the environment (for example kick the ball) and thereby render one's own action plans useless if too slow.

One of the frequent disappointments of robot competitions is that enormous research efforts are reduced to "show performance" in a particular event and cannot be appreciated adequately. Adapting from the home lab environment to the competition environment turns out to be quite tricky, and many programs are not as robust as their authors had hoped. On the other hand, the actual competitions are only one part of the event. Most competitions are part of conferences and encourage participants to present the research behind their competition entries, giving them the right forum to discuss related ideas.

Mobile robot competitions brought progress to the field by inspiring people and by continuously pushing the limits of what is possible. Through robot competitions, progress has been achieved in mechanics, electronics, and algorithms [Bräunl 1999].

CIIPS Glory
with local vision
on each robot

Note the colored
patches on top
of the Lilliputs
players. They
need them to
determine each
robot's position
and orientation
with global vision.

Figure 18.1: CIIPS Glory line-up and in play vs. Lilliputs (1998)

18.2 Team Structure

The CIIPS Glory robot soccer team (Figure 18.1) consists of four field players and one goal keeper robot [Bräunl, Graf 1999], [Bräunl, Graf 2000]. A local intelligence approach has been implemented, where no global sensing or control system is used. Each field player is equipped with the same control software, only the goal keeper – due to its individual design and task – runs a different program.

Different roles (left/right defender, left/right attacker) are assigned to the four field players. Since the robots have a rather limited field of view with their local cameras, it is important that they are always spread around the whole field. Therefore, each player's role is linked to a specific area of the field. When the ball is detected in a certain position, only the robot responsible for this area is meant to drive toward and play the ball. The robot which has detected the ball communicates the position of the ball to its team mates which try to find favorable positions on the field to be prepared to take over and play the ball as soon as it enters their area.

Situations might occur when no robot sees the ball. In that case, all robots patrol along specific paths in their assigned area of the field, trying to detect the ball. The goal keeper usually stays in the middle of the goal and only moves once it has detected the ball in a reasonably close position (Figure 18.2).

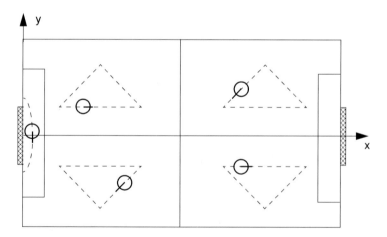

Figure 18.2: Robot patrolling motion

This approach appears to be quite efficient, especially since each robot acts individually and does not depend on any global sensing or communication system. For example, the communication system can be switched off without any major effects; the players are still able to continue playing individually.

18.3 Mechanics and Actuators

According to the RoboCup Small-Size League and FIRA RoboSot regulations the size of the SoccerBots has been restricted to 10cm by 15cm. The height is also limited, therefore the EyeCon controller is mounted on a mobile platform at an angle. To catch the ball, the robot has a curved front. The size of the curved area has been calculated from the rule that at least two-thirds of the ball's projected area must be outside the convex hull around the robot. With the ball having a diameter of approximately 4.5cm, the depth of the curved front must be no more than 1.5cm.

The robots are equipped with two motorized wheels plus two casters at the front and back of the vehicle. Each wheel is controlled separately, which enables the robots to drive forward, backward, as well as drive in curves or spin on the spot. This ability for quick movement changes is necessary to navigate successfully in rapidly changing environments such as during robot soccer competitions.

Two additional servo motors are used to activate a kicking device at the front of the robot and the movement of the on-board camera.

In addition to the four field players of the team, one slightly differing goal keeper robot has been constructed. To enable it to defend the goal successfully it must be able to drive sideways in front of the goal, but look and kick forward. For this purpose, the top plate of the robot is mounted at a 90° angle to the bottom plate. For optimal performance at the competition, the kicking device has been enlarged to the maximum allowed size of 18cm.

18.4 Sensing

Sensing a robot's environment is the most important part for most mobile robot applications, including robot soccer. We make use of the following sensors:

- Shaft encoders
- Infrared distance measurement sensors
- Compass module
- Digital camera

In addition, we use communication between the robots, which is another source of information input for each robot. Figure 18.3 shows the main sensors of a wheeled SoccerBot in detail.

Shaft encoders   The most basic feedback is generated by the motors' encapsulated shaft encoders. This data is used for three purposes:

- PI controller for individual wheel to maintain constant wheel speed.
- PI controller to maintain desired path curvature (i.e. straight line).
- Dead reckoning to update vehicle position and orientation.

The controller's dedicated timing processor unit (TPU) is used to deal with the shaft encoder feedback as a background process.

Figure 18.3: Sensors: shaft encoder, infrared sensors, digital camera

Infrared distance measurement

Each robot is equipped with three infrared sensors to measure the distance to the front, to the left, and to the right (PSD). This data can be used to:

• Avoid collision with an obstacle.

• Navigate and map an unknown environment.

• Update internal position in a known environment.

Since we are using low-cost devices, the sensors have to be calibrated for each robot and, due to a number of reasons, also generate false readings from time to time. Application programs have to take care of this, so a level of software fault tolerance is required.

Compass module

The biggest problem in using dead reckoning for position and orientation estimation in a mobile robot is that it deteriorates over time, unless the data can be updated at certain reference points. A wall in combination with a distance sensor can be a reference point for the robot's position, but updating robot orientation is very difficult without additional sensors.

In these cases, a compass module, which senses the earth's magnetic field, is a big help. However, these sensors are usually only correct to a few degrees and may have severe disturbances in the vicinity of metal. So the exact sensor placement has to be chosen carefully.

Digital camera

We use the EyeCam camera, based on a CMOS sensor chip. This gives a resolution of 60×80 pixels in 32bit color. Since all image acquisition, image processing, and image display is done on-board the EyeCon controller, there is no need to transmit image data. At a controller speed of 35MHz we achieve a frame capture rate of about 7 frames per second without FIFO buffer and up to 30 fps with FIFO buffer. The final frame rate depends of course on the image processing routines applied to each frame.

Robot-to-robot communication

While the wireless communication network between the robots is not exactly a sensor, it is nevertheless a source of input data to the robot from its environment. It may contain sensor data from other robots, parts of a shared plan, intention descriptions from other robots, or commands from other robots or a human operator.

18.5 Image Processing

Vision is the most important ability of a human soccer player. In a similar way, vision is the centerpiece of a robot soccer program. We continuously analyze the visual input from the on-board digital color camera in order to detect objects on the soccer field. We use color-based object detection since it is computationally much easier than shape-based object detection and the robot soccer rules define distinct colors for the ball and goals. These color hues are taught to the robot before the game is started.

The lines of the input image are continuously searched for areas with a mean color value within a specified range of the previously trained hue value and of the desired size. This is to try to distinguish the object (ball) from an area similar in color but different in shape (yellow goal). In Figure 18.4 a simplified line of pixels is shown; object pixels of matching color are displayed in gray, others in white. The algorithm initially searches for matching pixels at either end of a line (see region (a): first = 0, last = 18), then the mean color value is calculated. If it is within a threshold of the specified color hue, the object has been found. Otherwise the region will be narrowed down, attempting to find a better match (see region (b): first = 4, last = 18). The algorithm stops as soon as the size of the analyzed region becomes smaller than the desired size of the object. In the line displayed in Figure 18.4, an object with a size of 15 pixels is found after two iterations.

Figure 18.4: Analyzing a color image line

Distance estimation

Once the object has been identified in an image, the next step is to translate local image coordinates (x and y, in pixels) into global world coordinates (x′ and y′ in m) in the robot's environment. This is done in two steps:

- Firstly, the position of the ball as seen from the robot is calculated from the given pixel values, assuming a fixed camera position and orientation. This calculation depends on the *height* of the object in the image. The higher the position in the image, the further an object's distance from the robot.

- Secondly, given the robot's current position and orientation, the local coordinates are transformed into global position and orientation on the field.

Since the camera is looking down at an angle, it is possible to determine the object distance from the image coordinates. In an experiment (see Figure 18.5), the relation between pixel coordinates and object distance in meters has been determined. Instead of using approximation functions, we decided to use the faster and also more accurate method of lookup tables. This allows us to calculate the exact ball position in meters from the screen coordinates in pixels and the current camera position/orientation.

Measurement

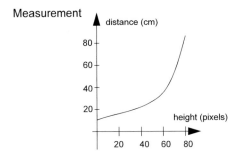

Schematic Diagram

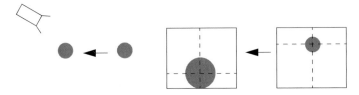

Figure 18.5: Relation between object height and distance

The distance values were found through a series of measurements, for each camera position and for each image line. In order to reduce this effort, we only used three different camera positions (up, middle, down for the tilting camera arrangement, or left, middle, right for the panning camera arrangement), which resulted in three different lookup tables.

Depending on the robot's current camera orientation, the appropriate table is used for distance translation. The resulting relative distances are then translated into global coordinates using polar coordinates.

An example output picture on the robot LCD can be seen in Figure 18.6. The lines indicate the position of the detected ball in the picture, while its global position on the field is displayed in centimeters on the right-hand side.

Figure 18.6: LCD output after ball detection

This simple image analysis algorithm is very efficient and does not slow down the overall system too much. This is essential, since the same controller doing image processing also has to handle sensor readings, motor control, and timer interrupts as well. We achieve a frame rate of 3.3 fps for detecting the ball when no ball is in the image and of 4.2 fps when the ball has been detected in the previous frame, by using coherence. The use of a FIFO buffer for reading images from the camera (not used here) can significantly increase the frame rate.

18.6 Trajectory Planning

Once the ball position has been determined, the robot executes an approach behavior, which should drive it into a position to kick the ball forward or even into the opponent's goal. For this, a trajectory has to be generated. The robot knows its own position and orientation by dead reckoning; the ball position has been determined either by the robot's local search behavior or by communicating with other robots in its team.

18.6.1 Driving Straight and Circle Arcs

The start position and orientation of this trajectory is given by the robot's current position, the end position is the ball position, and the end orientation is the line between the ball and the opponent's goal. A convenient way to generate a smooth trajectory for given start and end points with orientations are Hermite splines. However, since the robot might have to drive around the ball in order to kick it toward the opponent's goal, we use a case distinction to add "via-points" in the trajectory (see Figure 18.7). These trajectory points guide the robot around the ball, letting it pass not too close, but maintaining a smooth trajectory.

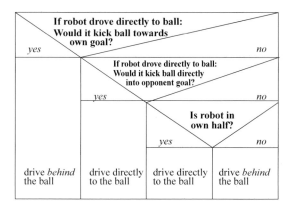

Figure 18.7: Ball approach strategy

In this algorithm, *driving directly* means to approach the ball without via-points on the path of the robot. If such a trajectory is not possible (for example for the ball lying between the robot and its own goal), the algorithm inserts a via-point in order to avoid an own goal. This makes the robot pass the ball on a specified side before approaching it. If the robot is in its own half, it is sufficient to drive to the ball and kick it toward the other team's half. When a player is already in the opposing team's half, however, it is necessary to approach the ball with the correct heading in order to kick it directly toward the opponent's goal.

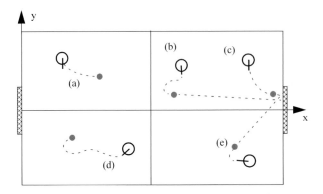

Figure 18.8: Ball approach cases

The different driving actions are displayed in Figure 18.8. The robot drives either directly to the ball (Figure 18.8 a, c, e) or onto a curve (either linear and circular segments or a spline curve) including via-points to approach the ball from the correct side (Figure 18.8 b, d).

Drive directly to the ball (Figure 18.8 a, b):
With *localx* and *localy* being the local coordinates of the ball seen from the robot, the angle to reach the ball can be set directly as:

$$\alpha = -\text{atan}\left(\frac{localy}{localx}\right)$$

With l being the distance between the robot and the ball, the distance to drive in a curve is given by:

$$d = l \cdot \alpha \cdot \sin(\alpha)$$

Drive around the ball (Figure 18.8 c, d, e):
If a robot is looking toward the ball but at the same time facing its own goal, it can drive along a circular path with a fixed radius that goes through the ball. The radius of this circle is chosen arbitrarily and was defined to be 5cm. The circle is placed in such a way that the tangent at the position of the ball also goes through the opponent's goal. The robot turns on the spot until it faces this

circle, drives to it in a straight line, and drives behind the ball on the circular path (Figure 18.9).

Compute turning angle γ for turning on the spot: $\quad\quad \gamma = \alpha + \beta$

Circle angle β between new robot heading and ball: $\quad \beta = \beta_1 + \beta_2$

Angle to be driven on circular path: $\quad\quad\quad\quad\quad 2{\cdot}\beta$

Angle $\beta 1$: goal heading from ball to x-axis:

$$\beta_1 = \text{atan}\left(\frac{\text{ball}_y}{\text{length-ball}_x}\right)$$

Angle β_2: ball heading from robot to x-axis:

$$\beta_2 = \text{atan}\left(\frac{\text{ball}_y\text{-robot}_y}{\text{ball}_x\text{-robot}_x}\right)$$

Angle α: from robot orientation to ball heading (φ is robot orientation):
$\alpha = -\varphi + \beta_2$

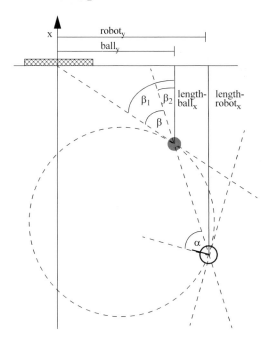

Figure 18.9: Calculating a circular path toward the ball

18.6.2 Driving Spline Curves

The simplest driving trajectory is to combine linear segments with circle arc segments. An interesting alternative is the use of splines. They can generate a smooth path and avoid turning on the spot, therefore they will generate a faster path.

Given the robot position P_k and its heading DP_k as well as the ball position P_{k+1} and the robot's destination heading DP_{k+1} (facing toward the opponent's goal from the current ball position), it is possible to calculate a spline which for every fraction u of the way from the current robot position to the ball position describes the desired location of the robot.

The Hermite blending functions H_0 .. H_3 with parameter u are defined as follows:

$$H_0 = 2u^3 - 3u^2 + 1$$

$$H_1 = -2u^3 + 3u^2$$

$$H_2 = u^3 - 3u^2 + u$$

$$H_3 = u^3 - u^2$$

The current robot position is then defined by:

$$P(u) = P_k H_0(u) + P_{k+1} H_1(u) + Dp_k H_2(u) + DP_{k+1} H_3(u)$$

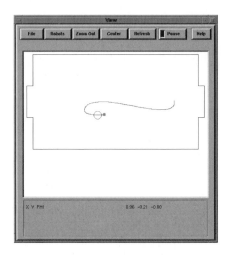
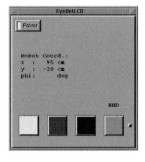

Figure 18.10: Spline driving simulation

A PID controller is used to calculate the linear and rotational speed of the robot at every point of its way to the ball, trying to get it as close to the spline curve as possible. The robot's speed is constantly updated by a background process that is invoked 100 times per second. If the ball can no longer be detected (for example if the robot had to drive around it and lost it out of sight), the robot keeps driving to the end of the original curve. An updated driving command is issued as soon as the search behavior recognizes the (moving) ball at a different global position.

This strategy was first designed and tested on the EyeSim simulator (see Figure 18.10), before running on the actual robot. Since the spline trajectory computation is rather time consuming, this method has been substituted by simpler drive-and-turn algorithms when participating in robot soccer tournaments.

18.6.3 Ball Kicking

After a player has successfully captured the ball, it can dribble or kick it toward the opponent's goal. Once a position close enough to the opponent's goal has been reached or the goal is detected by the vision system, the robot activates its kicker to shoot the ball into the goal.

The driving algorithm for the goal keeper is rather simple. The robot is started at a position of about 10cm in front of the goal. As soon as the ball is detected, it drives between the ball and goal on a circular path within the defense area. The robot follows the movement of the ball by tilting its camera up and down. If the robot reaches the corner of its goal, it remains on its position and turns on the spot to keep track of the ball. If the ball is not seen in a pre-defined number of images, the robot suspects that the ball has changed position and therefore drives back to the middle of the goal to restart its search for the ball.

Figure 18.11: CIIPS Glory versus Lucky Star (1998)

If the ball is detected in a position very close to the goalie, the robot activates its kicker to shoot the ball away.

Fair play is
obstacle
avoidance
"Fair Play" has always been considered an important issue in human soccer. Therefore, the CIIPS Glory robot soccer team (Figure 18.11) has also stressed its importance. The robots constantly check for obstacles in their way, and – if this is the case – try to avoid hitting them. In case an obstacle has been touched, the robot drives backward for a certain distance until the obstacle is out of reach. If the robot has been dribbling the ball to the goal, it turns quickly toward the opponent's goal to kick the ball away from the obstacle, which could be a wall or an opposing player.

18.7 References

ASADA, M. (Ed.) *RoboCup-98: Robot Soccer World Cup II*, Proceedings of the Second RoboCup Workshop, RoboCup Federation, Paris, July 1998

BALTES, J. *AllBotz*, in P. Stone, T. Balch, G. Kraetzschmar (Eds.), RoboCup-2000: Robot Soccer World Cup IV, Springer-Verlag, Berlin, 2001a, pp. 515-518 (4)

BALTES, J. *4 Stooges*, in P. Stone, T. Balch, G. Kraetzschmar (Eds.), RoboCup-2000: Robot Soccer World Cup IV, Springer-Verlag, Berlin, 2001b, pp. 519-522 (4)

BRÄUNL, T. *Research Relevance of Mobile Robot Competitions*, IEEE Robotics and Automation Magazine, vol. 6, no. 4, Dec. 1999, pp. 32-37 (6)

BRÄUNL, T., GRAF, B. *Autonomous Mobile Robots with Onboard Vision and Local Intelligence*, Proceedings of Second IEEE Workshop on Perception for Mobile Agents, Fort Collins, Colorado, 1999

BRÄUNL, T., GRAF, B. *Small robot agents with on-board vision and local intelligence*, Advanced Robotics, vol. 14, no. 1, 2000, pp. 51-64 (14)

CHO, H., LEE, J.-J. (Eds.) *Proceedings 2002 FIRA World Congress*, Seoul, Korea, May 2002

FIRA, *FIRA Official Website*, Federation of International Robot-Soccer Association, http://www.fira.net/, 2006

KITANO, H., ASADA, M., KUNIYOSHI, Y., NODA, I., OSAWA, E. *RoboCup: The Robot World Cup Initiative*, Proceedings of the First International Conference on Autonomous Agents (Agent-97), Marina del Rey CA, 1997, pp. 340-347 (8)

KITANO, H., ASADA, M., NODA, I., MATSUBARA, H. *RoboCup: Robot World Cup*, IEEE Robotics and Automation Magazine, vol. 5, no. 3, Sept. 1998, pp. 30-36 (7)

ROBOCUP FEDERATION, *RoboCup Official Site*, http://www.robocup.org, 2006

NEURAL NETWORKS

19

T he artificial neural network (ANN), often simply called neural network (NN), is a processing model loosely derived from biological neurons [Gurney 2002]. Neural networks are often used for classification problems or decision making problems that do not have a simple or straightforward algorithmic solution. The beauty of a neural network is its ability to learn an input to output mapping from a set of training cases without explicit programming, and then being able to generalize this mapping to cases not seen previously.

There is a large research community as well as numerous industrial users working on neural network principles and applications [Rumelhart, McClelland 1986], [Zaknich 2003]. In this chapter, we only briefly touch on this subject and concentrate on the topics relevant to mobile robots.

19.1 Neural Network Principles

A neural network is constructed from a number of individual units called neurons that are linked with each other via connections. Each individual neuron has a number of inputs, a processing node, and a single output, while each connection from one neuron to another is associated with a *weight*. Processing in a neural network takes place in parallel for all neurons. Each neuron constantly (*in an endless loop*) evaluates (*reads*) its inputs, calculates its local activation value according to a formula shown below, and produces (*writes*) an output value.

The activation function of a neuron $a(I, W)$ is the weighted sum of its inputs, i.e. each input is multiplied by the associated weight and all these terms are added. The neuron's output is determined by the output function $o(I, W)$, for which numerous different models exist.

In the simplest case, just thresholding is used for the output function. For our purposes, however, we use the non-linear "sigmoid" output function defined in Figure 19.1 and shown in Figure 19.2, which has superior characteristics for learning (see Section 19.3). This sigmoid function approximates the

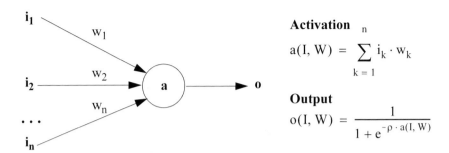

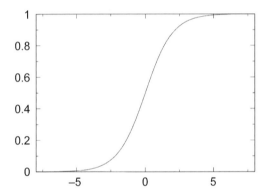

Activation

$$a(I, W) = \sum_{k=1}^{n} i_k \cdot w_k$$

Output

$$o(I, W) = \frac{1}{1 + e^{-\rho \cdot a(I, W)}}$$

Figure 19.1: Individual artificial neuron

Heaviside step function, with parameter ρ controlling the slope of the graph (usually set to 1).

Figure 19.2: Sigmoidal output function

19.2 Feed-Forward Networks

A neural net is constructed from a number of interconnected neurons, which are usually arranged in layers. The outputs of one layer of neurons are connected to the inputs of the following layer. The first layer of neurons is called the "input layer", since its inputs are connected to external data, for example sensors to the outside world. The last layer of neurons is called the "output layer", accordingly, since its outputs are the result of the total neural network and are made available to the outside. These could be connected, for example, to robot actuators or external decision units. All neuron layers between the input layer and the output layer are called "hidden layers", since their actions cannot be observed directly from the outside.

If all connections go from the outputs of one layer to the input of the next layer, and there are no connections within the same layer or connections from a later layer back to an earlier layer, then this type of network is called a "feed-forward network". Feed-forward networks (Figure 19.3) are used for the sim-

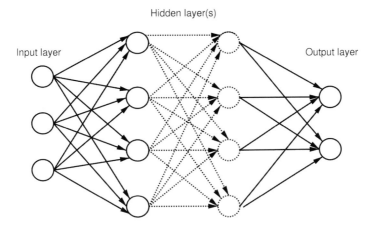

Hidden layer(s)

Input layer

Output layer

Figure 19.3: Fully connected feed-forward network

plest types of ANNs and differ significantly from feedback networks, which we will not look further into here.

For most practical applications, a single hidden layer is sufficient, so the typical NN for our purposes has exactly three layers:

- Input layer (for example input from robot sensors)
- Hidden layer (connected to input and output layer)
- Output layer (for example output to robot actuators)

Perceptron   Incidentally, the first feed-forward network proposed by Rosenblatt had only two layers, one input layer and one output layer [Rosenblatt 1962]. However, these so-called "Perceptrons" were severely limited in their computational power because of this restriction, as was soon after discovered by [Minsky, Papert 1969]. Unfortunately, this publication almost brought neural network research to a halt for several years, although the principal restriction applies only to two-layer networks, not for networks with three layers or more.

In the standard three-layer network, the input layer is usually simplified in the way that the input values are directly taken as neuron activation. No activation function is called for input neurons. The remaining questions for our standard three-layer NN type are:

- How many neurons to use in each layer?
- Which connections should be made between layer i and layer i + 1?
- How are the weights determined?

The answers to these questions are surprisingly straightforward:

- *How many neurons to use in each layer?*
  The number of neurons in the input and output layer are determined by the application. For example, if we want to have an NN drive a robot around a maze (compare Chapter 15) with three PSD sensors as input

and two motors as output, then the network should have three input neurons and two output neurons.

Unfortunately, there is no rule for the "right" number of hidden neurons. Too few hidden neurons will prevent the network from learning, since they have insufficient storage capacity. Too many hidden neurons will slow down the learning process because of extra overhead. The right number of hidden neurons depends on the "*complexity*" of the given problem and has to be determined through experimenting. In this example we are using six hidden neurons.

- *Which connections should be made between layer i and layer i + 1?*
  We simply connect *every output* from layer i to *every input* at layer i + 1. This is called a "fully connected" neural network. There is no need to leave out individual connections, since the same effect can be achieved by giving this connection a weight of zero. That way we can use a much more general and uniform network structure.

- *How are the weights determined?*
  This is the really tricky question. Apparently the whole *intelligence* of an NN is somehow encoded in the set of weights being used. What used to be a program (e.g. driving a robot in a straight line, but avoiding any obstacles sensed by the PSD sensors) is now reduced to a set of floating point numbers. With sufficient insight, we could just "program" an NN by specifying the correct (or let's say *working*) weights. However, since this would be virtually impossible, even for networks with small complexity, we need another technique.

  The standard method is supervised learning, for example through error backpropagation (see Section 19.3). The same task is repeatedly run by the NN and the outcome judged by a supervisor. Errors made by the network are backpropagated from the output layer via the hidden layer to the input layer, amending the weights of each connection.

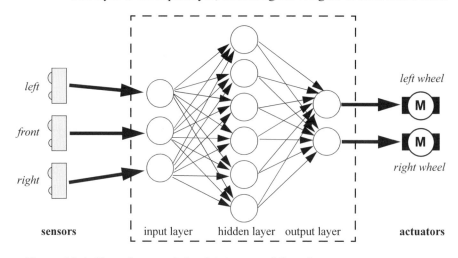

Figure 19.4: Neural network for driving a mobile robot

Evolutionary algorithms provide another method for determining the weights of a neural network. For example, a genetic algorithm (see Chapter 20) can be used to evolve an optimal set of neuron weights.

Figure 19.4 shows the experimental setup for an NN that should drive a mobile robot collision-free through a maze (for example left-wall following) with constant speed. Since we are using three sensor inputs and two motor outputs and we chose six hidden neurons, our network has $3 + 6 + 2$ neurons in total. The input layer receives the sensor data from the infrared PSD distance sensors and the output layer produces driving commands for the left and right motors of a robot with differential drive steering.

Let us calculate the output of an NN for a simpler case with $2 + 4 + 1$ neurons. Figure 19.5, top, shows the labelling of the neurons and connections in the three layers, Figure 19.5, bottom, shows the network with sample input values and weights. For a network with three layers, only two sets of connection weights are required:

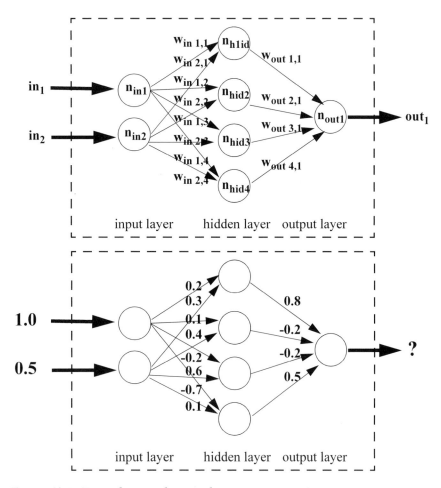

Figure 19.5: Example neural network

- Weights from the input layer to the hidden layer, summarized as matrix $w_{in\ i,j}$ (weight of connection from input neuron i to hidden neuron j).

- Weights from the hidden layer to the output layer, summarized as matrix $w_{out\ i,j}$ (weight of connection from hidden neuron i to output neuron j).

No weights are required from sensors to the first layer or from the output layer to actuators. These weights are just assumed to be always 1. All other weights are normalized to the range $[-1 .. +1]$.

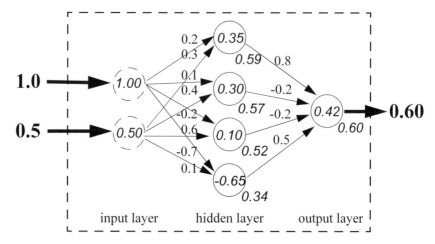

Figure 19.6: Feed-forward evaluation

Calculation of the output function starts with the input layer on the left and propagates through the network. For the input layer, there is one input value (sensor value) per input neuron. Each input data value is used directly as neuron activation value:

$$a(n_{in1}) = o(n_{in1}) = 1.00$$
$$a(n_{in2}) = o(n_{in2}) = 0.50$$

For all subsequent layers, we first calculate the activation function of each neuron as a weighted sum of its inputs, and then apply the sigmoid output function. The first neuron of the hidden layer has the following activation and output values:

$$a(n_{hid1}) = 1.00 \cdot 0.2 + 0.50 \cdot 0.3 = 0.35$$
$$o(n_{hid1}) = 1 / (1 + e^{-0.35}) = 0.59$$

The subsequent steps for the remaining two layers are shown in Figure 19.6 with the activation values printed in each neuron symbol and the output values below, always rounded to two decimal places.

Once the values have percolated through the feed-forward network, they will not change until the input values change. Obviously this is not true for networks with feedback connections. Program 19.1 shows the implementation of the feed-forward process. This program already takes care of two additional so-called "bias neurons", which are required for backpropagation learning.

Program 19.1: Feed-forward execution

```
1    #include <math.h>
2    #define NIN   (2+1)         // number of input neurons
3    #define NHID  (4+1)         // number of hidden neurons
4    #define NOUT 1              // number of output neurons
5    float w_in [NIN][NHID];     // in weights from 3 to 4 neur.
6    float w_out[NHID][NOUT];    // out weights from 4 to 1 neur.
7
8    float sigmoid(float x)
9    { return 1.0 / (1.0 + exp(-x));
10   }
11
12   void feedforward(float N_in[NIN], float N_hid[NHID],
13                    float N_out[NOUT])
14   { int i,j;
15     // calculate activation of hidden neurons
16     N_in[NIN-1] = 1.0; // set bias input neuron
17     for (i=0; i<NHID-1; i++)
18     { N_hid[i] = 0.0;
19       for (j=0; j<NIN; j++)
20         N_hid[i] += N_in[j] * w_in[j][i];
21       N_hid[i] = sigmoid(N_hid[i]);
22     }
23     N_hid[NHID-1] = 1.0; // set bias hidden neuron
24     // calculate activation and output of output neurons
25     for (i=0; i<NOUT; i++)
26     { N_out[i] = 0.0;
27       for (j=0; j<NHID; j++)
28         N_out[i] += N_hid[j] * w_out[j][i];
29       N_out[i] = sigmoid(N_out[i]);
30     }
31   }
```

19.3 Backpropagation

A large number of different techniques exist for learning in neural networks. These include supervised and unsupervised techniques, depending on whether a "teacher" presents the correct answer to a training case or not, as well as on-line or off-line learning, depending on whether the system evolves inside or outside the execution environment. Classification networks with the popular *backpropagation* learning method [Rumelhart, McClelland 1986], a supervised off-line technique, can be used to identify a certain situation from the network input and produce a corresponding output signal. The drawback of this method is that a complete set of all relevant input cases together with their solutions have to be presented to the NN. Another popular method requiring only incremental feedback for input/output pairs is *reinforcement learning* [Sutton, Barto 1998]. This on-line technique can be seen as either supervised or unsupervised, since the feedback signal only refers to the network's current performance and does not provide the desired network output. In the following, the backpropagation method is presented.

A feed-forward neural network starts with random weights and is presented a number of test cases called the *training set*. The network's outputs are compared with the known correct results for the particular set of input values and any deviations (error function) are propagated back through the net.

Having done this for a number of iterations, the NN hopefully has learned the complete training set and can now produce the correct output for each input pattern in the training set. The real hope, however, is that the network is able to *generalize*, which means it will be able to produce similar outputs corresponding to similar input patterns it has *not seen* before. Without the capability of generalization, no useful learning can take place, since we would simply store and reproduce the training set.

The backpropagation algorithm works as follows:

1. Initialize network with random weights.
2. For all training cases:

    a. Present training inputs to network and calculate output.
    b. For all layers (starting with output layer, back to input layer):

        i.   Compare network output with correct output (error function).
        ii. Adapt weights in current layer.

For implementing this learning algorithm, we do know what the correct results for the output layer should be, because they are supplied together with the training inputs. However, it is not yet clear for the other layers, so let us do this step by step.

Firstly, we look at the error function. For each output neuron, we compute the difference between the actual output value out_i and the desired output $d_{out\ i}$. For the total network error, we calculate the sum of square difference:

$$E_{out\ i} = d_{out\ i} - out_i$$

$$E_{total} = \sum_{i=0}^{num(n_{out})} E_{out\ i}^2$$

The next step is to adapt the weights, which is done by a gradient descent approach:

$$\Delta w = -\eta \cdot \frac{\partial E}{\partial w}$$

So the adjustment of the weight will be proportional to the contribution of the weight to the error, with the magnitude of change determined by constant

η. This can be achieved by the following formulas [Rumelhart, McClelland 1986]:

$$\text{diff}_{\text{out } i} = (o(n_{\text{out } i}) - d_{\text{out } i}) \cdot (1 - o(n_{\text{out } i})) \cdot o(n_{\text{out } i})$$

$$\Delta w_{\text{out } k,i} = -2 \cdot \eta \cdot \text{diff}_{\text{out } i} \cdot \text{input}_k(n_{\text{out } i})$$

$$= -2 \cdot \eta \cdot \text{diff}_{\text{out } i} \cdot o(n_{\text{hid } k})$$

Assuming the desired output d_{out1} of the NN in Figure 19.5 to be $d_{\text{out1}} = 1.0$, and choosing $\eta = 0.5$ to further simplify the formula, we can now update the four weights between the hidden layer and the output layer. Note that all calculations have been performed with full floating point accuracy, while only two or three digits are printed.

$$\text{diff}_{\text{out } 1} = (o(n_{\text{out } 1}) - d_{\text{out1}}) \cdot (1 - o(n_{\text{out } 1})) \cdot o(n_{\text{out } 1})$$

$$= (0.60 - 1.00) \cdot (1 - 0.60) \cdot 0.60 = -0.096$$

$$\Delta w_{\text{out } 1,1} = -\text{diff}_{\text{out1}} \cdot \text{input}_1(n_{\text{out1}})$$

$$= -\text{diff}_{\text{out1}} \cdot o(n_{\text{hid1}})$$

$$= -(-0.096) \cdot 0.59 = +0.057$$

$$\Delta w_{\text{out } 2,1} = 0.096 \cdot 0.57 = \quad +0.055$$

$$\Delta w_{\text{out } 3,1} = 0.096 \cdot 0.52 = \quad +0.050$$

$$\Delta w_{\text{out4},1} = 0.096 \cdot 0.34 = \quad +0.033$$

The new weights will be:

$$w'_{\text{out } 1,1} = w_{\text{out1},1} + \Delta w_{\text{out } 1,1} = 0.8 + 0.057 = \quad 0.86$$

$$w'_{\text{out } 2,1} = w_{\text{out2},1} + \Delta w_{\text{out } 2,1} = -0.2 + 0.055 = -0.15$$

$$w'_{\text{out } 3,1} = w_{\text{out3},1} + \Delta w_{\text{out } 3,1} = -0.2 + 0.050 = -0.15$$

$$w'_{\text{out } 4,1} = w_{\text{out4},1} + \Delta w_{\text{out } 4,1} = 0.5 + 0.033 = \quad 0.53$$

The only remaining step is to adapt the w_{in} weights. Using the same formula, we need to know what the desired outputs $d_{\text{hid } k}$ are for the hidden layer. We get these values by backpropagating the error values from the output layer multiplied by the activation value of the corresponding neuron in the hidden layer, and adding up all these terms for each neuron in the hidden layer. We could also use the difference values of the output layer instead of the error values, however we found that using error values improves convergence. Here, we use the old (unchanged) value of the connection weight, which again improves convergence. The error formula for the hidden layer (difference between desired and actual hidden value) is:

$$E_{\text{hid } i} = \sum_{k=1}^{\text{num}(n_{\text{out}})} E_{\text{out } k} \cdot w_{\text{out } i,k}$$

$$\text{diff}_{\text{hid } i} = E_{\text{hid } i} \cdot (1 - o(n_{\text{hid } i})) \cdot o(n_{\text{hid } i})$$

In the example in Figure 19.5, there is only one output neuron, so each hidden neuron has only a single term for its desired value. The value and difference values for the first hidden neuron are therefore:

$$E_{\text{hid } 1} = E_{\text{out } 1} \cdot w_{\text{out } 1,1}$$
$$= 0.4 \cdot 0.8 = 0.32$$

$$\text{diff}_{\text{hid } 1} = E_{\text{hid } 1} \cdot (1 - o(n_{\text{hid } 1})) \cdot o(n_{\text{hid } 1})$$
$$= 0.32 \cdot (1 - 0.59) \cdot 0.59 = 0.077$$

Program 19.2: Backpropagation execution

```
1    float backprop(float train_in[NIN], float train_out[NOUT])
2    /* returns current square error value */
3    { int i,j;
4      float err_total;
5      float N_out[NOUT],err_out[NOUT];
6      float diff_out[NOUT];
7      float N_hid[NHID], err_hid[NHID], diff_hid[NHID];
8
9      //run network, calculate difference to desired output
10     feedforward(train_in, N_hid, N_out);
11     err_total = 0.0;
12     for (i=0; i<NOUT; i++)
13     {  err_out[i] = train_out[i]-N_out[i];
14        diff_out[i]= err_out[i] * (1.0-N_out[i]) * N_out[i];
15        err_total += err_out[i]*err_out[i];
16     }
17
18     // update w_out and calculate hidden difference values
19     for (i=0; i<NHID; i++)
20     { err_hid[i] = 0.0;
21       for (j=0; j<NOUT; j++)
22       { err_hid[i]   += err_out[j] * w_out[i][j];
23         w_out[i][j] += diff_out[j] * N_hid[i];
24       }
25       diff_hid[i] = err_hid[i] * (1.0-N_hid[i]) * N_hid[i];
26     }
27
28     // update w_in
29     for (i=0; i<NIN; i++)
30       for (j=0; j<NHID; j++)
31         w_in[i][j] += diff_hid[j] * train_in[i];
32
33     return err_total;
34   }
```

The weight changes for the two connections from the input layer to the first hidden neuron are as follows. Remember that the input of the hidden layer is the output of the input layer:

$$\Delta w_{\text{in } k,i} = 2 \cdot \eta \cdot \text{diff}_{\text{hid } i} \cdot \text{input}_k(n_{\text{hid } i})$$
$$\text{for } \eta = 0.5$$
$$= \text{diff}_{\text{hid } i} \cdot o(n_{\text{in } k})$$

$$\Delta w_{in1,1} = diff_{hid\ 1} \cdot o(n_{in\ 1})$$
$$= 0.077 \cdot 1.0 = 0.077$$

$$\Delta w_{in2,1} = diff_{hid\ 1} \cdot o(n_{in\ 2})$$
$$= 0.077 \cdot 0.5 = 0.039$$

and so on for the remaining weights. The first two updated weights will therefore be:

$$w'_{in1,1} = w_{in1,1} + \Delta w_{in1,1} = 0.2 + 0.077 = 0.28$$
$$w'_{in2,1} = w_{in2,1} + \Delta w_{in2,1} = 0.3 + 0.039 = 0.34$$

The backpropagation procedure iterates until a certain termination criterion has been fulfilled. This could be a fixed number of iterations over all training patterns, or until sufficient convergence has been achieved, for example if the total output error over all training patterns falls below a certain threshold.

Bias neurons   Program 19.2 demonstrates the implementation of the backpropagation process. Note that in order for backpropagation to work, we need one additional input neuron and one additional hidden neuron, called "bias neurons". The activation levels of these two neurons are always fixed to 1. The weights of the connections to the bias neurons are required for the backpropagation procedure to converge (see Figure 19.7).

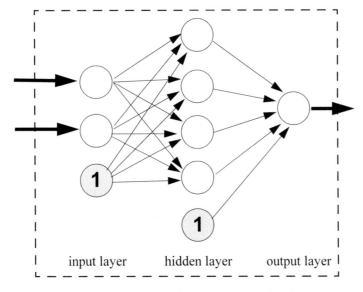

Figure 19.7: Bias neurons and connections for backpropagation

19.4 Neural Network Example

7-segment display A simple example for testing a neural network implementation is trying to learn the digits 0..9 from a seven-segment display representation. Figure 19.8 shows the arrangement of the segments and the numerical input and training output for the neural network, which could be read from a data file. Note that there are ten output neurons, one for each digit, 0..9. This will be much easier to learn than e.g. a four-digit binary encoded output (0000 to 1001).

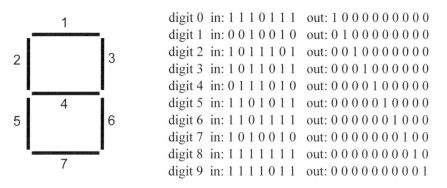

digit 0 in: 1 1 1 0 1 1 1 out: 1 0 0 0 0 0 0 0 0 0
digit 1 in: 0 0 1 0 0 1 0 out: 0 1 0 0 0 0 0 0 0 0
digit 2 in: 1 0 1 1 1 0 1 out: 0 0 1 0 0 0 0 0 0 0
digit 3 in: 1 0 1 1 0 1 1 out: 0 0 0 1 0 0 0 0 0 0
digit 4 in: 0 1 1 1 0 1 0 out: 0 0 0 0 1 0 0 0 0 0
digit 5 in: 1 1 0 1 0 1 1 out: 0 0 0 0 0 1 0 0 0 0
digit 6 in: 1 1 0 1 1 1 1 out: 0 0 0 0 0 0 1 0 0 0
digit 7 in: 1 0 1 0 0 1 0 out: 0 0 0 0 0 0 0 1 0 0
digit 8 in: 1 1 1 1 1 1 1 out: 0 0 0 0 0 0 0 0 1 0
digit 9 in: 1 1 1 1 0 1 1 out: 0 0 0 0 0 0 0 0 0 1

Figure 19.8: Seven-segment digit representation

Figure 19.9 shows the decrease of total error values by applying the back-propagation procedure on the complete input data set for some 700 iterations. Eventually the goal of an error value below 0.1 is reached and the algorithm terminates. The weights stored in the neural net are now ready to take on previously unseen real data. In this example the trained network could e.g. be tested against 7-segment inputs with a single defective segment (always on or always off).

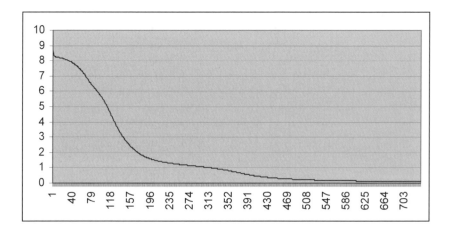

Figure 19.9: Error reduction for 7-segment example

19.5 Neural Controller

Control of mobile robots produces tangible actions from sensor inputs. A controller for a robot receives input from its sensors, processes the data using relevant logic, and sends appropriate signals to the actuators. For most large tasks, the ideal mapping from input to action is not clearly specified nor readily apparent. Such tasks require a control program that must be carefully designed and tested in the robot's operational environment. The creation of these control programs is an ongoing concern in robotics as the range of viable application domains expands, increasing the complexity of tasks expected of autonomous robots.

A number of questions need to be answered before the feed-forward ANN in Figure 19.4 can be implemented. Among them are:

How can the success of the network be measured?
The robot should perform a collision-free left-wall following.

How can the training be performed?
In simulation or on the real robot.

What is the desired motor output for each situation?
The motor function that drives the robot close to the wall on the left-hand side and avoids collisions.

Neural networks have been successfully used to mediate directly between sensors and actuators to perform certain tasks. Past research has focused on using neural net controllers to learn individual behaviors. Vershure developed a working set of behaviors by employing a neural net controller to drive a set of motors from collision detection, range finding, and target detection sensors [Vershure et al. 1995]. The on-line learning rule of the neural net was designed to emulate the action of Pavlovian classical conditioning. The resulting controller associated actions beneficial to task performance with positive feedback.

Adaptive logic networks (ALNs), a variation of NNs that only use boolean operations for computation, were successfully employed in simulation by Kube et al. to perform simple cooperative group behaviors [Kube, Zhang, Wang 1993]. The advantage of the ALN representation is that it is easily mappable directly to hardware once the controller has reached a suitable working state.

In Chapter 22 an implementation of a neural controller is described that is used as an arbitrator or selector of a number of behaviors. Instead of applying a learning method like backpropagation shown in Section 19.3, a genetic algorithm is used to evolve a neural network that satisfies the requirements.

19.6 References

GURNEY, K. *Neural Nets*, UCL Press, London, 2002

KUBE, C., ZHANG, H., WANG, X. *Controlling Collective Tasks with an ALN*, IEEE/RSJ IROS, 1993, pp. 289-293 (5)

MINSKY, M., PAPERT, S. *Perceptrons*, MIT Press, Cambridge MA, 1969

ROSENBLATT, F. *Principles of Neurodynamics*. Spartan Books, Washington DC, 1962

RUMELHART, D., MCCLELLAND, J. (Eds.) *Parallel Distributed Processing*, 2 vols., MIT Press, Cambridge MA, 1986

SUTTON, R., BARTO, A. *Reinforcement Learning: An Introduction*, MIT Press, Cambridge MA, 1998

VERSHURE, P., WRAY, J., SPRONS, O., TONONI, G., EDELMAN, G. *Multilevel Analysis of Classical Conditioning in a Behaving Real World Artifact*, Robotics and Autonomous Systems, vol. 16, 1995, pp. 247-265 (19)

ZAKNICH, A. *Neural Networks for Intelligent Signal Processing*, World Scientific, Singapore, 2003

GENETIC ALGORITHMS

20

E volutionary algorithms are a family of search and optimization techniques that make use of principles from Darwin's theory of evolution [Darwin 1859] to progress toward a solution. Genetic algorithms (GA) are a prominent part of this larger overall group. They operate by iteratively evolving a solution from a history of potential solutions, which are manipulated by a number of biologically inspired operations. Although only an approximation to real biological evolutionary processes, they have been proven to provide a powerful and robust means of solving problems.

The utility of genetic algorithms is their ability to be applied to problems without a deterministic algorithmic solution. Certain satisfiability problems in robotic control fall into this category. For example, there is no known algorithm to deterministically develop an optimal walking gait for a particular robot. An approach to designing a gait using genetic algorithms is to evolve a set of parameters controlling a gait generator. The parameters completely control the type of gait that is produced by the generator. We can assume there exists a set of parameters that will produce a suitable walk for the robot and environment – the problem is to find such a set. Although we do not have a way to obtain these algorithmically, we can use a genetic algorithm in a simulated environment to incrementally test and evolve populations of parameters to produce a suitable gait.

It must be emphasized that the effectiveness of using a genetic algorithm to find a solution is dependent on the problem domain, the existence of an optimal solution to the problem at hand, and a suitable fitness function. Applying genetic algorithms to problems that may be solved algorithmically is decidedly inefficient. They are best used for solving tasks that are difficult to solve, such as NP-hard problems. NP-hard problems are characterized by the difficulty of finding a solution due to a large solution search space, but being easy to verify once a candidate solution has been obtained.

For further reading see [Goldberg 1989] and [Langton 1995].

20.1 Genetic Algorithm Principles

In this section we describe some of the terminology used, and then outline the operation of a genetic algorithm. We then examine the components of the algorithm in detail, describing different implementations that have been employed to produce results.

Genotype and phenotype

Genetic algorithms borrow terminology from biology to describe their interacting components. We are dealing with *phenotypes*, which are possible solutions to a given problem (for example a simulated robot with a particular control structure), and *genotypes*, which are encoded representations of phenotypes. Genotypes are sometimes also called *chromosomes* and can be split into smaller chunks of information, called *genes* (Figure 20.1).

The genetic operators work only on genotypes (chromosomes), while it is necessary to construct phenotypes (individuals) in order to determine their fitness.

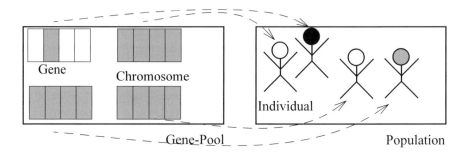

Figure 20.1: Terminology

GA execution

The basic operation of a genetic algorithm can be summarized as follows:

1. Randomly initialize a population of chromosomes.

2. While the terminating criteria have not been satisfied:

   a. Evaluate the fitness of each chromosome:

      i. Construct the *phenotype* (e.g. simulated robot) corresponding to the encoded *genotype* (chromosome).

      ii. Evaluate the phenotype (e.g. measure the simulated robot's walking abilities), in order to determine its fitness.

   b. Remove chromosomes with low fitness.

   c. Generate new chromosomes, using certain selection schemes and genetic operators.

The algorithm can start with either a set of random chromosomes, or ones that represent already approximate solutions to the given problem. The gene pool is evolved from generation to generation by a set of modifying operators and a selection scheme that depends on the fitness of each chromosome. The selection scheme determines which chromosomes should reproduce, and typically selects the highest-performing members for reproduction. Reproduction is achieved through various operators, which create new chromosomes from existing chromosomes. They effectively alter the parameters to cover the search space, preserving and combining the highest-performing parameter sets.

Each iteration of the overall procedure creates a new population of chromosomes. The total set of chromosomes at one iteration of the algorithm is known as a generation. As the algorithm iteration continues, it searches through the solution space, refining the chromosomes, until either it finds one with a sufficiently high fitness value (matching the desired criteria of the original problem), or the evolutionary progress slows down to such a degree that finding a matching chromosome is unlikely.

Fitness function   Each problem to be solved requires the definition of a unique fitness function describing the characteristics of an appropriate solution. The purpose of the fitness function is to evaluate the suitability of a solution with respect to the overall goal. Given a particular chromosome, the fitness function returns a numerical value corresponding to the chromosome's quality. For some applications the selection of a fitness function is straightforward. For example, in function optimization problems, fitness is evaluated by the function itself. However, in many applications there are no obvious performance measurements of the goal. In these cases a suitable fitness function may be constructed from a combination of desired factors characteristic of the problem.

Selection schemes   In nature, organisms that reproduce the most before dying will have the greatest influence on the composition of the next generation. This effect is employed in the genetic algorithm selection scheme that determines which individuals of a given population will contribute to form the new individuals for the next generation. "Tournament selection", "Random selection", and "Roulette wheel selection" are three commonly used selection schemes.

Tournament selection operates by selecting two chromosomes from the available pool, and comparing their fitness values when they are evaluated against each other. The better of the two is then permitted to reproduce. Thus, the fitness function chosen for this scheme only needs to discriminate between the two entities.

Random selection randomly selects the parents of a new chromosome from the existing pool. Any returned fitness value below a set operating point is instantly removed from the population. Although it would appear that this would not produce beneficial results, this selection mechanism can be employed to introduce randomness into a population that has begun to converge to a sub-optimal solution.

In *roulette wheel selection* (sometimes referred to as *fitness proportionate selection*) the chance for a chromosome to reproduce is proportional to the fitness of the entity. Thus, if the fitness value returned for one chromosome is twice as high as the fitness value for another, then it is twice as likely to reproduce. However, its reproduction is not guaranteed as in tournament selection.

Although genetic algorithms will converge to a solution if all chromosomes reproduce, it has been shown that the convergence rate can be significantly increased by duplicating unchanged copies of the fittest chromosomes for the next generation.

20.2 Genetic Operators

Genetic operators comprise the methods by which one or more chromosomes are combined to produce a new chromosome. Traditional schemes utilize only two operators: mutate and crossover [Beasley, Bull, Martin 1993a]. Crossover takes two individuals and divides the string into two portions at a randomly selected point inside the encoded bit string. This produces two "head" segments and two "tail" segments. The two tail segments for the chromosomes are then interchanged, resulting in two new chromosomes, where the bit string preceding the selected bit position belongs to one parent, and the remaining portion belongs to the other parent. This process is illustrated in Figure 20.2.

The mutate operator (Figure 20.3) randomly selects one bit in the chromosome string, and inverts the value of the bit with a defined probability. Historically, the crossover operator has been viewed as the more important of the two techniques for exploring the solution space; however, without the mutate operator portions of the solution space may not be searched, as the initial chromosomes may not contain all possible bit combinations [Beasley, Bull, Martin 1993b].

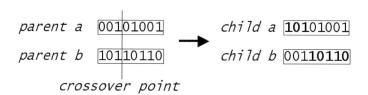

Figure 20.2: Crossover operator

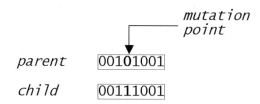

Figure 20.3: Mutate operator

There are a number of possible extensions to the set of traditional operators. The two-point crossover operates similarly to the single point crossover described, except that the chromosomes are now split into two positions rather than just one. The mutate operator can also be enhanced to operate on portions of the chromosome larger than just one bit, increasing the randomness that can be added to a search in one operation.

Further extensions rely on modifying the bit string under the assumption that portions of the bit string represent non-binary values (such as 8bit integer values or 32bit floating point values). Two commonly used operators that rely on this interpretation of the chromosome are the "Non-Binary Average" and "Non-Binary Creep" operators. Non-Binary Average interprets the chromosome as a string of higher cardinality symbols and calculates the arithmetic average of the two chromosomes to produce the new individual. Similarly, Non-Binary Creep treats the chromosomes as strings of higher cardinality symbols and increments or decrements a randomly selected value in these strings by a small randomly generated amount.

parent a   00101001

parent b   10110110

child   01100101

Figure 20.4: Non-Binary Average operator

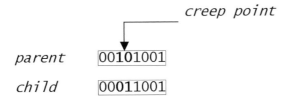

parent   00101001

child   00011001

Figure 20.5: Non-Binary Creep operator

The operation of the Non-Binary Average operator is illustrated in Figure 20.4. In the example shown the bit string is interpreted as a set of two bit symbols, and averaged using truncation. Thus, zero plus two averages to one, i.e.

$$(00 + 10) / 2 = 01$$

but two plus three will average to two, i.e.

$$(10 + 11) / 2 = 10$$

The Non-Binary Creep operator shown in Figure 20.5 also represents the bit string as two bit symbols and decrements the second symbol by a value of one.

Encoding   The encoding method chosen to transform the parameters to a chromosome can have a large effect on the performance of the genetic algorithm. A compact encoding allows the genetic algorithm to perform efficiently. There are two common encoding techniques applied to the generation of a chromosome. Direct encoding explicitly specifies every parameter within the chromosome, whereas indirect encoding uses a set of rules to reconstruct the complete

295

parameter space. Direct encoding has the advantage that it is a simple and powerful representation, but the resulting chromosome can be quite large. Indirect encoding is far more compact, but it often represents a highly restrictive set of the original structures.

20.3 Applications to Robot Control

Three applications of genetic algorithms to robot control are briefly discussed in the following sections. These topics are dealt with in more depth in the following chapters on behavior-based systems and gait evolution.

Gait generation   Genetic algorithms have been applied to the evolution of neural controllers for robot locomotion by numerous researchers [Ijspeert 1999], [Lewis Fagg, Bekey 1994]. This approach uses the genetic algorithm to evolve the weightings between interconnected neurons to construct a controller that achieves the desired gait. Neuron inputs are taken from various sensors on the robot, and the outputs of certain neurons are directly connected to the robot's actuators. [Lewis, Fagg, Bekey 1994] successfully generated gaits for a hexapod robot using a simple traditional genetic algorithm with one-point crossover and mutate. A simple neural network controller was used to control the robot, and the fitness of the individuals generated was evaluated by human designers. [Ijspeert 1999] evolved a controller for a simulated salamander using an enhanced genetic algorithm. The neural model employed was biologically based and very complex. However, the system developed was capable of operating without human fitness evaluators.

Schema-based navigation   Genetic algorithms have been used in a variety of different ways to newly produce or optimize existing behavioral controllers. [Ram et al. 1994] used a genetic algorithm to control the weightings and internal parameters of a simple reactive schema controller. In schema-based control, primitive motor and perceptual schemas do simple distributed processing of inputs (taken from sensors or other schemas) to produce outputs. Motor schemas asynchronously receive input from perceptual schemas to produce response outputs intended to drive an actuator. A schema arbitration controller produces output by summing contributions from independent schema units, each contributing to the final output signal sent to the actuators according to a weighting (Figure 20.6). These weightings are usually manually tuned to produce desired system behavior from the robot.

The approach taken by Ram et al. was to use a genetic algorithm to determine an optimal set of schema weightings for a given fitness function. By tuning the parameters of the fitness function, robots optimized for the qualities of safety, speed, and path efficiency were produced. The behavior of each of these robots was different from any of the others. This graphically demonstrates how behavioral outcomes may be easily altered by simple changes in a fitness function.

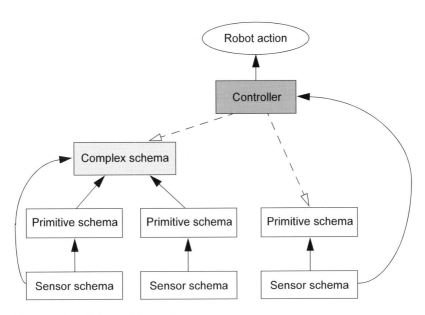

Figure 20.6: Schema hierarchy

Behavior selection   [Harvey, Husbands, Cliff 1993] used a genetic algorithm to evolve a robot neural net controller to perform the tasks of wandering and maximizing the enclosed polygonal area of a path within a closed space. The controller used sensors as its inputs and was directly coupled to the driving mechanism of the robot. A similar approach was taken in [Venkitachalam 2002] but the outputs of the neural network were used to control schema weightings. The neural network produces dynamic schema weightings in response to input from perceptual schemas.

20.4 Example Evolution

In this section we demonstrate a simple walk-through example. We will approach the problem of manually solving a basic optimization problem using a genetic algorithm, and suggest how to solve the same problem using a computer program. The fitness function we wish to optimize is a simple quadratic formula (Figure 20.7):

$$f(x) = -(x - 6)^2 \quad \text{for } 0 \le x \le 31$$

Using a genetic algorithm to search a solvable equation with a small search space like this is inefficient and not advised for reasons stated earlier in this chapter. In this particular example, the genetic algorithm is not significantly more efficient than a random search, and possibly worse. Rather, this problem has been chosen because its relatively small size allows us to examine the workings of the genetic algorithm.

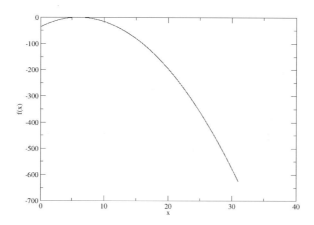

Figure 20.7: Graph of f(x)

The genetic algorithm we will use features a simple binary encoding, one-point crossover, elitism, and a decease rate. In this particular problem, the gene consists of a single integral number. Since there is only one parameter to be found, each chromosome consists only of a single gene. The rather artificial constraint on the function given above allows us to use a 5bit binary encoding for our genes. Hence chromosomes are represented by bit strings of length 5. We begin by producing a random population and evaluating each chromosome through the fitness function. If the terminating criteria are met, we stop the algorithm. In this example, we already know the optimal solution is x = 6 and hence the algorithm should terminate when this value is obtained. Depending on the nature of the fitness function, we may want to let the algorithm continue to find other (possibly better) feasible solutions. In these cases we may want to express the terminating criteria in relation to the rate of convergence of the top performing members of a population.

x	Bit String	f(x)	Ranking
2	00010	−16	1
10	01010	−16	2
0	00000	−36	3
20	10100	−196	4
31	11111	−625	5

Table 20.1: Initial population

The initial population, encodings, and fitnesses are given in Table 20.1. Note that chromosomes $x = 2$ and $x = 10$ have equal fitness values, hence their relative ranking is an arbitrary choice.

The genetic algorithm we use has a simple form of selection and reproduction. The top performing chromosome is reproduced and preserved for use in the next iteration of the algorithm. It replaces the lowest performing chromosome, which is removed from the population altogether. Hence we remove $x = 31$ from selection.

The next step is to perform crossover between the chromosomes. We randomly pair the top four ranked chromosomes and determine whether they are subject to crossover by a non-deterministic probability. In this example, we have chosen a crossover probability of 0.5, easily modeled by a coin toss. The random pairings selected are ranked chromosomes (1,4) and (2,3). Each pair of chromosomes will undergo a single random point crossover to produce two new chromosomes.

As described earlier, the single random point crossover operation selects a random point to perform the crossover. In this iteration, both pairs undergo crossover (Figure 20.8).

The resulting chromosomes from the crossover operation are as follows:

(1) $00|010 \downarrow\rightarrow 00100 = 4$
(4) $10|100 \uparrow\rightarrow 10010 = 18$

(2) $0|1010 \downarrow\rightarrow 00000 = 0$
(3) $0|0000 \uparrow\rightarrow 01010 = 10$

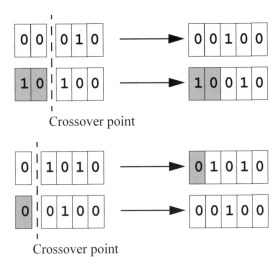

Figure 20.8: Crossover

Note that in the case of the second crossover, because the first bit is identical in both strings the resulting chromosomes are the same as the parents. This is effectively equivalent to no crossover operation occurring. After one iteration we can see the population has converged somewhat toward the optimal answer. We now repeat the evaluation process with our new population (Table 20.2).

x	Bit String	f(x)	Ranking
4	00100	−4	1
2	00010	−16	2
10	01010	−16	3
0	00000	−36	4
18	10010	−144	5

Table 20.2: Population after crossover

Again we preserve the *best* chromosome (x = 4) and remove the *worst* (x = 18). Our random pairings this time are ranked chromosomes (1, 2) and (3, 4). This time, only pair (3, 4) has been selected by a random process to cross over, and (1, 2) is selected for mutation. It is worth noting that the (1, 2) pair had the potential to produce the optimal solution x = 6 if it had undergone crossover. This missed opportunity is characteristic of the genetic algorithm's non-deterministic nature: the time taken to obtain an optimal solution cannot be accurately foretold. The mutation of (2), however, reintroduced some of the lost bit-string representation. With no mutate operator the algorithm would no longer be capable of representing odd values (bit strings ending with a one).

Mutation of (1) and (2)
(1) 00**1**00 → 00000 = 0
(2) 0001**0** → 00011 = 3

Crossover of pair (3, 4)
(3) 01|010 ↓→ 01000 = 8
(4) 00|000 ↑→ 00010 = 2

The results of the next population fitness evaluation are presented in Table 20.3.

As before, chromosome x = 0 is removed and x = 4 is retained. The selected pairs for crossover are (1, 3) and (1, 4), of which only (1, 4) actually undergoes crossover:

(1) 001|00 ↓→ 00110 = 6
(4) 000|10 ↑→ 00000 = 0

The optimal solution of x = 6 has been obtained. At this point, we can stop the genetic algorithm because we know this is the optimal solution. However, if we let the algorithm continue, it should eventually completely converge to

x	Bit String	f(x)	Ranking
4	00100	−4	1
8	01000	−4	2
3	00011	−9	3
2	00010	−16	4
0	00000	−36	5

Table 20.3: Connection between the input and output indices

x = 6. This is because the x = 6 chromosome is now persistent through subsequent populations due to its optimal nature. When another chromosome is set to x = 6 through crossover, the chance of it being preserved through populations increases due to its increased presence in the population. This probability is proportional to the presence of the x = 6 chromosome in the population, and hence given enough iterations the whole population should converge. The elitism operator, combined with the fact that there is only one maximum, ensures that the population will never converge to another chromosome.

20.5 Implementation of Genetic Algorithms

We have implemented a genetic algorithm framework in object-oriented C++ for the robot projects described in the following chapters. The base system consists of abstract classes `Gene`, `Chromosome`, and `Population`. These classes may be extended with the functionality to handle different data types, including the advanced operators described earlier and for use in other applications as required. The implementation has been kept simple to meet the needs of the application it was developed for. More fully featured third-party genetic algorithm libraries are also freely available for use in complex applications, such as GA Lib [GALib 2006] and OpenBeagle [Beaulieu, Gagné 2006]. These allow us to begin designing a working genetic algorithm without having to implement any infrastructure. The basic relationship between program classes in these frameworks tends to be similar.

Using C++ and an object-oriented methodology maps well to the individual components of a genetic algorithm, allowing us to represent components by classes and operations by class methods. The concept of inheritance allows the base classes to be extended for specific applications without modification of the original code.

The basic unit of the system is a child of the base Gene class. Each instance of a Gene corresponds to a single parameter. The class itself is completely abstract: there is no default implementation, hence it is more accurately described as an interface. The Gene interface describes a set of basic opera-

Program 20.1: Gene header

```
1    class Gene
2    {
3    //   Return our copy of data, suitable for reading
4         virtual void* getData(void) = 0;
5    //   Return new copy of data, suitable for manipulat.
6         virtual void* getDataCopy() = 0;
7    //   Copy the data from somewhere else to here
8         virtual void  setData(const void* data) = 0;
9    //   Copy data from another gene of same type to here
10        virtual void  setData(Gene& gene) = 0;
11   //   Set the data in this gene to a random value
12        virtual void  setRandom(void) = 0;
13   //   Mutate our data
14        virtual void  mutate(void) = 0;
15   //   Produce a new identical copy of this gene
16        virtual Gene& clone(void) = 0;
17   //   Return the unique type of this gene
18        virtual unsigned int type(void) = 0;
19   };
```

tions that all parameter types must implement so that they can be generically manipulated consistently externally. An excerpt of the Gene header file is given in Program 20.1.

Program 20.2: Chromosome header

```
1    class Chromosome
2    {
3    // Return the number of genes in this chromosome.
4         int getNumGenes();
5    // Set the gene at a specified index.
6         int setGene(int index, Gene* gene);
7    // Add a gene to the chromosome.
8         int addGene(Gene* gene);
9    // Get a gene at a specified index.
10        Gene* getGene(int index);
11   // Set fitness of chromosome as by ext. fitness function
12        void setFitness(double value);
13   // Retrieve the fitness of this chromosome
14        double getFitness(void);
15   // Perform single crossover with a partner chromosome
16        virtual void crossover(Chromosome* partner);
17   // Perform a mutation of the chromosome
18        virtual void mutate(void);
19   // Return a new identical copy of this chromosome
20        Chromosome& clone(void);
21   };
```

The chromosome class stores a collection of genes in a container class. It provides access to basic crossover and mutation operators. These can be over-ridden and extended with more complex operators as described earlier. An excerpt of the chromosome header file is given in Program 20.2.

Finally, the population class (Program 20.3) is the collection of Chromo-somes comprising a full population. It performs the iterative steps of the genetic algorithm, evolving its own population of Chromosomes by invoking their defined operators. Access to individual Chromosomes is provided, allow-ing evaluation of terminating conditions through an external routine.

Program 20.3: Population class

```
1    class Population
2    {// Initialise population with estimated no. chromosomes
3      Population(int   numChromosomes = 50,
4                 float deceaseRate    = 0.4f,
5                 float crossoverRate  = 0.5f,
6                 float mutationRate   = 0.05f );
7      ~Population();
8
9      void addChromosome(const Chromosome* c);
10
11     //  Set population parameters
12     void setCrossover(float rate);
13     void setMutation(float rate);
14     void setDecease(float rate);
15
16     // Create new pop. with selection, crossover, mutation
17     virtual void evolveNewPopulation(void);
18     int getPopulationSize(void);
19     Chromosome& getChromosome(int index);
20
21     //  Print pop. state, (chromosome vals & fitness stats)
22     void printState(void);
23
24     // Sort population according to fitness,
25     void sortPopulation(void);
26    };
```

As an example of using these classes, Program 20.4 shows an excerpt of code to solve our quadratic problem, using a derived integer representation class GeneInt and the Chromosome and Population classes described above.

Program 20.4: Main program

```
 1   int main(int argc, char *argv[])
 2   { int i;
 3
 4     GeneInt genes[5];
 5     Chromosome* chromosomes[5];
 6     Population population;
 7
 8     population.setCrossover(0.5f);
 9     population.setDecease(0.2f);
10     population.setMutation(0.0f);
11
12     //  Initialise genes and add them to our chromosomes,
13     //  then add the chromosomes to the population
14     for(i=0; i<5; i++) {
15       genes[i].setData((void*) rand()%32);
16       chromosomes[i].addGene(&genes[i]);
17       population.addChromosome(&chromosomes[i]);
18     }
19
20     //  Continually run the genetic algorithm until the
21     //  optimal solution is found by the top chromosome
22
23     i = 0;
24     do {
25       printf("Iteration %d", i++);
26       population.evolveNewPopulation();
27       population.printState();
28     } while((population.getChromosome(0)).getFitness()!=0);
29
30     //  Finished
31     return 0;
32   }
```

20.6 References

BEASLEY, D., BULL, D., MARTIN, R. *An Overview of Genetic Algorithms: Part 1, Fundamentals*, University Computing, vol. 15, no. 2, 1993a, pp. 58-69 (12)

BEASLEY, D., BULL, D., MARTIN, R. *An Overview of Genetic Algorithms: Part 2, Research Topics*, University Computing, vol. 15, no. 4, 1993b, pp. 170-181 (12)

BEAULIEU, J., GAGNÉ, C. *Open BEAGLE – A Versatile Evolutionary Computation Framework*, Département de génie électrique et de génie informatique, Université Laval, Québec, Canada, `http://www.gel.ulaval.ca/~beagle/`, 2006

DARWIN, C. *On the Origin of Species by Means of Natural Selection, or Preservation of Favoured Races in the Struggle for Life*, John Murray, London, 1859

GALIB *Galib – A C++ Library of Genetic Algorithm Components*, `http://lancet.mit.edu/ga/`, 2006

GOLDBERG, D. *Genetic Algorithms in Search, Optimization and Machine Learning*, Addison-Wesley, Reading MA, 1989

HARVEY, I., HUSBANDS, P., CLIFF, D. *Issues in Evolutionary Robotics*, in J. Meyer, S. Wilson (Eds.), From Animals to Animats 2, Proceedings of the Second International Conference on Simulation of Adaptive Behavior, MIT Press, Cambridge MA, 1993

IJSPEERT, A. *Evolution of neural controllers for salamander-like locomotion*, Proceedings of Sensor Fusion and Decentralised Control in Robotics Systems II, 1999, pp. 168-179 (12)

LANGTON, C. (Ed.) *Artificial Life – An Overview*, MIT Press, Cambridge MA, 1995

LEWIS, M., FAGG, A., BEKEY, G. *Genetic Algorithms for Gait Synthesis in a Hexapod Robot*, in Recent Trends in Mobile Robots, World Scientific, New Jersey, 1994, pp. 317-331 (15)

RAM, A., ARKIN, R., BOONE, G., PEARCE, M. *Using Genetic Algorithms to Learn Reactive Control Parameters for Autonomous Robotic Navigation*, Journal of Adaptive Behaviour, vol. 2, no. 3, 1994, pp. 277-305 (29)

VENKITACHALAM, D. *Implementation of a Behavior-Based System for the Control of Mobile Robots*, B.E. Honours Thesis, The Univ. of Western Australia, Electrical and Computer Eng., supervised by T. Bräunl, 2002

GENETIC PROGRAMMING

21

G enetic programming extends the idea of genetic algorithms discussed in Chapter 20, using the same idea of evolution going back to Darwin [Darwin 1859]. Here, the genotype is a piece of software, a directly executable program. Genetic programming searches the space of possible computer programs that solve a given problem. The performance of each individual program within the population is evaluated, then programs are selected according to their fitness and undergo operations that produce a new set of programs. These programs can be encoded in a number of different programming languages, but in most cases a variation of Lisp [McCarthy et al. 1962] is chosen, since it facilitates the application of genetic operators.

The concept of genetic programming was introduced by Koza [Koza 1992]. For further background reading see [Blickle, Thiele 1995], [Fernandez 2006], [Hancock 1994], [Langdon, Poli 2002].

21.1 Concepts and Applications

The main concept of genetic programming is its ability to create working programs without the full knowledge of the problem or the solution. No additional encoding is required as in genetic algorithms, since the executable program itself is the phenotype. Other than that, genetic programming is very similar to genetic algorithms. Each program is evaluated by running it and then assigning a fitness value. Fitness values are the base for selection and genetic manipulation of a new generation. As for genetic algorithms, it is important to maintain a wide variety of individuals (here: programs), in order to fully cover the search area.

Koza summarizes the steps in genetic programming as follows [Koza 1992]:

1. Randomly generate a combinatorial set of computer programs.

2. Perform the following steps iteratively until a termination criterion is satisfied (i.e. the program population has undergone the maximum number of generations, or the maximum fitness value has been reached, or the population has converged to a sub-optimal solution).

   a. Execute each program and assign a fitness value to each individual.

   b. Create a new population with the following steps:

      i. Reproduction: Copy the selected program unchanged to the new population.

      ii. Crossover: Create a new program by recombining two selected programs at a random crossover point.

      iii. Mutation: Create a new program by randomly changing a selected program.

3. The best sets of individuals are deemed the optimal solution upon termination.

Applications in robotics   The use of genetic programming is widely spread from evolving mathematical expressions to locating optimum control parameters in a PID controller. The genetic programming paradigm has become popular in the field of robotics and is used for evolving control architectures and behaviors of mobile robots.

[Kurashige, Fukuda, Hoshino 1999] use genetic programming as the learning method to evolve the motion planning of a six-legged walker. The genetic programming paradigm is able to use primitive leg-moving functions and evolve a program that performs robot walking with all legs moving in a hierarchical manner.

[Koza 1992] shows the evolution of a wall-following robot. He uses primitive behaviors of a subsumption architecture [Brooks 1986] to evolve a new behavior that lets the robot execute a wall-following pattern without prior knowledge of the hierarchy of behaviors and their interactions.

[Lee, Hallam, Lund 1997] apply genetic programming as the means to evolve a decision arbitrator on a subsumption system. The goal is to produce a high-level behavior that can perform box-pushing, using a similar technique to Koza's genetic programming.

[Walker, Messom 2002] use genetic programming and genetic algorithms to auto-tune a mobile robot control system for object tracking.

The initial population holds great importance for the final set of solutions. If the initial population is not diverse enough or strong enough, the optimal solution may not be found. [Koza 1992] suggests a minimum initial population size of 500 for robot motion control and 1,000 for robot wall-following (see Table 21.1).

Problem	Reference	Initial Pop. Size
Wall-following robot	[Koza 1992]	1,000
Box-moving robot	[Mahadevon, Connell 1991]	500
Evolving behavior primitives and arbitrators	[Lee, Hallam, Lund 1997]	150
Motion planning for six-legged robot	[Kurashige, Fukuda, Hoshino 1999]	2,000
Evolving communication agents	[Iba, Nonzoe, Ueda 1997]	500
Mobile robot motion control	[Walker, Messom 2002]	500

Table 21.1: Initial population sizes

21.2 Lisp

It is possible to formulate inductive programs in any programming language. However, evolving program structures such as C or Java are not straightforward. Therefore, Koza used the functional language Lisp ("List Processor") for genetic programming. Lisp was developed by McCarthy starting in 1958 [McCarthy et al. 1962], which makes it one of the oldest programming languages of all. Lisp is available in a number of implementations, among them the popular Common Lisp [Graham 1995]. Lisp is usually interpreted and provides only a single program and data structure: *the list*.

Lisp functions: atoms and lists

Every object in Lisp is either an *atom* (a constant, here: integer or a parameterless function name) or a *list* of objects, enclosed in parentheses.

Examples for atoms: `7, 123, obj_size`
Examples for lists: `(1 2 3), (+ obj_size 1), (+ (* 8 5) 2)`

S-Expression

Lists may be nested and are not only the representation for data structures, but also for program code as well. Lists that start with an operator, such as `(+ 1 2)`, are called *S-expressions*. An S-expression can be evaluated by the Lisp interpreter (Figure 21.1) and will be replaced by a result value (an atom or a list, depending on the operation). That way, a program execution in a procedural programming language like C will be replaced by a function call in Lisp:

`(+ (* 8 5) 2)` → `(+ 40 2)` → `42`

Lisp subset for robotics

Only a small subset of Lisp is required for our purpose of driving a mobile robot in a restricted environment. In order to speed up the evolutionary process, we use very few functions and constants (see Table 21.2).

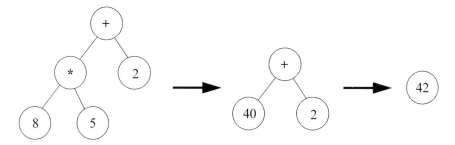

Figure 21.1: Tree structure and evaluation of S-expression

We deal only with integer data values. Our Lisp subset contains pre-defined constants `zero`, `low`, and `high`, and allows the generation of other integer constants by using the function `(INC v)`. Information from the robot's vision sensor can be obtained by calling `obj_size` or `obj_pos`. An evaluation of any of these two atoms will implicitly grab a new image from the camera and then call the color object detection procedure.

There are four atoms `psd_aaa` for measuring the distance between the robot and the nearest obstacle to the left, right, front, and back. Evaluating any of these atoms activates a measurement of the corresponding PSD (position sensitive device) sensor. These sensors are very useful for obstacle avoidance, wall-following, detecting other robots, etc.

There are four movement atoms remaining. Two for driving (forward and backward), and two for turning (left and right). When one of these is evaluated, the robot will drive (or turn, respectively) by a small fixed amount.

Finally, there are three program constructs for selection, iteration, and sequence. An "if-then-else" S-expression allows branching. Since we do not provide explicit relations, for example like `(< 3 7)`, the comparison operator "less" is a fixed part of the if-statement. The S-expression contains two integer values for the comparison, and two statements for the "then" and "else" branch. Similarly, the while-loop has a fixed "less" comparison operator as loop condition. The iteration continues while the first integer value is less than the second. The two integer arguments are followed by the iteration statement itself.

These are all constructs, atoms, and S-expression lists allowed in our Lisp subset. Although more constructs might facilitate programming, it might make evolution more complex and would therefore require much more time to evolve useful solutions.

Although Lisp is an untyped language, we are only interested in *valid* S-expressions. Our S-expressions have placeholders for either integer values or statements. So during genetic processing, only integers may be put into integer slots and only statements may be put into statement slots. For example, the first two entries following the keyword in a `WHILE_LESS`-list must be integer values, but the third entry must be a statement. An integer value can be either

Name	Kind	Semantics
zero	atom, int, constant	0
low	atom, int, constant	20
high	atom, int, constant	40
(INC v)	list, int, function	**Increment** v+1
obj_size	atom, int, image sensor	search image for color object, return height in pixels (0..60)
obj_pos	atom, int, image sensor	search image for color object, return x-position in pixels (0..80) or return −1 if not found
psd_left	atom, int, distance sensor	measure distance in mm to left (0..999)
psd_right	atom, int, distance sensor	measure distance in mm to right (0..999)
psd_front	atom, int, distance sensor	measure distance in mm to front (0..999)
psd_back	atom, int, distance sensor	measure distance in mm to back (0..999)
turn_left	atom, statem., act.	rotate robot 10° to the left
turn_right	atom, statem., act.	rotate robot 10° to the right
drive_straight	atom, statem., act.	drive robot 10cm forward
drive_back	atom, statem., act.	drive robot 10cm backward
(IF_LESS v1 v2 s1 s2)	list, statement, program construct	**Selection** if ($v_1 < v_2$) s_1; else s_2;
(WHILE_LESS v1 v2 s)	list, statement, program construct	**Iteration** while ($v_1 < v_2$) s;
(PROGN2 s1 s2)	list, statement, program construct	**Sequence** s_1; s_2;

The left brace spanning the first ten rows is labelled **value**; the brace spanning the last six rows is labelled **statement**.

Table 21.2: Lisp subset for genetic programming

an atom or an S-expression, for example low or INC(zero). In the same way, a statement can be either an atom or an S-expression, for example drive_straight or PROGN2(turn_left, drive_back).

We implemented the Lisp interpreter as a recursive C program (Lisp purists would have implemented it in Lisp) that is executed by the EyeSim simulator. Program 21.1 shows an extract of the main Lisp decoding routine.

Program 21.1: Lisp interpreter in C

```
1   int compute(Node n)
2   { int ret, return_val1, return_val2;
3     ...
4     CAMGetColFrame (&img, 0);
5     if (DEBUG) LCDPutColorGraphic(&img);
6     ret = -1; /* no return value */
7
8     switch(n->symbol) {
9     case PROGN2:
10      compute(n->children[0]);
11      compute(n->children[1]);
12      break;
13
14    case IF_LESS:
15      return_val1 = compute(n->children[0]);
16      return_val2 = compute(n->children[1]);
17      if (return_val1 <= return_val2) compute(n->children[2]);
18        else                        compute(n->children[3]);
19      break;
20
21    case WHILE_LESS:
22      do {
23        return_val1 = compute(n->children[0]);
24        return_val2 = compute(n->children[1]);
25        if (return_val1 <= return_val2) compute(n->children[2]);
26      } while (return_val1 <= return_val2);
27      break;
28
29    case turn_left:  turn_left(&vwhandle);
30      break;
31    case turn_right: turn_right(&vwhandle);
32      break;
33    ...
34    case obj_size: ColSearch2 (img, RED_HUE, 10, &pos, &ret);
35      break;
36    case obj_pos:  ColSearch2 (img, RED_HUE, 10, &ret, &val);
37      break;
38    case low:  ret = LOW;
39      break;
40    case high: ret = HIGH;
41      break;
42    default: printf("ERROR in compute\n");
43      exit(1);
44    }
45    return ret;
46  }
```

21.3 Genetic Operators

Similar to the genetic algorithm operators in Chapter 20, we have crossover and mutation. However, here they are applied directly to a Lisp program.

Crossover Crossover (sexual recombination) operation for genetic programming re-creates the diversity in the evolved population by combining program parts from two individuals:

1. Select two parent individuals from the current generation based on their fitness values.

2. Randomly determine a crossover point in each of the two parents. Both crossover points must match, i.e. they must both represent either a value or a statement.

3. Create the first offspring by using parent no. 1, replacing the sub-tree under its crossover point by the sub-tree under the crossover point from parent no. 2. Create the second offspring the same way, starting with parent no. 2.

Since we require the selected crossover points to match type, we have guaranteed that the two generated offspring programs will be valid and executable.

Crossover points can be external (a leaf node, i.e. replacing an atom) or internal (an internal tree node, i.e. replacing a function). External points may extend the program structure by increasing its depth. This occurs when one parent has selected an external point, and the other has selected an internal point for crossing over. An internal point represents a possibly substantial alteration of the program structure and therefore maintains the variety within the population.

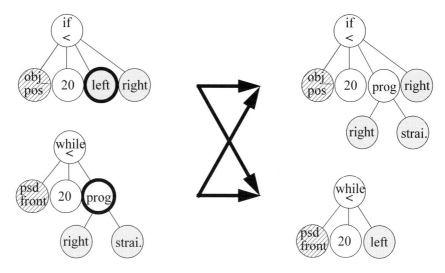

Figure 21.2: Crossover

The following shows an example with the crossover point marked in bold face:

```
1. (IF_LESS obj_pos low turn_left turn_right)
2. (WHILE_LESS psd_front (PROGN2 turn_right drive_straight))
```
⇒
```
1. (IF_LESS obj_pos low (PROGN2 turn_right drive_straight)
        turn_right)
2. (WHILE_LESS psd_front turn_left)
```

Both selected crossover points represent statements. The statement `turn-left` in parent no. 1 is replaced by the `PROGN2`-statement of parent no. 2, whereas the `PROGN2`-statement of parent no. 2 is replaced by statement `turn_left` as the new child program. Figure 21.2 presents the crossover operator graphically for this example.

Mutation   The mutation operation introduces a random change into an individual and thereby introduces diversity into the new individual and the next generation in general. While mutation is considered essential by some [Blickle, Thiele 1995], others believe it to be almost redundant [Koza 1992]. Mutation works as follows:

1. Select one parent from the current generation.

2. Select a mutation point.

3. Delete sub-tree at mutation point.

4. Replace sub-tree with randomly generated sub-tree.

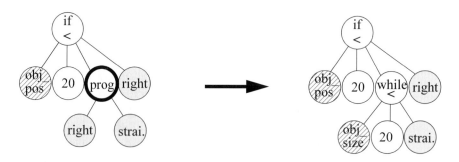

Figure 21.3: Mutation

The following shows an example with the mutation point marked in bold face:

```
(IF_LESS obj_pos low
    (PROGN2 drive_straight drive-straight) turn_right)
```
⇒
```
(IF_LESS obj_pos low
    (WHILE_LESS psd_front high drive_straight) turn_right)
```

The selected sub-tree (PROG2N2-sequence) is deleted from the parent program and subsequently replaced by a randomly generated sub-tree (here: a WHILE-loop construct containing a drive_straight statement). Figure 21.3 presents the mutation operator graphically.

21.4 Evolution

Initial population To start the evolutionary process, we first need an initial population. This consists of randomly generated individuals, which are random Lisp programs of limited tree depth. A large diversity of individuals improves the chances of locating the optimum solution after a set number of generations. [Koza 1992] suggests a number of methods to ensure a large diversity of different sizes and shapes in the initial population: full method, grow method, and ramped half-and-half (see below).

To ensure the validity and termination of each individual, the randomly generated Lisp programs must be sound and the root node must be a statement. All leaf nodes at the desired depth must be atoms.

A random program is initialized with a random statement for the root node. In case it is a function, the process continues recursively for all arguments until the maximum allowed depth is reached. For the leaf nodes, only atoms may be selected in the random process.

The "full method" requires the generated random tree to be fully balanced. That is, all leaves are at the same level, so there are no atoms at the inner level of the tree. This method ensures that the random program will be of the maximum allowed size.

The "grow method" allows the randomly generated trees to be of different shapes and heights. However, a maximum tree height is enforced.

The "ramped half-and-half method" is an often used mix between the grow method and the full method. It generates an equal number of grow trees and full trees of different heights and thereby increases variety in the starting generation. This method generates an equal number of trees of height 1, 2, ..., up to the allowed maximum height. For each height, half of the generated trees are constructed with the "full method" and half with the "grow method".

The initial population should be checked for duplicates, which have a rather high probability if the number of statements and values is limited. Duplicates should be removed, since they reduce the overall diversity of the population if they are allowed to propagate.

Evaluation and fitness Each individual (Lisp program) is now executed on the EyeSim simulator for a limited number of program steps. Depending on the nature of the problem, the program's performance is rated either continually or after it terminates (before or at the maximum allowed number of time steps). For example, the fitness of a wall-following program needs to be constantly monitored during execution, since the quality of the program is determined by the robot's distance to the wall at each time step. A search problem, on the other hand, only needs to check at the end of a program execution whether the robot has come sufficiently close to the desired location. In this case, the elapsed simulation time for achieving the goal will also be part of the fitness function.

Selection   After evaluating all the individuals of a population, we need to perform a selection process based on the fitness values assigned to them. The selection process identifies parents for generating the next generation with the help of previously described genetic operators. Selection plays a major role in genetic programming, since the diversity of the population is dependent on the choice of the selection scheme.

Fitness proportionate. The traditional genetic programming/genetic algorithm model selects individuals in the population according to their fitness value relative to the average of the whole population. However, this simple selection scheme has severe selection pressure that may lead to premature convergence. For example, during the initial population, an individual with the best fitness in the generation will be heavily selected, thus reducing the diversity of the population.

Tournament selection. This model selects n (e.g. two) individuals from the population and the best will be selected for propagation. The process is repeated until the number of individuals for the next generation is reached.

Linear rank selection. In this method, individuals are sorted according to their raw fitness. A new fitness value is then assigned to the individuals according to their rank. The ranks of individuals range from 1 to N. Now the selection process is identical to the proportionate schema. The advantage of the linear rank selection is that small differences between individuals are exploited and, by doing so, the diversity of the population is maintained.

Truncation selection. In this model, the population is first sorted according to its fitness values, and then from a certain point fitness value F, the poorer performing individuals below this value are cut off, only the better performing individuals remain eligible. Selection among these is now purely random; all remaining individuals have the same selection probability.

21.5 Tracking Problem

We chose a fairly simple problem to test our genetic programming engine, which can later be extended to a more complex scenario. A single robot is placed at a random position and orientation in a rectangular driving area enclosed by walls. A colored ball is also placed at a random position. Using its camera, the robot has to detect the ball, drive toward it, and stop close to it. The robot's camera is positioned at an angle so the robot can see the wall ahead from any position in the field; however, note that the ball will not always be visible.

Before we consider evolving a tracking behavior, we thoroughly analyze the problem by implementing a hand-coded solution. Our idea for solving this problem is shown below.

In a loop, grab an image and analyze it as follows:
- Convert the image from RGB to HSV.
- Use the histogram ball detection routine from Section 17.6 (this returns a ball position in the range [0..79] (left .. right) or no ball, and a ball size in pixels [0..60]).
- If the ball height is 20 pixels or more, then stop and terminate (the robot is then close enough to the ball).
- Otherwise:
  - if no ball is detected or the ball position is less than 20, turn slowly left.
  - if the ball position is between 20 and 40, drive slowly straight.
  - if the ball position is greater than 40, turn slowly right.

We experimentally confirm that this straightforward algorithm solves the problem sufficiently. Program 21.2 shows the main routine, implementing the algorithm described above. ColSearch returns the x-position of the ball (or −1 if not detected) and the ball height in pixels. The statement VWDriveWait following either a VWDriveTurn or a VWDriveStraight command suspends execution until driving or rotation of the requested distance or angle has finished.

Program 21.2: Hand-coded tracking program in C

```
1    do
2    { CAMGetColFrame(&c,0);
3      ColSearch(c, BALLHUE, 10, &pos, &val);   /* search image */
4
5      if (val < 20)   /* otherwise FINISHED */
6      { if (pos == -1 || pos < 20) VWDriveTurn(vw,  0.10, 0.4);/* left */
7          else if (pos > 60)        VWDriveTurn(vw, -0.10, 0.4);/* right*/
8          else                      VWDriveStraight(vw,  0.05, 0.1);
9        VWDriveWait(vw);   /* finish motion */
10     }
11   } while (val < 20);
```

Program 21.3: Hand-coded tracking program in Lisp

```
1    ( WHILE_LESS obj_size low
2      (IF_LESS obj_pos low rotate_left
3        (IF_LESS obj_pos high drive_straight
4          rotate_right )))
```

The next task is to hand-code the same solution in our Lisp notation. Program 21.3 shows the implementation as a Lisp string. This short program uses the following components:

Constants: low (20), high (40)

Sensor input: obj_size (0..60), obj_pos (−1, 0..79)
sensor values are evaluated from the image in each step

Constructs: (WHILE_LESS a b c)
C equivalent: while (a<b) c;

(IF_LESS a b c d)
C equivalent: if (a<b) c; else d;

Bearing in mind that obj_size and obj_pos are in fact calls to the image processing subroutine, the procedural translation of the Lisp program in Program 21.3 is almost identical to the hand-coded C solution in Program 21.2:

```
while (obj_size<20)
  if (obj_pos<20) rotate_left;
    else if (obj_pos<40) drive_straight;
      else rotate_right;
```

Figure 21.4 shows a graphical representation of the program tree structure. Of course the choice of the available constants and program constructs simplifies finding a solution and therefore also simplifies the evolution discussed below. Figure 21.5 and Figure 21.6 show the execution of the hand-coded solution from several different starting directions.

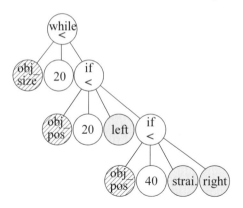

Figure 21.4: Lisp program tree structure

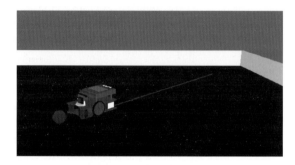

Figure 21.5: Execution of hand-coded solution in EyeSim simulator

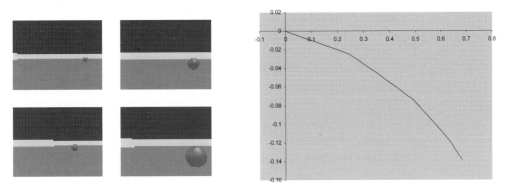

Figure 21.6: Robot's view and driving path in EyeSim simulator

21.6 Evolution of Tracking Behavior

Koza suggests the following steps for setting up a genetic programming system for a given problem [Koza 1992]:

1. Establish an objective.
2. Identify the terminals and functions used in the inductive programs.
3. Establish the selection scheme and its evolutionary operations.
4. Finalize the number of fitness cases.
5. Determine fitness function and hence the range of raw fitness values.
6. Establish the generation gap G, and the population M.
7. Finalize all control parameters.
8. Execute the genetic paradigm.

Our objective is to evolve a Lisp program that lets the robot detect a colored ball using image processing, drive toward it, and stop when it is close to it. Terminals are all statements that are also atoms, so in our case these are the four driving/turning routines from Table 21.2. Functions are all statements that are also lists, so in our case these are the three control structures sequence (PROGN2), selection (IF_LESS), and iteration (WHILE_LESS).

In addition, we are using a set of values, which comprises constants and implicit calls to image processing functions (obj_pos, obj_size) and to distance sensors (psd_aaa). Since we are not using PSD sensors for this experiment, we take these values out of the selection process. Table 21.3 shows all parameter choices for the experimental setup.

Figure 21.7 demonstrates the execution sequence for the evaluation procedure. The genetic programming engine generates a new generation of Lisp programs from the current generation. Each individual Lisp program is interpreted by the Lisp parser and run on the EyeSim simulator four times in order to determine its fitness. These fitness values are then in turn used as selection criteria for the genetic programming engine [Hwang 2002].

Control Parameters	Value	Description
Initial population	500	generated with ramp half-and-half method
Number of generations	50	[0..49]
Probability of crossover	90%	crossover operation is performed on 90% of selected individuals
Probability of reproduction	10%	copy operation is performed on 10% of the selected individual
Probability of mutation	0%	mutation is not used here
Probability of crossover point being a leaf node	10%	possibly extending program depth
Probability of crossover point being internal node	90%	possibly reducing program depth
Maximum tree height	5	maximum allowed program depth
Number of evaluations per individual	4	starting individual from different positions in the field
Maximum simulation time steps per trial	180	time limit for each individual to reach the ball

Table 21.3: Parameter settings

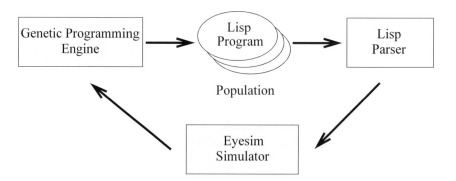

Population

Figure 21.7: Evaluation and simulation procedure

Fitness function   In order to determine the fitness for an individual program, we let each program run a number of times from different starting positions. Potentially more robust solutions can be achieved by choosing the starting positions and orientations at random. However, it is important to ensure that all individuals in a generation get the same starting *distances* for a run, otherwise, the fitness values will not be fair. In a second step, the ball position should also be made random.

Therefore, we run each individual in four *trials* with a random starting position and orientation, but with a trial-specific ball distance, i.e. the starting positions for all robots for a certain trial are located on a circle around the ball (Figure 21.8).

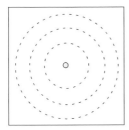

Figure 21.8: Robot starting positions

The fitness function is the difference between the initial and the final distance between robot and ball, added over four trials with different starting positions (the higher the fitness value, the better the performance):

$$f = \sum_{i=1}^{4} (dist_{i,0} - dist_{i,N})$$

Programs with a shorter execution time (fewer Lisp steps) are given a bonus, while all programs are stopped at the maximum number of time steps. Also, a bonus for programs with a lower tree height can be given.

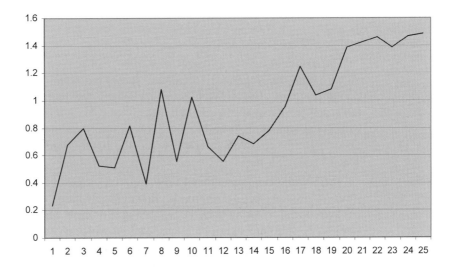

Figure 21.9: Maximum fitness over generations

Evaluation results The initial fitness diversity was quite large, which signifies a large search domain for the optimal solution. Figure 21.9 displays the maximum fitness of the population over 25 generations. Although we are retaining the best individuals of each generation unchanged as well as for recombination, the fitness function is not continuously increasing. This is due to the random robot starting positions in the experimental setup. A certain Lisp program may perform well "by accident", because of a certain starting position. Running the evolution over many generations and using more than four randomly selected starting positions per evaluation for each individual will improve this behavior.

The evolved Lisp programs and corresponding fitness values are shown in Program 21.4 for a number of generations.

Program 21.4: Optimal Lisp programs and fitness values

```
Generation  1, fitness 0.24
    (  IF_LESS obj-size obj-size turn-left move-forw )
Generation  6, fitness 0.82
    (  WHILE_LESS low obj-pos move-forw )
Generation 16, firness 0.95
    (  WHILE_LESS low high (  IF_LESS high low turn-left (  PROGN2 (
    IF_LESS low low move-back turn-left ) move-forw )  )  )
Generation 22, fitness 1.46
    (  PROGN2 (  IF_LESS low low turn-left (  PROGN2 (  PROGN2 (
    WHILE_LESS low obj-pos move-forw )  move-forw )  move-forw )  )  turn-
    left )
Generation 25, fitness 1.49
    (  IF_LESS low obj-pos move-forw (  PROGN2 move-back (  WHILE_LESS low
    high (  PROGN2 turn-right (  PROGN2 (  IF_LESS low obj-pos move-forw (
    PROGN2 turn-right (  IF_LESS obj-size obj-size (  PROGN2 turn-right
    move-back )  move-back )  )  )  move-forw )  )  )  )
```

The driving results of the evolved program can be seen in Figure 21.10, top. As a comparison, Figure 21.10, bottom, shows the driving results of the hand-coded program. The evolved program detects the ball and drives directly toward it in less than the allotted 180 time steps, while the robot does not exhibit any random movement. The evolved solution shows a similar performance to the hand-coded solution, while the hand-coded version still drives along a smoother path.

Speedup through parallelization The enormous computational time required by genetic programming is an inherent problem of this approach. However, evolution time can be significantly reduced by using parallel processing. For each generation, the population can be split into sub-populations, each of which evaluated in parallel on a workstation. After all sub-populations have finished their evaluation, the fitness results can be distributed among them, in order to perform a global selection operation for the next generation.

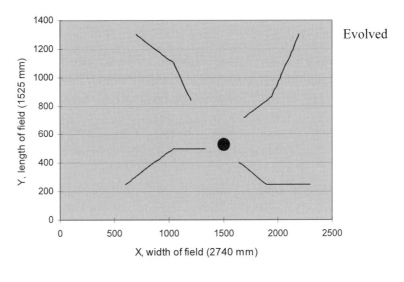

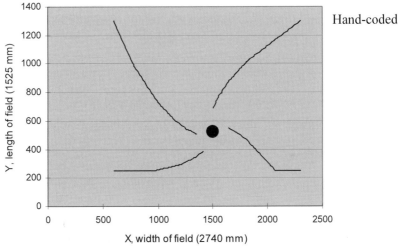

Figure 21.10: Evolved driving results versus hand-coded driving results

21.7 References

BLICKLE, T., THIELE, L. *A Comparison of Selection Schemes used in Genetic Algorithms*, Computer Engineering and Communication Networks Lab (TIK), Swiss Federal Institute of Technology/ETH Zürich, Report no. 11, 1995

BROOKS, R. *A Robust Layered Control System for a Mobile Robot*, IEEE Journal of Robotics and Automation, vol. 2, no. 1, March 1986, pp. 14-23 (10)

DARWIN, C. *On the Origin of Species by Means of Natural Selection, or Preservation of Favoured Races in the Struggle for Life*, John Murray, London, 1859

FERNANDEZ, J. *The GP Tutorial – The Genetic Programming Notebook*, `http://www.geneticprogramming.com/Tutorial/`, 2006

GRAHAM, P. *ANSI Common Lisp*, Prentice Hall, Englewood Cliffs NJ, 1995

HANCOCK, P. *An empirical comparison of selection methods in evolutionary algorithms*, in T. Fogarty (Ed.), Evolutionary Computing, AISB Workshop, Lecture Notes in Computer Science, no. 865, Springer-Verlag, Berlin Heidelberg, 1994, pp. 80-94 (15)

HWANG, Y. *Object Tracking for Robotic Agent with Genetic Programming*, B.E. Honours Thesis, The Univ. of Western Australia, Electrical and Computer Eng., supervised by T. Bräunl, 2002

IBA, H., NOZOE, T., UEDA, K. *Evolving communicating agents based on genetic programming*, IEEE International Conference on Evolutionary Computation (ICEC97), 1997, pp. 297-302 (6)

KOZA, J. *Genetic Programming – On the Programming of Computers by Means of Natural Selection*, The MIT Press, Cambridge MA, 1992

KURASHIGE, K., FUKUDA, T., HOSHINO, H. *Motion planning based on hierarchical knowledge for six legged locomotion robot*, Proceedings of IEEE International Conference on Systems, Man and Cybernetics SMC'99, vol. 6, 1999, pp. 924-929 (6)

LANGDON, W., POLI, R. *Foundations of Genetic Programming*, Springer-Verlag, Heidelberg, 2002

LEE, W., HALLAM, J., LUND, H. *Applying genetic programming to evolve behavior primitives and arbitrators for mobile robots*, IEEE International Conference on Evolutionary Computation (ICEC97), 1997, pp. 501-506 (6)

MAHADEVAN, S., CONNELL, J. *Automatic programming of behaviour-based robots using reinforcement learning*, Proceedings of the Ninth National Conference on Artificial Intelligence, vol. 2, AAAI Press/MIT Press, Cambridge MA, 1991

MCCARTHY, J., ABRAHAMS, P., EDWARDS, D., HART, T., LEVIN, M. *The Lisp Programmers' Manual*, MIT Press, Cambridge MA, 1962

WALKER, M., MESSOM, C. *A comparison of genetic programming and genetic algorithms for auto-tuning mobile robot motion control*, Proceedings of IEEE International Workshop on Electronic Design, Test and Applications, 2002, pp. 507-509 (3)

22

BEHAVIOR-BASED SYSTEMS

T raditional attempts at formulating complex control programs have been based on "Artificial Intelligence" (AI) theory. The dominant paradigm of this approach has been the sense–plan–act (SPA) organization: a mapping from perception, through construction of an internal world model, planning a course of action based upon this model, and finally execution of the plan in the real-world environment. Aspects of this method of robot control have been criticized, notably the emphasis placed on construction of a world model and planning actions based on this model [Agre, Chapman 1990], [Brooks 1986]. The computation time required to construct a symbolic model has a significant impact on the performance of the robot. Furthermore, disparity between the planning model and the actual environment may result in actions of the robot not producing the intended effect.

An alternative to this approach is described by behavior-based robotics. Reactive systems that do not use symbolic representation are demonstrably capable of producing reasonably complex behavior [Braitenberg 1984], see Section 1.1. Behavior-based robotic schemes extend the concept of simple reactive systems to combining simple concurrent behaviors working together.

22.1 Software Architecture

Often the importance of the software structure for mobile robots is stressed. Unfortunately, many published software structures are either too specialized for a particular problem or too general, so no advantage can be gained for the particular application at hand. Still, at least two standard models have emerged which we will discuss briefly.

The *classical model* (Figure 22.1, left) is known by many names: hierarchical model, functional model, engineering model, or three-layered model. It is a predictable software structure with top-down implementation. Names for the three levels in some systems are Pilot (lowest level), Navigator (intermediate level), and Planner (highest level), creating three levels of abstraction. Sensor data from the vehicle is pre-processed in two levels until the highest "intelligent" level takes the driving decisions. Execution of the actual driving (for example navigation and lower-level driving functions) is left to the layers below. The lowest layer is again the interface to the vehicle, transmitting driving commands to the robot's actuators.

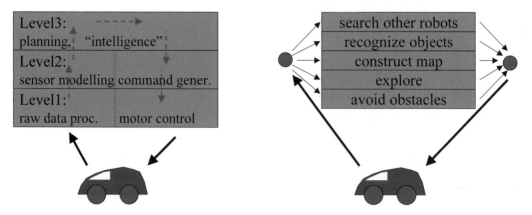

Figure 22.1: Software architecture models

The *behavior-based model* (Figure 22.1, right, [Brooks 1986]) is a bottom-up design that is not easily predictable. Instead of designing large blocks of code, each robot functionality has been encapsulated in a small self-contained module, here called a "behavior". All behaviors are executed in parallel, while explicit synchronization is not required. One of the goals of this design is to simplify extendability, for example for adding a new sensor or a new behavioral feature to the robot program. While all behaviors can access all vehicle sensors, a problem occurs at the reduction of the behaviors to produce a single output for the vehicle's actuators. The original "subsumption architecture" uses fixed priorities among behaviors, while modern implementations use more flexible selection schemes (see Chapter 22).

Subsumption architecture

22.2 Behavior-Based Robotics

The term "behavior-based robotics" is broadly applicable to a range of control approaches. Concepts taken from the original subsumption design [Brooks 1986] have been adapted and modified by commercial and academic research groups, to the point that the nomenclature has become generic. Some of the most frequently identified traits of behavior-based architectures are [Arkin 1998]:

- Tight coupling of sensing and action
  At some level, all behavioral robots are reactive to stimuli with actions that do not rely upon deliberative planning. Deliberative planning is eschewed in favor of computationally simple modules that perform a simple mapping from input to action, facilitating a rapid response. Brooks succinctly expressed this philosophy with the observation that *"Planning is just a way of avoiding figuring out what to do next"* [Brooks 1986].

- Avoiding symbolic representation of knowledge
  Rather than construct an internal model of the environment to perform planning tasks, the world is used as "its own best model" [Brooks 1986]. The robot derives its future behavior directly from observations of its environment, instead of trying to produce an abstract representation of the world that can be internally manipulated and used as a basis for planning future actions.

- Decomposition into contextually meaningful units
  Behaviors act as situation–action pairs, being designed to respond to certain situations with a definite action.

- Time-varying activation of concurrent relevant behaviors
  A control scheme is utilized to change the activation level of behaviors during run-time to accommodate the task that is trying to be achieved.

Behavior selection   In a behavior-based system, a certain number of behaviors run as parallel processes. While each behavior can access all sensors (*read*), only one behavior can have control over the robot's actuators or driving mechanism (*write*). Therefore, an overall controller is required to coordinate behavior selection or behavior activation or behavior output merging at appropriate times to achieve the desired objective.

Early behavior-based systems such as [Brooks 1986] used a fixed priority ordering of behaviors. For example, the wall avoidance behavior always has priority over the foraging behavior. Obviously such a rigid system is very restricted in its capabilities and becomes difficult to manage with increasing system complexity. Therefore, our goal was to design and implement a behavior-based system that employs an adaptive controller. Such a controller uses machine learning techniques to develop the correct selection response from the specification of desired outcomes. The controller is the "intelligence" behind the system, deciding from sensory and state input which behaviors to activate at any particular time. The combination of a reactive and planning (adaptive controller) component produces a hybrid system.

Hybrid systems combine elements of deliberative and reactive architectures. Various hybrid schemes have been employed to mediate between sensors and motor outputs to achieve a task. Perhaps the most appealing aspect of combining an adaptive controller with a hybrid architecture is that the system learns to perform the task from only the definition of criteria favoring task completion. This shifts the design process from specifying the system itself to defining outcomes of a working system. Assuming that the criteria for successful task completion are easier to define than a complete system specification, this would significantly reduce the work required of the system designer.

A more advanced and more flexible method for behavior selection is to use a neural network controller (see Chapter 19, Figure 22.2), as we did for some of the following applications. The neural network will receive input from all sensors (including pre-computed high-level sensor data), a clock, plus status lines from each behavior and will generate output to select the currently active behavior, and thereby cause a robot action. The structure of the network is itself developed with a genetic algorithm designed to optimize a fitness function describing the task criteria (see Chapter 20).

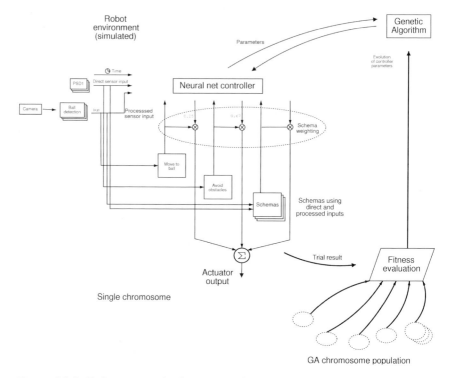

Figure 22.2: Behaviors and selection mechanism in robot environment

Emergent functionality The terms *emergent functionality, emergent intelligence* or *swarm intelligence* (if multiple robots are involved) are used to describe the manifestation of an overall behavior from a combination of smaller behaviors that may not have been designed for the original task [Moravec 1988], [Steels, Brooks 1995]. The appearance of this behavior can be attributed to the complexity of *interactions* between simple tasks instead of the tasks themselves. Behavior is generally classified as emergent if the response produced was outside the analysis of system design but proves to be beneficial to system operation.

Arkin argues that the coordination between simpler sub-units does not explain emergence completely [Arkin 1998]. As coordination of a robot is achieved by a deterministic algorithm, a sufficiently sophisticated analysis should be able to perfectly predict the behavior of a robot. Rather, the emergent phenomenon is attributed to the non-deterministic nature of real-world environments. These can never be modeled completely accurately, and so there is always a margin of uncertainty in system design that could cause unexpected behavior to be exhibited.

22.3 Behavior-Based Applications

Typical behavior-based applications involve a group of interacting robots mimicking some animal behavior pattern and thereby exhibiting some form of *swarm intelligence*. Communication between the robots can be either direct (e.g. wireless) or indirect via changes in the shared environment (*stigmergy*). Depending on the application, communication between individuals can range from essential to not required [Balch, Arkin 1994]. Some prominent applications are:

- Foraging
  One or more robots search an area for "food" items (usually easy to detect objects, such as colored blocks), collect them and bring them to a "home" area. Note that this is a very broad definition and also applies to tasks such as collecting rubbish (e.g. cans).

- Predator-Prey
  Two or more robots interact, with at least one robot in the role of the predator, trying to catch one of the prey robots, who are in turn trying to avoid the predator robots.

- Clustering
  This application mimics the social behavior of termites, which individually follow very simple rules when building a mound together.

Emergence   In the mound building process, each termite places new building material to the largest mound so far in its vicinity – or at a random position when starting. The complex mound structure resulting from the interaction of each of the colony's terminates is a typical example for *emergence*.

Cube clustering   This phenomenon can be repeated by either computer simulations or real robots interacting in a similar way [Iske, Rückert 2001], [Du, Bräunl 2003]. In our implementation (see Figure 22.3) we let a single or multiple robots search an area for red cubes. Once a robot has found a cube, it pushes it to the position of the largest collection of red cubes it has seen previously – or, if this is the first cube the robot has encountered, it uses this cube's coordinates for the start of a new cluster.

Over time, several smaller clusters will appear, which will eventually be merged to a single large cluster, containing all cubes from the robots' driving area. No communication between the robots is required to accomplish this task, but of course the process can be sped up by using communication. The number of robots used for this task also affects the average completion time. Depending on the size of the environment, using more and more robots will result in a faster completion time, up to the point where too many robots encumber each other (e.g. stopping to avoid collisions or accidentally destroying each other's cluster), resulting in an increasing completion time [Du, Bräunl 2003].

Figure 22.3: Cube clustering with real robots and in simulation

22.4 Behavior Framework

The objective of a behavior framework is to simplify the design and implementation of behavior-based programs for a robot platform such as the EyeBot. At its foundation is a programming interface for consistently specified behaviors.

We adapt the convention of referring to simple behaviors as schemas and extend the term to encompass any processing element of a control system. The specification of these schemas is made at an abstract level so that they may be generically manipulated by higher-level logic and/or other schemas without specific knowledge of implementation details.

Schemas may be recursively combined either by programming or by generation from a user interface. Aggregating different schemas together enables more sophisticated behaviors to be produced. The mechanism of arbitration between grouped schemas is up to the system designer. When combined with coordination schemas to select between available behaviors, the outputs of the contributing modules can be directed to actuator schemas to produce actual robot actions. A commonly used technique is to use a weighted sum of all schemas that drive an actuator as the final control signal.

Behavior design   The framework architecture was inspired by AuRA's reactive component [Arkin, Balch 1997], and takes implementation cues from the TeamBots environment realization [Balch 2006].

The basic unit of the framework is a schema, which may be perceptual (for example a sensor reading) or behavioral (for example move to a location). A schema is defined as a unit that produces an output of a pre-defined type. In our implementation, the simplest types emitted by schemas are integer, floating point, and boolean scalar values. More complex types that have been implemented are the two-dimensional floating point vector and image types. The floating point vector may be used to encode any two-dimensional quantity commonly used by robot schemas, such as velocities and positions. The image type corresponds to the image structure used by the RoBIOS image processing routines.

Schemas may optionally embed other schemas for use as inputs. Data of the pre-defined primitive types is exchanged between schemas. In this way behaviors may be recursively combined to produce more complex behaviors.

In a robot control program, schema organization is represented by a processing tree. Sensors form the leaf nodes, implemented as embedded schemas. The complexity of the behaviors that embed sensors varies, from simple movement in a fixed direction to ball detection using an image processing algorithm. The output of the tree's root node is used every processing cycle to determine the robot's next action. Usually the root node corresponds to an actuator output value. In this case output from the root node directly produces robot action.

Behavior implementation   The behavioral framework has been implemented in C++, using the RoBIOS API to interface with the Eyebot. These same functions are simulated and available in EyeSim (see Chapter 13), enabling programs created with the framework to be used on both the real and simulated platforms.

The framework has been implemented with an object-oriented methodology. There is a parent `Node` class that is directly inherited by type-emitting schema classes for each pre-defined type. For example, the `NodeInt` class represents a node that emits an integer output. Every schema inherits from a node child class, and is thus a type of node itself.

All schema classes define a `value(t)` function that returns a primitive type value at a given time `t`. The return type of this function is dependent on the class – for example, schemas deriving from `NodeInt` return an integer type. Embedding of schemas is by a recursive calling structure through the schema tree. Each schema class that can embed nodes keeps an array of pointers to the embedded instances. When a schema requires an embedded node value, it iterates through the array and calls each embedded schema's respective `value(t)` function. This organization allows invalid connections between schemas to be detected at compile time: when a schema embedding a node of an invalid type tries to call the value function, the returned value will not be of the required type. The compiler checks that the types from connected emitting and embedding nodes are the same at compilation time and will flag any mismatch to the programmer.

The hierarchy of schema connections forms a tree, with actuators and sensors mediated by various schemas and schema aggregations. Time has been discretized into units, Schemas in the tree are evaluated from the lowest level (sensors) to the highest from a master clock value generated by the running program.

Schemas A small working set of schemas using the framework was created for use in a neural network controller design task. The set of schemas with a short description of each is listed in Table 22.1. The schemas shown are either perceptual (for example Camera, PSD), behavioral (for example Avoid), or generic (for example Fixed vector). Perceptual schemas only emit a value of some type that is used by the behavioral schemas. Behavioral schemas transform their input into an egocentric output vector that would fulfill its goal.

A front-end program has been created to allow point-and-click assemblage of a new schema from pre-programmed modules [Venkitachalam 2002]. The representation of the control program as a tree of schemas maps directly to the interface presented to the user (Figure 22.4).

For a schema to be recognized by the user interface, the programmer must "tag" the header file with a description of the module. A sample description block is shown in Program 22.1. The graphical user interface then parses the header files of a schema source directory to determine how to present the modules to the user.

Schema	Description	Output
Camera	Camera perceptual schema	Image
PSD	PSD sensor perceptual schema	Integer
Avoid	Avoid obstacles based on PSD reading	2D vector
Detect ball	Detects ball position in image by hue analysis	2D vector
Fixed vector	Fixed vector representation	2D vector
Linear movement	Moves linearly from current position to another point	2D vector
Random	Randomly directed vector of specified size	2D vector

Table 22.1: Behavior schemas

The header block specifies how a particular schema interconnects with other schemas. It includes a description of typed initialization parameters for the module, a list of ports that can be used for emitting or embedding other modules, and various meta-information.

From the interconnection of the visual modules, the user interface generates appropriate code to represent the tree specified by the user. The structure of the program is determined by analyzing the behavior tree and translating it into a series of instantiations and embedding calls. The uniform nature of the behavioral API facilitates a simple code generation algorithm.

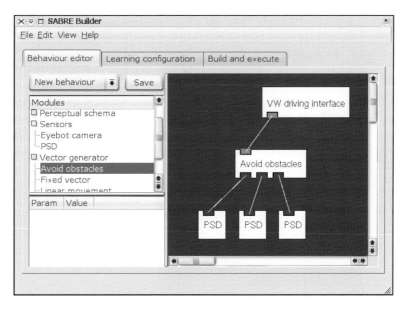

Figure 22.4: Graphical user interface for assembling schemas

Program 22.1: Schema header descriptor block

```
1       NAME            "Avoid obstacles"
2       CATEGORY        "Vector generator"
3       ORG             "edu.uwa.ciips"
4
5       DESC            "Produces vector to avoid obstacle\"
6       DESC            "based on three PSD readings"
7       INIT INT        "Object detection range in mm"
8       INIT DOUBLE     "Maximum vector magnitude"
9
10      EMIT VEC2       "Avoidance vector"
11      EMBED INT       "PSD Front reading"
12      EMBED INT       "PSD Left reading"
13      EMBED INT       "PSD Right reading"
```

22.5 Adaptive Controller

The adaptive controller system used in this system consists of two parts: a neural network controller (see Chapter 19) and a genetic algorithm learning system (see Chapter 20). The role of the neural network controller is to transform inputs to control signals that activate the behaviors of the robot at the appropriate time. The structure of the neural network determines the functional transformation from input to output. As a consequence of the neural network topology, this is effectively determined by changing the weights of all the network arcs. Evolution of the structure to achieve an objective task is performed by the genetic algorithm. The set of parameters describing the neural network arc weights is optimized to produce a controller capable of performing the task.

Our implementation of a genetic algorithm uses a direct binary encoding scheme to encode the numeric weights and optionally the thresholds of the controller's neural network. A single, complete, neural network controller configuration is encoded into a chromosome. The chromosome itself is a concatenation of individual floating point genes. Each gene encodes a single weight of the neural network. The population consists of a number of chromosomes, initially evaluated for fitness and then evolved through numerous iterations. Population evolution is achieved by a single point crossover operation at gene boundaries on a set percentage of the population. This is supplemented by mutation operations (random bit-wise inversion) on a small percentage of the population, set as a user parameter. The top performing members of a population are preserved between iterations of the algorithm (elitism). The lowest performing are removed and replaced by copies of the highest performing chromosomes. In our trials we have used population sizes of between 100 and 250 chromosomes.

The simulator environment was built around an early version of EyeSim 5 [Waggershauser 2002]. EyeSim is a sophisticated multi-agent simulation of the Eyebot hardware platform set in a virtual 3D environment. As well as simulating standard motor and hardware sensors, the environment model allows realistic simulation of image capture by an on-board camera sensor (Figure 22.5). This allows for complete testing and debugging of programs and behaviors using image processing routines.

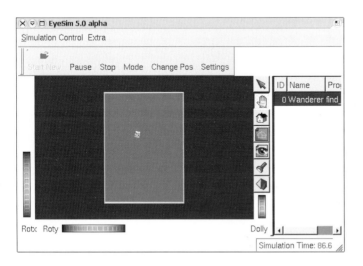

Figure 22.5: EyeSim 5 simulator screen shot

Because we run our programs in a simulated environment, we can obtain records of the positions and orientations of all objects in the environment with perfect accuracy. The logging calls determine positions during execution from the simulator's internal world model. The control program of the robot calls these functions after it completes execution and writes them in a suitable file format for reading by the separate evolutionary algorithm. The results are not used by the robot to enhance its performance while running. The final output of the program to the logfile is analyzed after termination to determine how well the robot performed its task.

Program 22.2: Schema example definition

```
1    #include "Vec2.h"
2    #include "NodeVec2.h"
3    class v_Random : public NodeVec2
4    { public:
5      v_Random(int seed);
6      ~v_Random();
7      Vec2* value(long timestamp);
8
9      private:
10     Vec2*    vector;
11     long     lastTimestamp;
12   };
```

Program 22.3: Schema example implementation

```
1    v_Random::v_Random(double min = 0.0f, double max = 1.0f,
2                       int seed = 5)
3    { double phi, r;
4      srand( (unsigned int) seed);
5      phi = 2*M_PI / (rand() % 360);
6      r = (rand() % 1000) / 1000.0;
7      vector = new Vec2(phi, r);
8      lastTimestamp = -1;
9    }
10
11   v_Random::~v_Random()
12   { if(vector)
13     delete vector;
14   }
15
16   Vec2* v_Random::value(long timestamp)
17   { if(timestamp > lastTimestamp)
18     { lastTimestamp = timestamp;
19       //  Generate a new random vector for this timestamp
20       vector->setx(phi = 2*M_PI / (rand() % 360));
21       vector->sety(r = (rand() % 1000) / 1000.0);
22     }
23     return vector;
24   }
```

As an example of a simple motor schema we will write a behavior to move in a random direction. This does not take any inputs so does not require behavior embedding. The output will be a 2D vector representing a direction and distance to move to. Accordingly, we subclass the `NodeVec2` class, which is the base class of any schemas that produce 2D vector output. Our class definition is shown in Program 22.2.

The constructor specifies the parameters with which the schema is initialized, in this case a seed value for our random number generator (Program 22.3). It also allocates memory for the local vector class where we store our output, and produces an initial output. The destructor for this class frees the memory allocated for the 2D vector.

The most important method is value, where the output of the schema is returned each processing cycle. The value method returns a pointer to our produced vector; had we subclassed a different type (for example `NodeInt`), it would have returned a value of the appropriate type. All value methods should take a timestamp as an argument. This is used to check if we have already computed an output for this cycle. For most schemas, we only want to produce a new output when the timestamp is incremented.

Program 22.4: Avoid schema

```
 1   v_Avoid_iii::v_Avoid_iii(int sensitivity, double maxspeed)
 2   { vector = new Vec2();  //  Create output vector
 3     initEmbeddedNodes(3);  //  Allocate space for nodes
 4     sense_range = sensitivity;  //  Initialise sensitivity
 5     this->maxspeed = maxspeed;
 6   }
 7   Vec2* v_Avoid_iii::value(long timestamp)
 8   { double front, left, right;
 9     if(timestamp != lastTimestamp) {
10     // Get PSD readings
11     frontPSD = (NodeInt*) embeddedNodes[0];
12     leftPSD  = (NodeInt*) embeddedNodes[1];
13     rightPSD = (NodeInt*) embeddedNodes[2];
14     front = frontPSD->value(timestamp);
15     left  = leftPSD->value(timestamp);
16     right = rightPSD->value(timestamp);
17     //  Calculate avoidance vector
18     // Ignore object if out of range
19     if (front >= sense_range) front = sense_range;
20     if (left >= sense_range) left = sense_range;
21     if (right >= sense_range) right = sense_range;
22     ...
23
```

Schemas that embed a node (i.e. take the output of another node as input) must allocate space for these nodes in their constructor. A method to do this is already available in the base class (`initEmbeddedNodes`), so the schema only needs to specify how many nodes to allocate. For example, the avoid schema embeds three integer schemas; hence the constructor calls `initEmbedded-Nodes` shown in Program 22.4. The embedded nodes are then accessible in an array `embeddedNodes`. By casting these to their known base classes and calling their value methods, their outputs can be read and processed by the embedding schema.

22.6 Tracking Problem

The evolved controller task implemented in this project is to search an enclosed space to find a colored ball. We began by identifying the primitive schemas that could be combined to perform the task. These are selected by the evolved controller during program execution to perform the overall task. A suitable initial fitness function for the task was constructed and then an initial random population generated for refinement by the genetic algorithm.

Primitive schemas — We identified the low-level motor schemas that could conceivably perform this task when combined together. Each schema produces a single normalized 2D vector output, described in Table 22.2.

Behavior	Normalized Vector Output
Move straight ahead	In the direction the robot is facing
Turn left	Directed left of the current direction
Turn right	Directed right of the current direction
Avoid detected obstacles	Directed away from detected obstacles

Table 22.2: Primitive schemas

The "avoid detected obstacles" schema embeds PSD sensor schemas as inputs, mounted on the front, left, and right of the robot (Figure 22.6). These readings are used to determine a vector away from any close obstacle (see Figure 22.6). Activation of the "avoid detected obstacles" schema prevents collisions with walls or other objects, and getting stuck in areas with a clear exit.

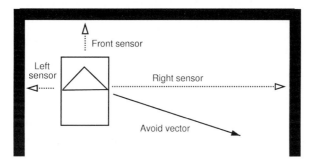

Figure 22.6: Avoidance schema

Ball detection is achieved by a hue recognition algorithm that processes images captured from the Eyebot camera (Figure 22.5) and returns ball position in the x-direction and ball height as "high-level sensor signals". The system should learn to activate the "turn left" behavior whenever the ball drifts toward the left image border and the "turn right" behavior whenever the balls drifts to the right. If the sensors detect the ball roughly in the middle, the system should learn to activate the "drive straight" behavior.

At the moment, only one behavior can be active at a time. However, as a future extension, one could combine multiple active behaviors by calculating a weighted sum of their respective vector outputs.

22.7 Neural Network Controller

The role of the neural network controller is to select the currently active behavior from the primitive schemas during each processing cycle. The active behavior will then take control over the robot's actuators and drive the robot in the direction it desires. In principle, the neural network receives information from all sensor inputs, status inputs from all schemas, and a clock value to determine the activity of each of the schemas. Inputs may be in the form of raw sensor readings or processed sensor results such as distances, positions, and pre-processed image data. Information is processed through a number of hidden layers and fed into the output layer. The controller's output neurons are responsible for selecting the active schema.

An additional output neuron is used to have the controller learn when it has finished the given task (here driving close to the ball and then stop). If the controller does not stop at a maximal number of time steps, it will be terminated and the last state is analyzed for calculating the fitness function. Using these fitness values, the parameters of the neural network controller are evolved by the genetic algorithm as described in Chapter 20.

We decided to use an off-line learning approach for the following reasons:

- Generation of ideal behavior
  There is the possibility of the system adapting to a state that fulfills some but not all of the task's fitness criteria. This typically happens when the method of learning relies on gradient descent and becomes stuck in a local fitness maxima [Gurney 2002]. Off-line evolution allows the application of more complex (and hence processor intensive) optimization techniques to avoid this situation.

- Time to convergence
  The time a robot takes to converge to an appropriate controller state reduces the robot's effective working time. By evolving suitable parameters off-line, the robot is in a suitable working state at run-time.

- System testing
  Evolution of behavior in a real environment limits our ability to test the controller's suitability for a task. Off-line evolution enables extensive testing of the system in simulation before actual use.

- Avoid physical damage to system
  While the controller is evolving, its response may cause damage to the physical robot until it learns to perform a task safely. Evolving the controller in simulation allows such responses to be modified before real harm can be done to expensive hardware.

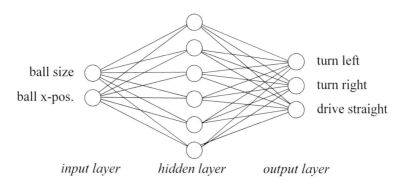

Figure 22.7: Neural network structure used

Figure 22.7 shows the neural network structure used. Image processing is done in an "intelligent sensor", so the ball position and ball size in the image are determined by image processing for all frames and directly fed into the network's input layer. The output layer has three nodes, which determine the robot's action, either turning left, turning right, or driving forward. The neuron with the highest output value is selected in every cycle.

Each chromosome holds an array of floating point numbers representing the weights of the neural network arbitrator. The numbers are mapped first-to-last in the neural network as is demonstrated for a simpler example network in Figure 22.8.

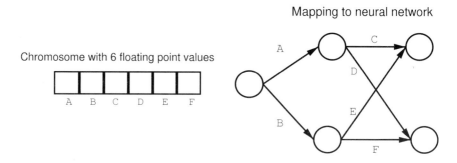

Figure 22.8: Chromosome encoding

22.8 Experiments

The task set for the evolution of our behavior-based system was to make a robot detect a ball and drive toward it. The driving environment is a square area with the ball in the middle and the robot placed at a random position and orientation. This setup is similar to the one used in Section 21.5.

The evolution has been run with minimal settings, namely 20 generations with 20 individuals. In order to guarantee a fair evaluation, each individual is run three times with three different original distances from the ball. The same three distance values are used for the whole population, while the individual robot placement is still random, i.e. along a circle around the ball.

Program 22.5: Fitness function for ball tracking

```
1    fitness = initDist - b_distance();
2    if (fitness < 0.0) fitness = 0.0;
```

The fitness function used is shown in Program 22.5. We chose only the improvement in distance toward the ball for the fitness function, while negative values are reset to zero. Note that only the desired outcome of the robot getting close to the ball has been encoded and not the robot's performance during the driving. For example, it would also have been possible to increase an individual's fitness whenever the ball is in its field of view – however, we did not want to favor this selection through hard-coding. The robot should discover this itself through evolution. Experiments also showed that even a simpler neural network with 2×4×3 nodes is sufficient for evolving this task, instead of 2×6×3.

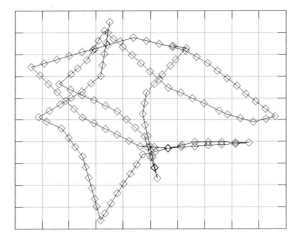

Figure 22.9: Robot driving result

This elementary fitness function worked surprisingly well. The robots learned to detect the ball and drive toward it. However, since there are no incentives to stop once the ball has been approached, most high-scoring robots continued pushing and chasing the ball around the driving environment until the maximum simulation time ran out. Figure 22.9 shows typical driving results obtained from the best performing individual after 11 generations. The robot is able to find the ball by rotating from its starting position until it is in its field of view, and can then reliably track the ball while driving toward it, and will continue to chase the ball that is bouncing off the robot and off the walls.

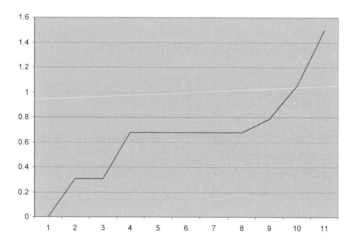

Figure 22.10: Fitness development over generations

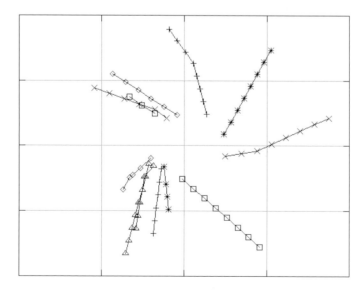

Figure 22.11: Individual runs of best evolved behavioral controller

Figure 22.10 shows the development of the maximum fitness over 10 generations. The maximum fitness increases consistently and finally reaches a level of acceptable performance.

This experiment can be extended if we want to make the robot stop in front of the ball, or change to a different behavioral pattern (for example goal kicking). What needs to be done is to change the fitness function, for example by adding a bonus for stopping in a time shorter than the maximum allowed simulation time, and to extend the neural network with additional output (and hidden) nodes. Care needs to be taken that only robots with a certain fitness for approaching the ball get the time bonus, otherwise "lazy" robots that do not move and stop immediately would be rewarded. Figure 22.11 shows several runs of the best evolved behavioral controller. This state was reached after 20 generations; the simulation is halted once the robot gets close to the ball.

22.9 References

AGRE, P., CHAPMAN, D. *What are plans for?*, Robotics and Autonomous Systems, vol. 6, no. 1-2, 1990, pp. 17-34 (18)

ARKIN, R. *Behavior Based Robotics*, MIT Press, Cambridge MA, 1998

ARKIN, R., BALCH, T. *AuRA: Principles and Practice in Review*, Journal of Experimental and Theoretical Artificial Intelligence, vol. 9, no. 2-3, 1997, pp. 175-189 (15)

BALCH, T., ARKIN, R., *Communication in Reactive Multiagent Robotic Systems*, Autonomous Robots, vol. 1, no. 1, 1994, pp. 27-52 (26)

BALCH T. *TeamBots simulation environment*, available from http://www.teambots.org, 2006

BRAITENBERG, V. *Vehicles, experiments in synthetic psychology*, MIT Press, Cambridge MA, 1984

BROOKS, R. *A Robust Layered Control System For A Mobile Robot*, IEEE Journal of Robotics and Automation, vol. 2, no.1, 1986, pp. 14-23 (7)

DU J., BRÄUNL, T. *Collaborative Cube Clustering with Local Image Processing*, Proc. of the 2nd Intl. Symposium on Autonomous Minirobots for Research and Edutainment, AMiRE 2003, Brisbane, Feb. 2003, pp. 247-248 (2)

GURNEY, K. *Neural Nets*, UCL Press, London, 2002

ISKE, B., RUECKERT, U. *Cooperative Cube Clustering using Local Communication*, Autonomous Robots for Research and Edutainment - AMiRE 2001, Proceedings of the 5th International Heinz Nixdorf Symposium, Paderborn, Germany, 2001, pp. 333-334 (2)

MORAVEC, H. *Mind Children: The Future of Robot and Human Intelligence*, Harvard University Press, Cambridge MA, 1988

STEELS, L., BROOKS, R. *Building Agents out of Autonomous Behavior Systems*, in L. Steels, R. Brooks (Eds.), The Artificial Life Route to AI: Building Embodied, Situated Agents, Erlbaum Associates, Hillsdale NJ, 1995

VENKITACHALAM, D. *Implementation of a Behavior-Based System for the Control of Mobile Robots*, B.E. Honours Thesis, The Univ. of Western Australia, Electrical and Computer Eng., supervised by T. Bräunl, 2002

WAGGERSHAUSER, A. *Simulating small mobile robots*, Project Thesis, Univ. Kaiserslautern / The Univ. of Western Australia, supervised by T. Bräunl and E. von Puttkamer, 2002

23

Evolution of Walking Gaits

D esigning or optimizing control systems for legged locomotion is a complex and time consuming process. Human engineers can only produce and evaluate a limited number of configurations, although there may be numerous competing designs that should be investigated. Automation of the controller design process allows the evaluation of thousands of competing designs, without requiring prior knowledge of the robot's walking mechanisms [Ledger 1999]. Development of an automated approach requires the implementation of a control system, a test platform, and an adaptive method for automated design of the controller. Thus, the implemented control system must be capable of expressing control signals that can sufficiently describe the desired walking pattern. Furthermore, the selected control system should be simple to integrate with the adaptive method.

One possible method for automated controller design is to utilize a spline controller and evolve its control parameters with a genetic algorithm [Boeing, Bräunl 2002], [Boeing, Bräunl 2003]. To decrease the evolution time and remove the risk of damaging robot hardware during the evolution, a dynamic mechanical simulation system can be employed.

23.1 Splines

Splines are a set of special parametric curves with certain desirable properties. They are piecewise polynomial functions, expressed by a set of control points. There are many different forms of splines, each with their own attributes [Bartels, Beatty, Barsky 1987]; however, there are two desirable properties:

- **Continuity**, so the generated curve smoothly connects its parts.
- **Locality** of the control points, so the influence of a control point is limited to a neighborhood region.

The Hermite spline is a special spline with the unique property that the curve generated from the spline passes through the control points that define the spline. Thus, a set of pre-determined points can be smoothly interpolated by simply setting these points as control points for the Hermite spline. Each segment of the curve is dependent on only a limited number of the neighboring control points. Thus, a change in the position of a distant control point will not alter the shape of the entire spline. The Hermite spline can also be constrained so as to achieve C^{K-2} continuity.

The function used to interpolate the control points, given starting point p_1, ending point p_2, tangent values t_1 and t_2, and interpolation parameter s, is shown below:

$$f(s) = h_1 p_1 + h_2 p_2 + h_3 t_1 + h_4 t_2$$

where

$$h_1 = 2s^3 - 3s^2 + 1$$
$$h_2 = -2s^3 + 3s^2$$
$$h_3 = s^3 - 2s^2 + s$$
$$h_4 = s^3 - s^2$$

for $0 \leq s \leq 1$

Program 23.1 shows the routine utilized for evaluating splines. Figure 23.1 illustrates the output from this function when evaluated with a starting point at one, with a tangent of zero, and an ending point of zero with a tangent of zero. The Hermite_Spline function was then executed with s ranging from zero to one.

Program 23.1: Evaluating a simple cubic Hermite spline section

```
1    float Hermite_Spline(float s) {
2      float ss=s*s;
3      float sss=s*ss;
4      float h1 =   2*sss -  3*ss +1;   // calculate basis funct. 1
5      float h2 = -2*sss +  3*ss;       // calculate basis funct. 2
6      float h3 =     sss - 2*ss + s;   // calculate basis funct. 3
7      float h4 =     sss -   ss;       // calculate basis funct. 4
8      float value =   h1*starting_point_location
9                    + h2*ending_point_location
10                   + h3*tangent_for_starting_point
11                   + h4*tangent_for_ending_point;
12     return value;
13   }
```

23.2 Control Algorithm

Using splines for modeling robot joint motions   Larger, more complex curves can be achieved by concatenating a number of cubic Hermite spline sections. This results in a set of curves that are capable of expressing the control signals necessary for legged robot locomotion. The

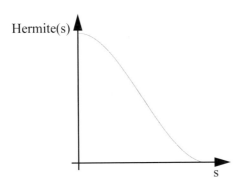

Figure 23.1: Cubic Hermite spline curve

spline controller consists of a set of joined Hermite splines. The first set contains robot initialization information, to move the joints into the correct positions and enable a smooth transition from the robot's starting state to a traveling state. The second set of splines contains the cyclic information for the robot's gait. Each spline can be defined by a variable number of control points, with variable degrees of freedom. Each pair of a start spline and a cyclic spline corresponds to the set of control signals required to drive one of the robot's actuators.

An example of a simple spline controller for a robot with three joints (three degrees of freedom) is illustrated in Figure 23.2. Each spline indicates the controller's output value for one actuator.

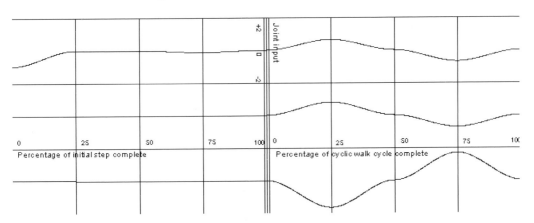

Figure 23.2: Spline joint controller

There are a number of advantages offered by Hermite spline controllers. Since the curve passes through all control points, individual curve positions can be pre-determined by a designer. This is especially useful in situations where the control signal directly corresponds to angular, or servo, positions. Program 23.2 provides a simplified code snippet for calculating the position values for a one-dimensional spline.

Program 23.2: Evaluating a concatenated Hermite spline

```
1    Hspline hs[nsec]; //A spline with nsec sections
2
3    float SplineEval(float s) {
4        int sect;      //what section are we in?
5        float z;       //how far into that section are we?
6        float secpos;
7        secpos=s*(nsec-1);
8        sect=(int)floorf(secpos);
9        z=fmodf(secpos,1);
10       return hs[sect].Eval(z);
11   }
```

There is a large collection of evidence that supports the proposition that most gaits for both animals and legged robots feature synchronized movement [Reeve 1999]. That is, when one joint alters its direction or speed, this change is likely to be reflected in another limb. Enforcing this form of constraint is far simpler with Hermite splines than with other control methods. In order to force synchronous movement with a Hermite spline, all actuator control points must lie at the same point in cycle time. This is because the control points represent the critical points of the control signal when given default tangent values.

23.3 Incorporating Feedback

Most control methods require a form of feedback in order to correctly operate (see Chapter 10). Spline controllers can achieve walking patterns without the use of feedback; however, incorporating sensory information into the control system allows a more robust gait. The addition of sensory information to the spline control system enabled a bipedal robot to maneuver on uneven terrain.

In order to incorporate sensor feedback information into the spline controller, the controller is extended into another dimension. The extended control points specify their locations within both the gait's cycle time and the feedback value. This results in a set of control surfaces for each actuator. Extending the controller in this form significantly increases the number of control points required. Figure 23.3 illustrates a resulting control surface for one actuator.

The actuator evaluates the desired output value from the enhanced controller as a function of both the cycle time and the input reading from the sensor. The most appropriate sensory feedback was found to be an angle reading from an inclinometer (compare Section 2.8.3) placed on the robot's central body (torso). Thus, the resultant controller is expressed in terms of the percentage cycle time, the inclinometer's angle reading, and the output control signal.

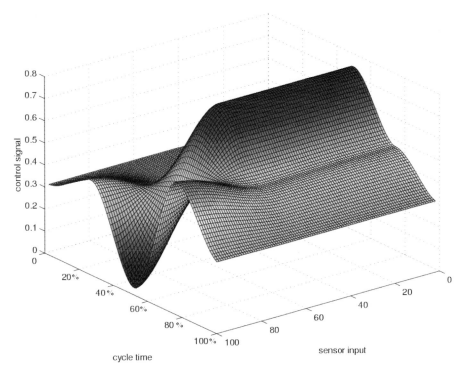

Figure 23.3: Generic extended spline controller

23.4 Controller Evolution

Genetic algorithms can be applied to automate the design of the control system. To achieve this, the parameters for the control system need to be encoded in a format that can be evolved by the genetic algorithm. The parameters for the spline control system are simply the position and tangent values of the control points that are used to describe the spline. Thus, each control point has three different values that can be encoded:

- Its position in the cycle time
  (i.e. position along the x-axis)

- The value of the control signal at that time
  (i.e. position along the y-axis)

- The tangent value

To allow these parameters to evolve with a genetic algorithm in minimal time, a more compact format of representing the parameters is desired. This can be achieved by employing fixed point values.

For example, if we wanted to encode the range [0..1] using 8bit fixed point values, then the 8 bits can represent any integer value from 0 to 255. By simply dividing this value by 255, we can represent any number ranging from 0 to 1, with an accuracy of 0.004 (1/256).

The curve shown in Figure 23.1 was generated by a one-dimensional spline function, with the first control point (s = 0) at position 1 with tangent value of 0, and the second control point (s = 1) at position 0 with tangent value of 0. If an encoding which represented each value as an 8bit fixed point number from 0 to 1 is used, then the control parameters in this case would be represented as a string of 3 bytes with values of [0, 255, 0] for the first control point's position and tangent, and [255, 0, 0] for the second control point's position and tangent.

Thus, the entire spline controller can be directly encoded using a list of control point values for each actuator. An example structure to represent this information is shown in Program 23.3.

Program 23.3: Full direct encoding structures

```
 1   struct encoded_controlpoint {
 2      unsigned char x,y,tangent;
 3   };
 4
 5   struct encoded_splinecontroller {
 6      encoded_controlpoint
 7        initialization_spline[num_splines][num_controlpoints];
 8      encoded_controlpoint
 9        cyclic_spline          [num_splines][num_controlpoints];
10   };
```

Staged evolution   There are a number of methods for optimizing the performance of the genetic algorithm. One method for increasing the algorithm's performance is *staged evolution*. This concept is an extension to "Behavioural Memory", and was first applied to controller evolution by [Lewis, Fagg, Bekey 1994]. Staged evolution divides a problem task into a set of smaller, manageable challenges that can be sequentially solved. This allows an early, approximate solution to the problem to be solved. Then, incrementally increasing the complexity of the problem provides a larger solution space for the problem task and allows for further refinements of the solution. Finally, after solving all the problem's sub-tasks, a complete solution can be determined. Solving the sequence of sub-tasks is typically achieved in less time than required if the entire problem task is tackled without decomposition.

This optimization technique can also be applied to the design of the spline controller. The evolution of the controller's parameters can be divided into the following three phases:

1. Assume that each control point is equally spaced in the cycle time. Assume the tangent values for the control points are at a default value. Only evolve the parameters for the control points' output signal (y-axis).

2. Remove the restriction of equidistant control points, and allow the control points to be located at any point within the gait time (x-axis).

3. Allow final refinement of the solution by evolving the control point tangent values.

To evolve the controller in this form, a staged encoding method is required. Table 23.1 indicates the number of control points required to represent the controller in each phase. In the case of an encoding where each value is represented as an 8 bit fixed-point number, the encoding complexity directly corresponds to the number of bytes required to describe the controller.

Evolution Phase	Encoding Complexity
Phase 1	$a(s + c)$
Phase 2	$2a(s + c)$
Phase 3	$3a(2 + c)$

with

 a number of actuators
 s number of initialization control points, and
 c number of cyclic control points

Table 23.1: Encoding complexity

23.5 Controller Assessment

In order to assign a fitness value to each controller, a method for evaluating the generated gait is required. Since many of the generated gaits result in the robot eventually falling over, it is desirable to first simulate the robot's movement in order to avoid damaging the actual robot hardware. There are many different dynamic simulators available that can be employed for this purpose.

One such simulator is DynaMechs, developed by McMillan [DynaMechs 2006]. The simulator implements an optimized version of the Articulated Body algorithm, and provides a range of integration methods with configurable step sizes. The package is free, open source, and can be compiled for a variety of operating systems (Windows, Linux, Solaris). The simulator provides information about an actuator's location, orientation, and forces at any time, and this information can be utilized to determine the fitness of a gait.

A number of fitness functions have been proposed to evaluate generated gaits. Reeve proposed the following sets of fitness measures [Reeve 1999]:

- FND (forward not down): The average speed the walker achieves minus the average distance of the center of gravity below the starting height.

- DFND (decay FND): Similar to the FND function, except it uses an exponential decay of the fitness over the simulation period.

- DFNDF (DFND or fall): As above, except a penalty is added for any walker whose body touches the ground.

Fitness function   These fitness functions do not consider the direction or path that is desired for the robot to walk along. Thus, more appropriate fitness functions can be employed by extending the simple FND function to include path information, and including terminating conditions [Boeing, Bräunl 2002]. The terminating conditions assign a very low fitness value to any control system which generates a gait that results in:

- A robot's central body coming too close to the ground. This termination condition ensures that robots do not fall down.

- A robot that moves too far from the ground. This removes the possibility of robots achieving high fitness values early in the simulation by propelling themselves forward through the air (jumping).

- A robot's head tilting too far forward. This ensures the robots are reasonably stable and robust.

Thus, the overall fitness function is calculated, taking into account the distance the robot moves along the desired path, plus the distance the robot deviates from the path, minus the distance the robot's center of mass has lowered over the period of the walk, as well as the three terminating conditions.

23.6 Evolved Gaits

This system is capable of generating a wide range of gaits for a variety of robots. Figure 23.4 illustrates a gait for a simple bipedal robot. The robot moves forward by slowly lifting one leg by rotating the hip forward and knee backward, then places its foot further in front, straightens its leg, and repeats this process. The gait was evolved within 12 hours on a 500MHz AMD Athlon PC. The genetic algorithm typically requires the evaluation of only 1,000 individuals to evolve an adequate forward walking pattern for a bipedal robot.

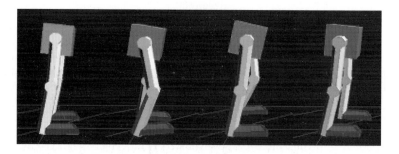

Figure 23.4: Biped gait

Figure 23.5 illustrates a gait generated by the system for a tripod robot. The robot achieves forward motion by thrusting its rear leg toward the ground, and lifting its forelimbs. The robot then gallops with its forelimbs to produce a dynamic gait. This illustrates that the system is capable of generating walking patterns for legged robots, regardless of the morphology and number of legs.

Figure 23.5: Tripod gait

The spline controller also evolves complex dynamic movements. Removing the termination conditions allows for less stable and robust gaits to be evolved. Figure 23.6 shows a jumping gait evolved for an android robot. The resultant control system depicted was evolved within 60 generations and began convergence toward a unified solution within 30 generations. However, the gait was very unstable, and the android could only repeat the jump three times before it would fall over.

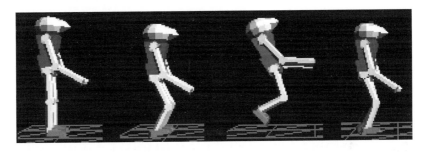

Figure 23.6: Biped jumping

The spline controller utilized to create the gait depicted in Figure 23.4 was extended to include sensory information from an inclinometer located in the robot's torso. The inclinometer reading was successfully interpreted by the control system to provide an added level of feedback capable of sustaining the generated gait over non-uniform terrain. An example of the resultant gait is

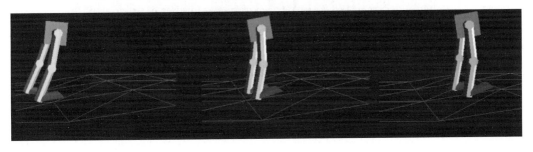

Figure 23.7: Biped walking over uneven terrain

illustrated in Figure 23.7. The controller required over 4 days of computation time on a 800MHz Pentium 3 system, and was the result of 512 generations of evaluation.

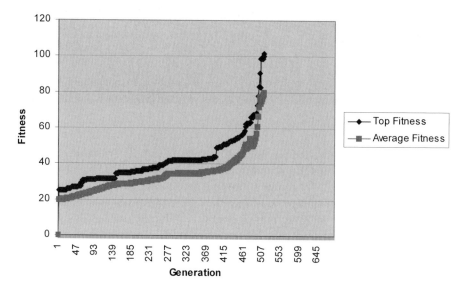

Figure 23.8: Fitness versus generation for extended spline controller

The graph in Figure 23.8 demonstrates the increase in fitness value during the evolution of the extended controller depicted in Figure 23.7. A rapid increase in fitness values can clearly be observed at around 490 generations. This corresponds to the convergence point where the optimal solution is located. The sharp increase is a result of the system managing to evolve a controller that was capable of traversing across flat, rising, and lowering terrains.

This chapter presented a flexible architecture for controller evolution, and illustrated a practical robotics application for genetic algorithms. The control system was shown to describe complex dynamic walking gaits for robots with differing morphologies. A similar system can be employed to control any robot consisting of multiple actuators, and the present system could be extended to evolve the robot's morphology in unison with the controller. This would enable the robot's design to be improved, such that the robot's structure was optimally designed to suit its desired purpose. Further extensions of this could be to automatically construct the designed robots using 3D printing technology, removing the human designer completely from the robot design process [Lipson, Pollack 2006].

23.7 References

BARTELS, R,. BEATTY, J., BARSKY, B. *An Introduction to Splines for Use in Computer Graphics and Geometric Models*, Morgan Kaufmann, San Francisco CA, 1987

BOEING, A., BRÄUNL, T. *Evolving Splines: An alternative locomotion controller for a bipedal robot*, Proceedings of the Seventh International Conference on Control, Automation, Robotics and Vision (ICARV 2002), CD-ROM, Nanyang Technological University, Singapore, Dec. 2002, pp. 1-5 (5)

BOEING, A., BRÄUNL, T. *Evolving a Controller for Bipedal Locomotion*, Proceedings of the Second International Symposium on Autonomous Minirobots for Research and Edutainment, AMiRE 2003, Brisbane, Feb. 2003, pp. 43-52 (10)

DYNAMECHS, *Dynamics of Mechanisms: A Multibody Dynamic Simulation Library*, http://dynamechs.sourceforge.net, 2006

LEDGER, C. *Automated Synthesis and Optimization of Robot Configurations*, Ph.D. Thesis, Carnegie Mellon University, 1999

LEWIS, M., FAGG, A., BEKEY, G. *Genetic Algorithms for Gait Synthesis in a Hexapod Robot*, in Recent Trends in Mobile Robots, World Scientific, New Jersey, 1994, pp. 317-331 (15)

LIPSON, H., POLLACK, J. *Evolving Physical Creatures*, http://citeseer.nj.nec.com/523984.html, 2006

REEVE, R. *Generating walking behaviours in legged robots*, Ph.D. Thesis, University of Edinburgh, 1999

OUTLOOK

24

I n this book we have presented the application of embedded systems for small autonomous mobile robots. We looked at embedded systems in general, their interfacing to sensors and actuators, basic operation system functions, device drivers, multitasking, and system tools. A number of detailed programming examples were used to aid understanding of this practical subject area.

Of course, time does not stand still. In the decade of development of the EyeBot robots and the EyeCon controller we have already seen quite a remarkable development in components.

A whole new generation of image sensors has emerged. CMOS sensors are slowly overtaking CCD sensors, because of their lower production cost and larger brightness range. Image sensor resolution has increased in a similar fashion to the processing power of microprocessors. However, a higher resolution is not always desirable in small robot systems, because there is a trade-off between image resolution versus frame rate, and for most of our applications a higher frame rate is more important than a higher resolution. The required processing time usually grows much faster than linearly with the number of image pixels.

Also, the development of microcontrollers has not kept up with the increased processing speeds of microprocessors, most likely because of insufficient industrial demand for fast microcontrollers. In general, the latest-generation embedded systems are about an order of magnitude slower than high-end PCs or workstations. On the other hand, commercial embedded systems meet additional requirements such as an extended temperature range and electromagnetic compatibility (EMC). That means these systems must be able to function in a harsh environment, at cold or hot temperatures, and in the presence of electromagnetic noise, while their own level of electromagnetic emission is strictly limited.

With this rapid development in processor and image sensor chips, advances in motors, gearboxes, and battery technology seem slower. However, one should not forget that improvements in the resolution of image sensors and in the speed of processor chips are mainly a consequence of miniaturization – a technique that cannot easily be applied to other components.

The biggest challenge and the largest effort, however, remains software development. One can easily overlook how many person-years of software development are required for a project like EyeBot/RoBIOS. This includes operating system routines, compiler adaptations, system tools, simulation systems, and application programs. Especially time-consuming are all low-level device drivers, most of them written in assembly language or incompatible C/C++ code. Every migration to a different CPU or sensor chip requires the redevelopment of this code.

We are still far away from intelligent reasoning robots, integrated into our human environment. However, extrapolating the achievements of the past, and projecting exponential increase, maybe the RoboCup/FIRA vision for the year 2050 will become a reality.

APPENDICES

PROGRAMMING TOOLS A

A.1 System Installation

We are using the "GNU" cross-compiler tools [GNU 2006] for operating system development as well as for compiling user programs. GNU stands for "*Gnu's not Unix*", representing an independent consortium of worldwide distributed software developers that have created a huge open-source software collection for Unix systems. The name, however, seems to be a relic from the days when proprietary Unix implementations had a larger market share.

Supported operating systems for EyeCon are Windows (from DOS to XP) and Unix (Linux, Sun Solaris, SGI Unix, etc.).

Windows System installation in Windows has been made extremely simple, by providing an installer script, which can be executed by clicking on:

```
rob65win.exe
```

This executable will run an installer script and install the following components on a Windows system:

- GNU cross-compiler for C/C++ and assembly
- RoBIOS libraries, include-files, hex-files and shell-scripts
- Tools for downloading, sound conversion, remote control, etc.
- Example programs for real robot and simulator

Unix For installation under Unix, several pre-compiled packages are available for the GNU cross-compiler. For Linux Red-Hat users, "rpm" packages are available as well. Because a number of different Unix systems are supported, the cross-compiler and the RoBIOS distribution have to be installed separately, for example:

- gcc68-2.95.3-linux.rpm   cross-compiler for Linux
- rob65usr.tgz   complete RoBIOS distribution

The cross-compiler has to be installed in a directory that is contained in the command path, to ensure the Unix operating system can execute it (when using "rpm" packages, a standard path is being chosen). The RoBIOS distribution can be installed at an arbitrary location. The following lists the required steps:

- `>setenv ROBIOS /usr/local/robios/`
  Set the environment variable `ROBIOS` to the chosen installation path.

- `>setenv PATH "${PATH}:/usr/local/gnu/bin:${ROBIOS}/cmd"`
  Include both the cross-compiler binaries and the RoBIOS commands in the Unix command path, to make sure they can be executed.

Example program library Besides the compiler and operating system, a huge EyeBot/RoBIOS example program library is available for download from:

> `http://robotics.ee.uwa.edu.au/eyebot/ftp/EXAMPLES-ROB/`
> `http://robotics.ee.uwa.edu.au/eyebot/ftp/EXAMPLES-SIM/`

or in compressed form:

> `http://robotics.ee.uwa.edu.au/eyebot/ftp/PARTS/`

The example program library contains literally hundreds of well-documented example programs from various application areas, which can be extremely helpful for familiarizing oneself with a particular aspect or application of the controller or robot.

After installing and unpacking the examples (and after installing both the cross-compiler and RoBIOS distribution), they can be compiled all at once by typing:

> `make`

(In Windows first open a console window by double-clicking on "`start-rob.bat`".) This will compile all C and assembly files and generate corresponding hex-files that can subsequently be downloaded to the controller and run.

RoBIOS upgrade Upgrading to a newer RoBIOS version or updating a hardware description file (HDT) with new sensors/actuators is very simple. Simple downloading of the new binary file is required. RoBIOS will automatically detect the system file and prompt the user to authorize overwriting of the flash-ROM. Only in the case of a corrupted flash-ROM is the background debugger required to re-install RoBIOS (see Section A.4). Of course, the RoBIOS version installed on the local host system has to match the version installed on the EyeCon controller.

A.2 Compiler for C and C++

The GNU cross-compiler [GNU 2006] supports C, C++, and assembly language for the Motorola 68000 family. All source files have specific endings that determine their type:

- .c   C program
- .cc or .cpp   C++ program
- .s   Assembly program
- .o   Object program (compiled binary)
- a.out   Default generated executable
- .hex   Hex-file, downloadable file (ASCII)
- .hx   Hex-file, downloadable file (compressed binary)

Hello World   Before discussing the commands (shell-scripts) for compiling a C or C++ source program, let us have a look at the standard "hello world" program in Program A.1. The standard "hello world" program runs on the EyeCon in the same way as on an ordinary PC (note that ANSI C requires `main` to be of type `int`). Library routine `printf` is used to write to the controller's LCD, and in the same way, `getchar` can be used to read key presses from the controller's menu keys.

Program A.1: "Hello World" program in C

```
1   #include <stdio.h>
2   int main ()
3   { printf ("Hello !\n");
4     return 0;
5   }
```

Program A.2 shows a slightly adapted version, using RoBIOS-specific commands that can be used in lieu of standard Unix `libc`-commands for printing to the LCD and reading the menu keys. Note the inclusion of `eyebot.h` in line 1, which allows the application program to use all RoBIOS library routines listed in Appendix B.5.

Program A.2: Extended C program

```
1   #include "eyebot.h"
2   int main ()
3   { LCDPrintf ("Hello !\n");
4     LCDPrintf ("key %d pressed\n", KEYGet ());
5     return 0;
6   }
```

Assuming one of these programs is stored under the filename `hello.c`, we can now compile the program and generate a downloadable binary:

```
>gcc68 hello.c -o hello.hex
```

This will compile the C (or C++) source file, print any error messages, and – in case of an error-free source program – generate the downloadable output file `hello.hex`. This file can now be downloaded (see also Section A.5) with the following command from the host PC to the EyeCon controller via a serial cable or a wireless link:

```
>dl hello.hex
```

On the controller, the program can now be executed by pressing "RUN" or stored in ROM.

Optionally, it is possible to compress the generated hex-file to the binary hx-format by using the utility srec2bin as shown in the command below. This reduces the file size and therefore shortens the time required for transmitting the file to the controller.

```
>srec2bin hello.hex hello.hx
```

The gcc GNU C/C++ compiler has a large number of options, which all are available with the script gcc68 as well. For details see [GNU 2006]. For compilation of larger program systems with many source files, the Makefile utility should be used. See [Stallman, McGrath 2002] for details. Note that if the output clause is omitted if during compilation (see below), then the default C output filename a.out is assumed:

```
>gcc68 hello.c
```

A.3 Assembler

Since the same GNU cross-compiler that handles C/C++ can also translate Motorola 68000 assembly programs, we do not need an additional tool or an additional shell-script. Let us first look at an assembly version of the "hello world" program (Program A.3).

Program A.3: Assembly demo program

```
1           .include "eyebot.i"
2           .section .text
3           .globl main
4
5   main:   PEA hello, -(SP)      | put parameter on stack
6           JSR LCDPutString      | call RoBIOS routine
7           ADD.L 4,SP            | remove param. from stack
8           RTS
9
10          .section .data
11  hello: .asciz "Hello !"
```

We include eyebot.i as the assembly equivalent of eyebot.h in C. All program code is placed in assembly section text (line 2) and the only label visible to the outside is main, which specifies the program start (equivalent to main in C).

The main program starts by putting all required parameters on the stack (LCDPutString only has one: the start address of the string). Then the

RoBIOS routine is called with command JSR (jump subroutine). After returning from the subroutine, the parameter entry on the stack has to be cleared,
which is simply done by adding 4 (all basic data types int, float, char, as
well as addresses, require 4 bytes). The command RTS (return from subroutine)
terminates the program. The actual string is stored in the assembly section
data with label hello as a null-terminated string (command asciz).

For further details on Motorola assembly programming, see [Harman
1991]. However, note that the GNU syntax varies in some places from the
standard Motorola assembly syntax:

- Filenames end with ". s".

- Comments start with "|".

- If the length attribute is missing, WORD is assumed.

- Prefix "0x" instead of "$" for hexadecimal constants.

- Prefix "0b" instead of "%" for binary constants.

As has been mentioned before, the command for translating an assembly
file is identical to compiling a C program:

```
>gcc68 hello.s -o hello.hex
```

Combining C and
Assembly

It is also possible to combine C/C++ and assembly source programs. The
main routine can be either in assembly or in the C part. Calling a C function
from assembly is done in the same way as calling an operating system function
shown in Program A.3, passing all parameters over the stack. An optional
return value will be passed in register D0.

Program A.4: Calling assembly from C

```
1   #include "eyebot.h"
2   int fct(int); /* define ASM function prototype */
3
4   int main (void)
5   { int x=1,y=0;
6       y = fct(x);
7       LCDPrintf("%d\n", y);
8       return 0;
9   }
```

```
1   .globl   fct
2   fct:      MOVE.L 4(SP), D0   | copy parameter x in register
3             ADD.L #1,D0        | increment x
4             RTS
```

The more common way of calling an assembly function from C is even
more flexible. Parameters can be passed on the stack, in memory, or in registers. Program A.4 shows an example, passing parameters over the stack.

From the C program (top of Program A.4) the function call does not look
any different from calling a C function. All parameters of a function are
implicitly passed via the stack (here: variable x). The assembly function (bot-

tom of Program A.4) can then copy all its parameters to local variables or registers (here: register D0).

Note that an assembly routine called from a C function can freely use data registers D0, D1 and address registers A0, A1. Using any additional registers requires storing their original contents on the stack at the beginning of the routine and restoring their contents at the end of the routine.

After finishing all calculations, the function result (here: x+1) is stored in register D0, which is the standard register for returning a function result to the calling C routine. Compiling the two source files (assuming filenames main.c and fct.s) into one binary output file (demo.hex) can be done in a single command:

```
>gcc68 main.c fct.s -o demo.hex
```

A.4 Debugging

The debugging system BD32 (Figure A.1) is a free program for DOS (also running under Windows) utilizing the M68332 controller's built-in "background debugger module" (BDM). This means it is a true hardware debugger that can stop the CPU, display memory and register contents, disassemble code, upload programs, modify memory, set breakpoints, single-step, and so on. Currently, BD32 is only available for DOS and only supports debugging at assembly level. However, it may be possible to integrate BDM with a Unix source-level debugger for C, such as gdb.

```
 BD32                                                              _□×
┌──────────────────────────────────────────────────────────────────┐
│ 8 x 12 ▾   □ ▣ ▣  ⊞  ▣▣  A                                         │
├────────────────────────── CPU Registers ──────────────────────────┤
│ D0-7 00000000 0000000A 00000000 0000007F 0000001F FFFFFFFF 00004E20 0000FFFF │
│ A0-7 0000004B 00000009 00013694 00013E74 0000D558 00000400 0001FFA8 0001FF6C │
│   PC 00020010      USP 24047108      SFC 00000000      SR 10S--210---XNZVC │
│  VBR 00000000      SSP 0001FF6C      DFC 00000000         0010000000000000 │
├────────────────────────── Memory Display ─────────────────────────┤
│   00020000 4EF9 0002 000A            JMP      $2000A.L              │
│   00020006 0002 F814                 ORI.B    #$14,D2               │
│ B*0002000A 23CF 0002 F80C            MOVE.L   A7,$2F80C.L           │
│   00020010 23FC 0003 7930            MOVE.L   #$37930,$2F810.L      │
│   0002001A 0CB9 0002 0000            CMPI.L   #$20000,$2F810.L      │
│   00020024 6E00 0014                 BGT      $2003A.W              │
│   00020028 2E7C 0001 FFF8            MOVEA.L  #$1FFF8,A7            │
│   0002002E 21FC 0000 0000            MOVE.L   #$0,$3C4.W            │
│   00020036 6000 0012                 BRA      $2004A.W              │
│   0002003A 2038 03C0                 MOVE.L   $3C0.W,D0             │
│   0002003E 5180                      SUBQ.L   #8,D0                 │
│   00020040 2E40                      MOVEA.L  D0,A7                 │
├─────────────────────────── Command Line ──────────────────────────┤
│ BD32->br $2000a                                                    │
│ BD32->go $20000                                                    │
│ BD32->s                                                            │
│ BD32->                                                             │
├────────────────────────────────────────────────────────────────────┤
│ MCU: BRKPNT  Port: LPT1 Halt: ? Reset: 1      V1.22 (C) 1990-94 Scott Howard │
└──────────────────────────────────────────────────────────────────┘
```

Figure A.1: Background debugger

Whenever the debugger is used, the EyeCon controller has to be connected to the parallel port of a Windows-PC using a BDM-cable. The actual debugging hardware interface is included on the EyeCon controller, so the BDM-cable contains no active components. The main uses for the BD32 debugger are:

- Debugging an assembly program.
- Rewriting a corrupted flash-ROM.

Debugging   When debugging an assembly program, the program first has to be loaded in memory using the button sequence `Usr/Ld` on the controller. Then, the BD32 debugger is started and the CPU execution is halted with the command `STOP`.

The user program is now located at the hex address $20000 and can be viewed with the disassemble debugger command:

```
dasm $20000
```

To go through a program step by step, use the following commands:

`window on`	Continuously display registers and memory contents.
`br $20a44`	Set breakpoint at desired address.
`s`	"Single-step", execute program one command at a time, but skip over subroutine calls at full speed.
`t`	"Trace", execute program one command at a time, including subroutine calls.

Detailed information on the background debugger can be found at:

```
http://robotics.ee.uwa.edu.au/eyebot/
```

Restoring the   Under normal conditions, rewriting the EyeCon's on-board flash-ROM is
flash-ROM   handled by the RoBIOS operating system, requiring no user attention. Whenever a new RoBIOS operating system or a new HDT is downloaded through the serial port, the operating system detects the system file and asks the user for authorization to overwrite the flash-ROM. In the same way, the user area of the flash-ROM can be overwritten by pressing the corresponding buttons for storing a downloaded program in flash-ROM.

Unfortunately, there are cases when the EyeCon's on-board flash-ROM can be corrupted, for example through a power failure during the write cycle or through a misbehaving user program. If this has happened, the EyeCon can no longer boot (start) and no welcome screen is printed on power-up. Since the operating system that normally takes care of the flash-ROM writing has been wiped out, trying to download the correct operating system does not work. While simpler controllers require the flash-ROM chip to be replaced and externally reprogrammed, the EyeCon has an on-board reprogramming capability using the processor's BDM interface. This allows restoration of the flash-ROM without having to remove it.

Similar to the debugging procedure, the controller has to be connected to a Windows-PC and its execution stopped before issuing the rewrite command via the BDM. The command sequence is:

`stop`	Stop processor execution; if EyeCon does not halt, press the reset button.
`do mapcs`	Initialize chip select lines.
`flash 11000000 rob52f.hex 0`	
	Delete RoBIOS in flash-ROM, overwrite with new version (bit string `11111111` can be used instead, to delete all sectors in the flash-ROM, including user programs). This process takes a few minutes.
`flash 00000000 hdt-std.hex $1c000`	
	Without deleting any flash-ROM sectors, write the HDT file at offset $1c000.

The parameters of the `flash` command are:

- Deletion of individual sectors:
  Each flash-ROM has eight sectors; specifying a "1" means delete, specifying a "0" means keep.

- Filename of hex-file to be written to flash-ROM.

- Address-offset:
  RoBIOS starts at address 0 in the ROM, the HDT starts at $1c000.

Note that because of the flash-ROM sector structure, only complete sectors can be deleted and overwritten. In the case of a corrupted RoBIOS, both RoBIOS and HDT need to be restored. In the case of a corrupted HDT and intact RoBIOS, the HDT can be restored by flashing it to the to the first user program slot at offset $20000. During restart, RoBIOS will detect the updated HDT and re-flash it as part of the operating system ROM sector:

```
flash 00100000 hdt-std.hex $20000
```

After rewriting the flash-ROM, the EyeCon needs to be reset of switched off and on again. It will then start with the normal greeting screen.

A.5 Download and Upload

Download For downloading a program, the EyeCon controller needs to be connected to a host PC via a standard serial cable (nine-pin RS232). Data transmission is possible at a number of different baud rates with default value 115,200 Baud. Executable programs can be transmitted as ASCII ".hex" files following the Motorola S-record format, or faster as compressed binary ".hx" files. The RoBIOS system tool `srec2bin` transforms `hex`-files to `hx`-files and vice versa.

To start a user program download from the host PC to the EyeCon, the data transfer has to be initialized on both sides:

- **On the EyeCon:**
  Press `Usr` / `Ld`
  (The LCD screen will indicate that the controller is ready to receive data. Download progress is indicated graphically and in the number of bytes transmitted.)

- **On the host PC:**
  Use the command `dl` for download:

  ```
  >dl userprog.hx
  ```

Upload   Besides downloading executable programs, it is also possible to transfer data under program control either from the PC to the EyeCon or from the Eye-Con to the PC. For uploading a block of data to the PC, the shell-script `ul` can be used instead of `dl`. A number of more elaborate example programs are available on the web to illustrate this procedure, for example for uploading images or measurement data [Bräunl 2006].

Turn-key system   A turn-key system can be created if the uploaded program name is either `startup.hex` or `startup.hx` (for compressed programs). The program has to be stored under this name in one of the three ROM slots. At start-up, RoBIOS will then bypass the standard monitor program and directly execute the user program. If the user program terminates, the RoBIOS monitor program will become active.

In case of a user program error like an endless loop, it would seem impossible to return to the monitor program in order to undo the turn-key setting and delete the user program, unless resorting to the background debugger. In order to solve this problem, it is possible to hold down one of the user buttons during start-up. In this case, the turn-key system will be temporarily deactivated and the regular RoBIOS monitor program will start.

A.6 References

BRÄUNL, T., *EyeBot Online Documentation*,
`http://robotics.ee.uwa.edu.au/eyebot/`, 2006

GNU. *GNU Compiler*, `http://www.delorie.com/gnu/docs/`, 2006

HARMAN, T. *The Motorola MC68332 Microcontroller - Product Design, Assembly Language Programming, and Interfacing*, Prentice Hall, Englewood Cliffs NJ, 1991

STALLMAN, R., MCGRATH, R. *Make: A Program for Directed Compilation*, GNU Press, Free Software Foundation, Cambridge MA, 2002

RoBIOS
OPERATING SYSTEM

B.1 Monitor Program

On power-up of the EyeCon controller, RoBIOS is booted and automatically starts a small monitor program which presents a welcome screen on the LCD and plays a small tune. This monitor program is the control interface for RoBIOS. The user can navigate via the four keys through numerous information and settings pages that are displayed on the LCD. In particular, the monitor program allows the user to change all basic settings of RoBIOS, test every single system component, receive and run user programs, and load or store them in flash-ROM.

Following the welcome screen, the monitor program displays the RoBIOS status screen with information on operating system version and controller hardware version, user-assigned system name, network ID, supported camera type, selected CPU frequency, RAM and ROM size with usage, and finally the current battery charge status (see Figure B.1).

All monitor pages (and most user programs) use seven text lines for displaying information. The eighth or bottom display line is reserved for menus that define the current functionality of the four user keys (*soft keys*). The pages that can be reached by pressing buttons from the main status page will be discussed in the following.

B.1.1 Information Section

The information screen displays the names of people that have contributed to the EyeBot project. On the last page a timer is perpetually reporting the elapsed time since the last reset of the controller board.

371

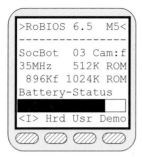

Figure B.1: RoBIOS status page and user keys

By pressing the REG-labelled key, a mask is displayed that shows the serial number of the controller and allows the user to enter a special keyword to unlock the wireless communication library of RoBIOS (see Chapter 6). This key will be saved in the flash-ROM so that it has to be entered only once for a controller, even if RoBIOS is being updated.

B.1.2 Hardware Settings

The hardware screens allow the user to monitor, modify, and test most of the on-board and off-board sensors, actuators, and interfaces. The first page displays the user-assigned HDT version number and a choice for three more submenus.

The setup menu (Set) offers two sections that firstly (Ser) deal with the settings of the serial port for program downloads and secondly (Rmt) with settings of the remote control feature. All changes that are made in those two pages are valid only as long as the controller is supplied with power. The default values for the power-up situation can be set in the HDT as described in Section B.3.

For download, the interface port, baud rate, and transfer protocol can be selected. There are three different transfer protocols available that all just differ in the handling of the RTS and CTS handshake lines of the serial port:

- NONE   Completely disregard handshaking.
- RTS/CTS   Full support for hardware handshaking.
- IrDA   No handshaking but the handshake lines are used to select different baud rates on an infrared module.

For wireless communication, the interface port and the baud rate can be selected in the same manner. In addition, specific parameters for the remote control protocol can be set. These are the network unique id-number between 0 and 255, the image quality, and the protocol. The protocol modes (to be set in the HDT file) are:

- RADIO_BLUETOOTH Communication via a serial Bluetooth module.
- RADIO_WLAN   Communication via a serial WLAN module.

- `RADIO_METRIX`   Communication via a serial transceiver module.

The image quality modes are:

- `Off`   No images are sent.
- `Reduced`   Images are sent in reduced resolution and color depth.
- `Full`   Images are sent in full resolution and color depth.

The second sub-menu (`HDT`) of the hardware settings page displays a list of all devices found in the HDT that can be tested by RoBIOS. For each device type, the number of registered instances and their associated names are shown. Currently nine different device types can be tested:

- **Motor**
  The corresponding test function directly drives the selected motor with user-selectable speed and direction. Internally it uses the `MOTORDrive` function to perform the task.

- **Encoder**
  The encoder test is an extension of the motor test. In the same manner the speed of the motor linked to the encoder is set. In addition, the currently counted encoder ticks and the derived speed in ticks per second are displayed by internally calling the `QUADRead` function.

- **vω Interface**
  This test is somewhat more "high level" since it utilizes the vω interface for differential drives, which is based upon the motor and encoder drivers. Wheel distance and encoder IDs are shown, as stored in the HDT. By pressing the `Tst`-labelled key, vω commands to drive a straight line of 40cm, turn 180° on the spot, come back in a straight line, and turn 180° again are issued.

- **Servo**
  In analogy to the motor test, an angular value between 0 and 255 can be entered to cause an attached servo to take the corresponding position by using the `SERVOSet` function.

- **PSD**
  The currently measured distance value from the selected PSD is displayed graphically in a fast scrolling fashion. In addition, the numeric values of raw and calibrated sensor data (through a lookup table in the HDT) are shown by using functions `PSDGetRaw` and `PSDGet`.

- **IR**
  The current binary state of the selected sensor is permanently sampled by calling `IRRead` and printed on the LCD. With this test, any binary sensor that is connected to an HDT-assigned TPU channel and entered in the HDT can be monitored.

- **Bumper**
  The precise transition detection driver is utilized here. Upon detection of a signal edge (predefined in the HDT) on the selected TPU channel

the corresponding time of a highly accurate TPU timer is captured and posted for 1s on the LCD before restarting the process. The applied function is BUMPCheck.

- **Compass**
  A digital compass can be calibrated and its read-out displayed. For the calibration process, the compass first has to be placed in a level position aligned to a virtual axis. After acknowledging this position, the compass has to be turned in the opposite direction followed by another confirmation. The calibration data is permanently stored in the compass module so that no further calibration should be required. In the read-out mode, a graphical compass rose with an indicator for the north direction and the corresponding numerical heading in degrees (from function COMPASSGet) is displayed.

- **IRTV**
  The currently received infrared remote control code is displayed in numerical form. All the necessary parameters for the different remote control types have to be defined in the HDT before any valid code will be displayed. This test is very useful to find out which code each button of the remote control will deliver upon calling the IRTVPressed function, so these codes can be used in the software.

If any of these tests shows an unsatisfactory result, then the parameters in the corresponding HDT structure should be checked first and modified where appropriate before any conclusions about the hardware status are drawn. All of these tests and therefore the RoBIOS drivers solely rely upon the stored values in the HDT, which makes them quite universal, but they depend on correct settings.

The third sub-menu (IO) of the hardware settings page deals with the status of the on-board I/O interfaces. Three different groups are distinguished here. These are the input and output latches (Dig), the parallel port interface (Parl), and the analog input channels (AD). In the latch section, all eight bits of the input latch can be monitored and each of the eight bits of the output latch can be modified. In the parallel port section the port can be handled as an input port to monitor the eight data pins plus the five incoming status pins or as an output port to set any of the eight data pins plus the four outgoing control pins. Finally in the analog input section, the current readings of the eight available A/D converter (ADC) channels can be read with a selectable refresh rate.

B.1.3 Application Programs

The application program screens are responsible for the download of all RoBIOS-related binaries, their storage in the flash-ROM, or the program execution from RAM. In the first screen, the program name together with the file-size and, if applicable, the uncompressed size of an application in RAM are displayed. From here, there is a choice between three further actions: Ld, Run, or ROM.

1. **Load**

   The display shows the current settings for the assigned download port and RoBIOS starts to monitor this port for any incoming data. If a special start sequence is detected, the subsequent data in either binary or S-record format is received. Download progress is displayed as either a graphical bar (for binary format) or byte counter (for S-record). If the cyclic redundancy check (crc) reveals no error, the data type is being checked. If the file contains a new RoBIOS or HDT, the user will be prompted for storing it in ROM. If it contains a user application, the display changes back to the standard download screen.

   Auto-download   There is an alternative method to enter the download screen. If in the HDT info-structure, the "auto_download" member is set to "AUTOLOAD" or "AUTOLOADSTART", RoBIOS will perform the scanning of the download port during the status screen that appears at power-up. If data is being downloaded, the system jumps directly to the download screen. In the "AUTOLOADSTART" case, it even automatically executes the downloaded application. This mode comes in handy if the controller is fixed in a difficult-to-reach assembly, where the LCD may not be visible or even attached, or none of the four keys can be reached.

2. **Run**

   If there is a user program ready in RAM, either by downloading or copying from ROM, it can be executed by choosing this option. If the program binary is compressed RoBIOS will decompress it before execution. Program control is completely transferred to the user application rendering the monitor program inactive during the application's run-time. If the user program terminates, control is passed back to the monitor program. It will display the overall run-time of the application before showing the Usr screen again. The application can be restarted, but one has to be aware that any *global variables* that are not initialized from the main program will still contain the *old values* of the last run. Global declaration initializations such as:

   Explicitly initialize global variables

   ```
   int x = 7;
   ```

   will not work a second time in RAM!

   The application in RAM will survive a reset, so any necessary reset during the development phase of an application will not make it necessary to reload the application program.

3. **ROM**

   In order to store user programs permanently, they need to be saved to the flash-ROM. Currently, there are three program slots of 128KB each available. By compressing user programs before downloading, larger applications can be stored. The ROM screen displays the name of the current program in RAM and the names of the three stored programs or NONE if empty. With the Sav key, the program currently in RAM will be saved to the selected ROM slot. This will only be performed if the program size does not exceed the 128KB limit and the program in RAM has not yet been executed. Otherwise programs could be stored that have already modified their global variable initializations or are already decompressed. With the corresponding Ld key, a stored program is copied from flash-ROM to RAM, either for execution or for copying to a different ROM slot.

Demo programs in ROM

There are two reserved names for user applications that will be treated in a special way. If a program is called "demos.hex" or "demos.hx" (compressed program), it will be copied to RAM and executed if the Demo key is pressed in the main menu of the monitor program (see Section B.1.4 below). The second exception is that a program stored as "startup.hex" or

Turn-key system in ROM

"startup.hx" will automatically be executed on power-up or reset of the controller without any keys being pressed. This is called a "turn-key" system and is very useful in building a complete embedded application with an EyeCon controller. To prevent RoBIOS from automatically starting such an application, any key can be pressed at boot time.

B.1.4 Demo Programs

As described above, if a user program with the name "demos.hex" or "demos.hx" is stored in ROM, it will be executed when the Demo key is pressed in the main screen. The standard demo program of RoBIOS includes some small demonstrations: Camera, Audio, Network, and Drive.

In the camera section three different demos are available. The Gry demo captures grayscale camera images and lets the user apply up to four image processing filters on the camera data before displaying them with the effective frame rate in frames per second (fps). The Col demo grabs color images and displays the current red, green, and blue values of the center pixel. By pressing Grb, the color of the center pixel is memorized so that a subsequent press of Tog can toggle between the normal display and showing only those pixels in black that have a similar RGB color value to the previously stored value. The third camera demo FPS displays color images and lets the user vary the frame rate. Camera performance at various frame rates can be tested depending on image resolution and CPU speed. At too high a frame rate the image will start to roll through. Recorded images can be sent via serial port 1 to a PC by pressing the Upl key in PPM format. Also, the vω interface can be started in order to check image processing while slowly driving the robot.

In the audio section, a simple melody or a voice sample can be played. Also, the internal microphone can be monitored or used to record and play back a sample sound.

In the network section, the radio module on serial port 2 can be tested. The first test Tst simply sends five messages of 1,000 characters via the radio module. The next test requires two controllers with a radio module. One EyeCon acts as the sender by pressing Snd, while the other acts as the receiver by pressing Rcv. The sender now permanently sends a short changing message that the receiver will print on its LCD.

The last section drive performs the same task as described for the vω inter-face HDT test function in Section B.1.2. In addition to this, driving can be per-formed with the camera activated, showing captured images while driving.

B.2 System Function and Device Driver Library

The RoBIOS binary contains a large library of system functions and device drivers to access and control all on-board and off-board hardware and to utilize the operating system's services. The advantages of placing those functions as a shared library in the operating system code instead of distributing them as a static library that is linked to all user programs are obvious. Firstly, the user programs are kept small in size so that they can be downloaded faster to the controller and naturally need less space in the case of being stored in ROM. Secondly, if the function library is updated in ROM, every user program can directly benefit from the new version without the need of being re-compiled. Lastly, the interaction between the library functions and the operating system internal functions and structures is straightforward and efficient since they are integrated in the same code segment. Any user program running under RoBIOS can call these library functions. Only the `eyebot.h` header file needs to be included in the program source code.

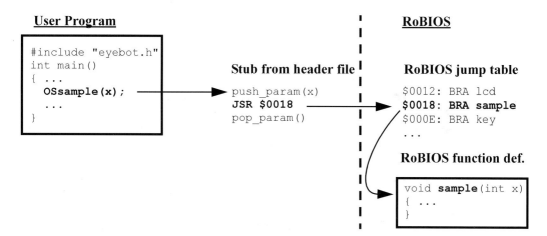

Figure B.2: RoBIOS function call

A special mechanism takes place to redirect a system call from a user pro-gram to the appropriate RoBIOS library function. The header file only con-tains so-called "function stubs", which are simple macro definitions handling parameter passing via stack or registers and then calling the "real" RoBIOS functions via a jump address table. With this mechanism, any RoBIOS func-tion call from a user program will be replaced by a function stub that in turn calls the RAM address of the matching RoBIOS function. Since the order of the current RoBIOS functions in this lookup table is static, no user program has to be re-compiled if a new version of RoBIOS is installed on the EyeCon controller (see Figure B.2).

The library functions are grouped in the following categories:

- Image Processing — *A small set of sample image processing functions for demonstration purposes*
- Key Input — *Reading the controller's user keys*
- LCD Output — *Printing text of graphics to the controller's LCD screen*
- Camera — *Camera drivers for several grayscale and color camera modules*
- System Functions — *Low-level system functions and interrupt handling*
- Multi-Tasking — *Thread system with semaphore synchronization*
- Timer — *Timer, wait, sleep functions as well as real-time clock*
- Serial Communication — *Program and data download/upload via RS232*
- Audio — *Sound recording and playback functions, tone and wave-format playing functions*
- Position Sensitive Devices — *Infrared distance sensor functions with digital distance values*
- Servos and Motors — *Driving functions for model servos and DC motors with encoders*
- vω Driving Interface — *High-level vehicle driving interface with PI controller for linear and angular velocity*
- Bumper+Infrared Sensors — *Routines for simple binary sensors (on/off switches)*
- Latches — *Access routines for digital I/O ports of the controller*
- Parallel Port — *Reading/writing data from/to standard parallel port, setting/reading of port status lines*
- Analog-Digital Converter — *Access routines for A/D converter, including*
  · microphone input (analog input 0)
  · battery status (analog input 1)
- Radio Communication — *Wireless communication routines for virtual token ring of nodes (**requires enabling**)*
- Compass — *Device driver for digital compass sensor*
- IR Remote Control — *Reading a standard infrared TV remote as user interface*

All library functions are described in detail in Section B.5.

B.3 Hardware Description Table

The EyeCon controller was designed as a core component for the large EyeBot family of mobile robots and numerous external robot projects that implement very different kinds of locomotion. Among those are wheeled, tracked, legged, and flying robots. They all have in common that they utilize the same RoBIOS library functions to control the attached motors, servos, and other supported devices.

Therefore, the RoBIOS operating system is not committed to one hardware design or one locomotion type only. This makes the operating system more open toward different hardware applications but also complicates software integration of the diverse hardware setups. Without any system support, a user program would have to know exactly which hardware ports are used by all the used actuators and sensors and what their device characteristics are. For instance, even motors of the same type may have different performance curves that have to be individually measured and compensated for in software. Due to the same reasons another problem emerges: a piece of software that was written for a particular target will not show exactly the same performance on a similar model, unless adapted for any differences in the hardware characteristics.

To overcome those deficiencies a hardware abstraction layer (called the "Hardware Description Table", HDT) has been introduced to RoBIOS. The idea is that, for each controller, the characteristics and connection ports of all attached devices are stored in a simple internal database. Each entry is associated with a unique keyword that reflects the semantics of the device. Thus, the application programs only need to pass the desired semantics to the corresponding RoBIOS driver to gain control. The driver searches the database for the corresponding entry and reads out all necessary configurations for this device. With this abstraction layer, a user program becomes portable not only between robots of the same model, but also between electronically and mechanically different robots.

If, for example, an application requests access to the left and right motor of a vehicle, it simply calls the motor driver with the pre-defined semantics constants (a kind of "device name", see definition file `htd_sem.h`) MOTOR_LEFT and MOTOR_RIGHT, without having to know where the motors are connected to and what characteristic performance curves they have. By using the high level vω interface, an application can even issue commands like "drive 1m forward" without having to know what kind of locomotion system the robot is actually based on. Furthermore, a program can dynamically adapt to different hardware configurations by trying to access multiple devices through a list of semantics and only cope with those that respond positively. This can be used for sensors like the PSD distance sensors or IR binary sensors that help to detect surrounding obstacles, so that the software can adapt its strategy on the basis of the available sensors and their observed area or direction.

The HDT not only incorporates data about the attached sensors and actuators, but also contains a number of settings for the internal controller hardware (including CPU frequency, chip-access waitstates, serial port settings and I/O-latch configuration) and some machine-dependent information (for example radio network ID, robot name, start-up melody, and picture).

As already noted in Section 1.4, the HDT is stored separately in the flash-ROM, so modifications can easily be applied and downloaded to the controller without having to reload RoBIOS as well. The size of the HDT is limited to 16KB, which is more than enough to store information about a fully equipped and configured controller.

B.3.1 HDT Component List

The HDT primarily consists of an array of component structures. These structures carry information about the object they are referring to (see Program B.1).

Program B.1: Component structure of HDT array

```
1   typedef struct
2   { TypeID              type_id;
3     DeviceSemantics     semantics;
4     String6             device_name;
5     void*               data_area;
6   } HDT_entry_type;
```

- type_id: This is the unique identifier (listed in hdt.h) of the category the described object belongs to. Examples are MOTOR, SERVO, PSD, COMPASS, etc. With the help of this category information, RoBIOS is able to determine the corresponding driver and hence can count how many entries are available for each driver. This is used among others in the HDT section of the monitor program to display the number of candidates for each test.

- semantics: The abstraction of a device from its physical connection is primarily achieved by giving it a meaningful name, such as MOTOR_RIGHT, PSD_FRONT, L_KNEE, etc., so that a user program only needs to pass such a name to the RoBIOS driver function, which will in turn search the HDT for the valid combination of the according TypeID and this name (DeviceSemantics). The list of already assigned semantics can be found in hdt_sem.h. It is strongly recommended to use the predefined semantics in order to support program portability.

- device_name: This is a string representation of the numerical *semantics* parameter with a maximum of six letters. It is only used for testing purposes, to produce a readable semantics output for the HDT test functions of the monitor program.

- `data_area`: This is a typeless pointer to the different category-dependent data structures that hold type-specific information for the assigned drivers. It is a typeless pointer, since no common structure can be used to store the diversity of parameters for all the drivers.

The array of these structures has no predefined length and therefore requires a special end marker to prevent RoBIOS from running past the last valid entry. This final entry is denoted as:

```
{END_OF_HDT,UNKNOWN_SEMANTICS,"END",(void *)0}
```

Apart from this marker, two other entries are mandatory for all HDTs:

- `WAIT`: This entry points to a list of waitstate values for the different chip-access times on the controller platform, which are directly derived from the chosen CPU frequency.

- `INFO`: This entry points to a structure of numerous basic settings, like the CPU frequency to be used, download and serial port settings, network ID, robot name, etc.

Program B.2 is an example of the shortest valid HDT.

Program B.2: Shortest valid HDT file

```
1   #include "robios.h"
2   int     magic = 123456789;
3   extern HDT_entry_type HDT[];
4   HDT_entry_type   *hdtbase = &HDT[0];
5
6   /* Info: EyeBot summary */
7   info_type roboinfo  = {0, VEHICLE, SER115200, RTSCTS,
8    SERIAL1, 0, 0, AUTOBRIGHTNESS, BATTERY_ON, 35, 1.0,
9    "Eye-M5",1};
10  /* waitstates for: ROM, RAM, LCD, IO, UART */
11  waitstate_type waitstates = {0,3,1,2,1,2};
12
13  HDT_entry_type HDT[] =
14  {   {WAIT,WAIT,"WAIT",(void *)&waitstates},
15      {INFO,INFO,"INFO",(void *)&roboinfo},
16      {END_OF_HDT,UNKNOWN_SEMANTICS,"END",(void *)0}
17  };
```

The descriptions of all the different HDT data structures can be found in Appendix C. Together with the array of component structures, the used data structures build up the complete source code for an HDT binary. To obtain a downloadable binary image the HDT source code has to be compiled with the special HDT batch commands provided with the RoBIOS distribution. For example:

```
gcchdt myhdt.c -o myhdt.hex
```

The HDT code is compiled like a normal program except for a different linker file that tells the linker not to include any start-up code or `main()` function, since only the data part is needed. During the download of an HDT binary to the controller, the "magic number" in the HDT header is recognized by RoBIOS and the user is prompted to authorize updating the HDT in flash-ROM.

B.3.2 HDT Access Functions

There are five internal functions in RoBIOS to handle the HDT. They are mainly used by hardware drivers to find the data structure corresponding to a given semantics or to iterate through all assigned data structures with the same type identifier:

```
int HDT_Validate(void)
```
This function is used by RoBIOS to check the magic number of the HDT and to initialize the global HDT access data structure.

```
void *HDTFindEntry(TypeID typeid,DeviceSemantics semantics)
```
With the help of this function the address of the data structure that corresponds to the given type identifier and semantics is found. This is the only function that can also be called from a user program to obtain more detailed information about a specific device configuration or characteristic.

```
DeviceSemantics HDT_FindSemantics(TypeID typeid, int x)
```
This is the function that is needed to iterate through all available entries of the same type. By calling this function in a loop with increasing values for x until reaching UNKNOWN_SEMANTICS, it is possible to inspect all instances of a specific category. The return value is the semantics of the corresponding instance of this type and might be used in calling HDT_FindEntry() or the device driver initialization function.

```
int HDT_TypeCount(TypeID typeid)
```
This function returns the number of entries found for a specific type identifier.

```
char *HDT_GetString(TypeID typeid,DeviceSemantics semantics)
```
This function returns the readable name found in the entry associated with the given type and semantics.

Normally, an application program does not need to bother with the internal structure of the HDT. It can simply call the driver functions with the defined semantics as shown in an example for the motor driver functions in Program B.3. For details of all HDT entries see Appendix C.

Program B.3: Example of HDT usage

```
1    /* Step1: Define handle variable as a motor reference */
2    MotorHandle leftmotor;
3
4    /* Step2: Initialize handle with the semantics (name) of
5             chosen motor. The function will search the HDT
6             for a MOTOR entry with given semantics and, if
7             successful, initialize motor hardware and return
8             the corresponding handle */
9    leftmotor = MOTORInit(LEFTMOTOR);
10
11   /* Step3: Use a motor command to set a certain speed.
12             Command would fail if handle was not initial. */
13   MOTORDrive (leftmotor,50);
14
15   /* Step4: Release motor handle when motor is no longer
16             needed */
17   MOTORRelease (leftmotor);
```

B.4 Boot Procedure

The time between switching on the EyeCon controller and the display of the RoBIOS user interface is called the boot phase. During this time numerous actions are performed to bring the system up to an initialized and well-defined state.

In the beginning, the CPU is trying to fetch the start address of an executable program from memory location $000004. Since the RAM is not yet initialized, the default memory area for the CPU is the flash-ROM, which is activated by the hardware chip-select line CSBOOT' and therefore is internally mapped to address $000000. As shown in Figure 1.11, RoBIOS starts at exactly that memory location, so the CPU will start executing the RoBIOS bootstrap loader, which precedes the compressed RoBIOS binary. This code initializes the CPU chip-select signals for the RAM chips, so that the compressed RoBIOS can later be unpacked into RAM. Furthermore, the address mapping is changed so that after the unpacking the RAM area will start at address $000000, while the ROM area will start at $C00000.

It seems to be a waste of RAM space to have RoBIOS in ROM and in RAM, but this offers a number of advantages. First, RAM access is about three times faster than ROM access because of different waitstates and the 16bit RAM bus compared to the 8bit ROM bus. This increases RoBIOS performance considerably. Secondly, it allows storage of the RoBIOS image in compressed form in ROM, saving ROM space for user programs. And finally, it allows the use of self-modifying code. This is often regarded as bad programming style, but can result in higher performance, e.g for time consuming tasks like frame grabbing or interrupt handling. On the other hand, a RAM location has the disadvantage of being vulnerable to memory modifications caused by user programs, which can temporarily lead to an unexpected system behavior or a total crash. However, after a reset everything will work fine again, since a new RoBIOS copy will be read from the protected flash-ROM.

After the RAM chips and all other required chip-select pins have been initialized, the start-up code copies a small decompression algorithm to a CPU-local RAM area (TPU-RAM), where it can be executed with zero waitstates, which speeds up the unpacking of the RoBIOS binary to external RAM. Finally, after having placed the executable RoBIOS image in the address area from $000000 to $020000, the start-up code jumps into the first line of the now uncompressed RoBIOS in RAM, where the remaining initialization tasks are performed. Among those are the test for additional mounted RAM chips and the subsequent calculation of the actual RAM size.

In the same manner the ROM size is checked, to see if it exceeds the minimum of 128KB. If so, RoBIOS knows that user programs can be stored in ROM. Now it is checked if a new HDT is located at the first user ROM slot. In this case, a short message is printed on the LCD that the re-programming of the flash-ROM will take place before the system continues booting. Now that an HDT is available, RoBIOS checks it for integrity and starts extracting information from it like the desired CPU clock rate, or the waitstate settings for different chip-select lines. Finally, the interrupt handlers, the 100Hz system timer, and some basic drivers, for example for serial interface, ADC, in/out-latches and audio, are started just before the welcome screen is shown and a melody is played.

Before displaying the standard monitor status screen, RoBIOS has to check whether a program called "startup.hex" or "startup.hx" is stored in ROM. If this is the case, a turn-key system has been created and the application program will be loaded and started immediately, unless a button is being pressed. This is very useful for an embedded application of the EyeCon controller or in the case when no LCD is mounted, which obviously would make a manual user program start difficult.

B.5 RoBIOS Library Functions

This section describes the RoBIOS operating system library routines in version 6.5 (2006). Newer versions of the RoBIOS software may differ from the functionality described below – see the latest software documentation. The following libraries are available in ROM for programming in C.

In application files use:

```
#include "eyebot.h"
```

The following libraries are available in ROM for programming in C and are automatically linked when calling "gcc68" and the like (using librobi.a).

Note that there are also a number of libraries available which are *not listed here*, since they are not in ROM but in the *EyeBot* distribution (e.g. elaborate image processing library). They can also be linked with an application program, as shown in the demo programs provided.

Return Codes

Unless specifically noted otherwise, all routines return 0 when successful, or a value !=0 when an error has occurred. Only very few routines support multiple return codes.

B.5.1 Image Processing

A few basic image processing functions are included in RoBiOS. A larger collection of image processing functions is contained in the "image processing library", which can be linked to an application program.

```
Data Types:
        /* image is 80x60 but has a border of 1 pixel */
        #define imagecolumns 82
        #define imagerows 62

        typedef BYTE image[imagerows][imagecolumns];
        typedef BYTE colimage[imagerows][imagecoulmns][3];
```

```
int IPLaplace (image *src, image *dest);
        Input:          (src) source b/w image
        Output:         (dest) destination b/w image
        Semantics:      The Laplace operator is applied to the source image
                        and the result is written to the destination image
```

```
int IPSobel (image *src, image *dest);
        Input:          (src) source b/w image
        Output:         (dest) destination b/w image
        Semantics:      The Sobel operator is applied to the source image
                        and the result is written to the destination image
```

```
int IPDither (image *src, image *dest);
        Input:          (src) source b/w image
        Output:         (dest) destination b/w image
        Semantics:      The Dithering operator with a 2x2 pattern is applied
                        to the source image and the result is written to the
                        destination image
```

```
int IPDiffer (image *current, image *last, image *dest);
        Input:          (current) the current b/w image
                        (last) the last read b/w image
        Output:         (dest) destination b/w image
        Semantics:      Calculate the grey level difference at each pixel
                        position between current and last image, and
                        store the result in destination.
```

```
int IPColor2Grey (colimage *src, image *dest);
        Input:          (src) source color image
        Output:         (dest) destination b/w image
        Semantics:      Convert RGB color image given as source to 8-bit
                        grey level image and store the result in
                        destination.
```

Advanced image processing functions are available as library "improc.a".
For detailed info see the Improv web-page:
  http://robotics.ee.uwa.edu.au/improv/

B.5.2 Key Input

Using the standard Unix "libc" library, it is possible to use standard C "scanf" commands to read key "characters" from the "keyboard".

int **KEYGetBuf** (char *buf);
Input: (buf) a pointer to one character
Output: (buf) the keycode is written into the buffer
                  Valid keycodes are: KEY1,KEY2,KEY3,KEY4 (keys
                  from left to right)
Semantics: Wait for a keypress and store the keycode into
                  the buffer

int **KEYGet** (void);
Input: NONE
Output: (returncode) the keycode of a pressed key is returned
                  Valid keycodes are: KEY1,KEY2,KEY3,KEY4 (keys
                  from left to right)
Semantics: Wait for a keypress and return keycode

int **KEYRead** (void);
Input: NONE
Output: (returncode) the keycode of a pressed key is
                  returned or 0 if no key is pressed.
                  Valid keycodes are: KEY1,KEY2,KEY3,KEY4 (keys
                  from left to right) or 0 for no key.
Semantics: Read keycode and return it. Function does not wait.

int **KEYWait** (int excode);
Input: (excode) the code of the key expected to be pressed
                  Valid keycodes are: KEY1,KEY2,KEY3,KEY4 (keys
                  from left to right) or ANYKEY.
Output: NONE
Semantics: Wait for a specific key

B.5.3 LCD Output

Using the standard Unix "libc" library, it is possible to use standard C "printf" commands to print on the LCD "screen". E.g. the "hello world" program works:

   printf("Hello, World!\n");

The following routines can be used for specific output functions:

int **LCDPrintf** (const char format[], ...);
Input: format string and parameters
Output: NONE
Semantics: Prints text or numbers or combination of both
                  onto LCD. This is a simplified and smaller
                  version of standard Clib "printf".

int **LCDClear** (void);
Input: NONE
Output: NONE
Semantics: Clear the LCD

int **LCDPutChar** (char char);
Input: (char) the character to be written
Output: NONE
Semantics: Write the given character to the current cursor

```
                            position and increment cursor position

        int LCDSetChar (int row, int column, char char);
        Input:              (char) the character to be written
                            (column) the number of the column
                            Valid values are: 0-15
                            (row) the number of the row
                            Valid values are: 0-6
        Output:             NONE
        Semantics:          Write the given character to the given display position

        int LCDPutString (char *string);
        Input:              (string) the string to be written
        Output:             NONE
        Semantics:           Write the given string to the current cursor pos.
                            and increment cursor position

        int LCDSetString (int row, int column, char *string);
        Input:              (string) the string to be written
                            (column) the number of the column
                            Valid values are: 0-15
                            (row) the number of the row
                            Valid values are: 0-6
        Output:             NONE
        Semantics:          Write the given string to the given display position

        int LCDPutHex (int val);
        Input:              (val) the number to be written
        Output:             NONE
        Semantics:          Write the given number in hex format at current
                            cursor position

        int LCDPutHex1 (int val);
        Input:              (val) the number to be written (single byte 0..255)
        Output:             NONE
        Semantics:          Write the given number as 1 hex-byte at current
                            cursor position

        int LCDPutInt (int val);
        Input:              (val) the number to be written
        Output:             NONE
        Semantics:          Write the given number as decimal at current
                            cursor position

        int LCDPutIntS (int val, int spaces);
        Input:              (val)    the number to be written
                            (spaces) the minimal number of print spaces
        Output:             NONE
        Semantics:          Write the given number as decimal at current
                                cursor position using extra spaces in front if necessary

        int LCDPutFloat (float val);
            Input:              (val) the number to be written
            Output:             NONE
            Semantics:          Write the given number as floating point number
                                at current cursor position

        int LCDPutFloatS (float val, int spaces, int decimals);
            Input:              (val)      the number to be written
                                (spaces)   the minimal number of print spaces
                                (decimals) the number of decimals after the point
            Output:             NONE
            Semantics:          Write the given number as a floating point number
                                at current cursor position using extra spaces in
                                front if necessary and with specified number of
```

387

```
                          decimals

int LCDMode (int mode);
Input:            (mode) the display mode you want
                         Valid values are: (NON)SCROLLING|(NO)CURSOR
Output:           NONE
Semantics:        Set the display to the given mode
                  SCROLLING: the display will scroll up one
                  line, when the right bottom corner is
                  reached and the new cursor position
                  will be the first column of the now
                  blank bottom line
                  NONSCROLLING: display output will resume in
                      the top left corner when the bottom
                      right corner is reached
                  NOCURSOR: the blinking hardware cursor is not
                      displayed at the current cursor position
                  CURSOR: the blinking hardware cursor is
                      displayed at the current cursor position

int LCDSetPos (int row, int column);
Input:            (column) the number of the column
                  Valid values are: 0-15
                  (row) the number of the row
                  Valid values are: 0-6
Output:           NONE
Semantics:        Set the cursor to the given position

int LCDGetPos (int *row, int *column);
Input:             (column) pointer to the storing place for current
                  column.
                      (row) pointer to the storing place for current row.
Output:           (*column) current column
                  Valid values are: 0-15
                  (row) current row
                  Valid values are: 0-6
Semantics:        Return the current cursor position

int LCDPutGraphic (image *buf);
Input:            (buf) pointer to a grayscale image (80*60 pixel)
Output:           NONE
Semantics:        Write the given graphic b/w to the display
                      it will be written starting in the top left corner
                  down to the menu line.  Only 80x54 pixels will
                  be written to the LCD, to avoid destroying the
                  menu line.

int LCDPutColorGraphic (colimage *buf);
        Input:            (buf) pointer to a color image (80*60 pixel)
        Output:           NONE
        Semantics:        Write the given graphic b/w to the display
                          it will be written starting in the top left corner
                          down to the menu line.  Only 80x54 pixels will
                          be written to the LCD, to avoid destroying the
                          menu line.  Note: The current implementation
                          destroys the image content.

int LCDPutImage (BYTE *buf);
Input:            (buf) pointer to a b/w image (128*64 pixel)
Output:           NONE
Semantics:        Write the given graphic b/w to the hole display.

int LCDMenu (char *string1, char *string2, char *string3,char *string4);
Input:            (string1) menu entry above key1
                  (string2) menu entry above key2
```

```
                   (string3) menu entry above key3
                   (string4) menu entry above key4
                   Valid Values are:
                   - a string with max 4 characters, which
                     clears the menu entry and writes the new one
                   - "" : leave the menu entry untouched
                   - " " : clear the menu entry
Output:            NONE
Semantics:         Fill the menu line with the given menu entries

int LCDMenuI (int pos, char *string);
Input:             (pos) number of menu entry to be exchanged (1..4)
                   (string) menu entry above key <pos> a string
                   with max 4 characters
Output:            NONE
Semantics:         Overwrite the menu line entry at position pos with
                   the given string

int LCDSetPixel (int row, int col, int val);
Input:             (val) pixel operation code
                   Valid codes are:       0 = clear pixel
                                          1 = set pixel
                                          2 = invert pixel
                   (column) the number of the column
                   Valid values are: 0-127
                   (row) the number of the row
                   Valid values are: 0-63
Output:            NONE

Semantics:         Apply the given operation to the given pixel
                   position.  LCDSetPixel(row, col, 2) is the
                   same as LCDInvertPixel(row, col).

int LCDInvertPixel (int row, int col);
Input:             (column) the number of the column
                   Valid values are: 0-127
                   (row) the number of the row
                   Valid values are: 0-63
Output:            NONE
Semantics:         Invert the pixel at the given pixel position.
                   LCDInvertPixel(row, col) is the same as
                   LCDSetPixel(row, col, 2).

int LCDGetPixel (int row, int col);
Input:             (column) the number of the column
                   Valid values are: 0-127
                   (row) the number of the row
                   Valid values are: 0-63
Output:            (returncode) the value of the pixel
                   Valid values are:      1 for set pixel
                                          0 for clear pixel
Semantics:         Return the value of the pixel at the given
                   position

int LCDLine (int x1, int y1, int x2, int y2, int col)
Input:             (x1,y1) (x2,y2) and color
Output:            NONE
Semantics:         Draw a line from (x1,y1) to (x2,y2) using Bresenham Algorithm
                   top left  is   0, 0
                   bottom right is 127,63
                   color: 0 white
                          1 black
                          2 negate image contents

int LCDArea (int x1, int y1, int x2, int y2, int col)
```

```
Input:   (x1,y1) (x2,y2) and color
Output: NONE
Semantics: Fill rectangular area from (x1,y1) to (x2,y2) it must
           hold: x1 < x2 AND y1 < y2
                 top    left  is   0, 0
                 bottom right is 127,63
                 color: 0 white
                        1 black
                        2 negate image contents
```

B.5.4 Camera

The following functions handle initializing and image reading from either gray-scale or color camera:

int **CAMInit** (int mode);

```
      Input:            (mode) camera initialization mode
                        Valid Values are: NORMAL
      Output:           (return code) Camera version or Error code
                        Valid values: 255     = no camera connected
                               200..254= camera init error (200 + cam. code)
                                      0      = QuickCam V1 grayscale
                                      16     = QuickCam V2 color
                                      17     = EyeCam-1 (6300), res.  82x 62 RGB
                                      18     = EyeCam-2 (7620), res. 320x240 Bayer
                                      19     = EyeCam-3 (6620), res. 176x144 Bayer
      Semantics:    Reset and initialize connected camera
      Notes:        [Previously used camera modes for Quickcam:
                     WIDE,NORMAL,TELE]
                    The maximum camera speed is determined by processor speed
                    and any background tasks. E.g. when using v-omega motor
                    control as a background task, set the camera speed to:
                    CAMSet (FPS1_875, 0, 0);
```

int **CAMRelease** (void);
```
      Input:        NONE
      Output:       (return code) 0 = success
                                 -1 = error
      Semantics:    Release all resources allocated using CAMInit().
```

int **CAMGetFrame** (image *buf);
```
      Input:        (buf) a pointer to a gray scale image
      Output:       NONE
      Semantics:    Read an image size 62x82 from gray scale camera.
                    Return 8 bit gray values 0 (black) .. 255 (white)
```

int **CAMGetColFrame** (colimage *buf, int convert);
```
      Input:        (buf) a pointer to a color image
                    (convert) flag if image should be reduced to 8 bit gray
                           0 = get 24bit color image
                           1 = get 8bit grayscale image
      Output:       NONE
      Semantics:    Read an image size 82x62 from color cam and reduce it
                    if required to 8 bit gray scale.
      Note:         - buf needs to be a pointer to 'image'
                    - enable conversion like this:
                         image buffer;
                         CAMGetColFrame((colimage*)&buffer, 1);
```

int **CAMGetFrameMono** (BYTE *buf);
```
      Note:         This function works only for EyeCam
      Input:        (buf) pointer to image buffer of full size (use CAMGet)
```

```
        Output:         (return code) 0 = success
                                      -1 = error (camera not initialized)
        Semantics:      Reads one full gray scale image
                        (e.g. 82x62, 88x72, 160x120) dep. on camera module


int CAMGetFrameRGB (BYTE *buf);
        Note:           This function works only for EyeCam
        Input:          (buf) pointer to image buffer of full size (use CAMGet)
        Output:         (return code) 0 = success
                                      -1 = error (camera not initialized)
        Semantics:      Reads full color image in RBG format, 3 bytes per pixel,
                        (e.g. 82x62*3, 88x72*3, 160x120*3)
                        depending on camera module


int CAMGetFrameBayer (BYTE *buf);
        Note:           This function works only for EyeCam
        Input:          (buf) pointer to image buffer of full size (use CAMGet)
        Output:         (return code) 0 = success
                                      -1 = error (camera not initialized)
        Semantics:      Reads full color image in Bayer format, 4 bytes per pix,
                        (e.g. 82x62*4, 88x72*4, 160x120*4)
                        depending on camera module


int CAMSet (int para1, int para2, int para3);
        Note:           parameters have different meanings for different cameras
        Input:QuickCam  (para1) camera brightness
                        (para2) camera offset (b/w camera) / hue (color camera)
                        (para3) contrast (b/w camera) / saturation (color camera)
                        Valid values are: 0-255
                        --------------------------------------------------
              EyeCam    (para1) frame rate in frames per second
                        (para2) not used
                        (para3) not used
                        Valid values are: FPS60, FPS30, FPS15,
                         FPS7_5, FPS3_75, FPS1_875, FPS0_9375, and FPS0_46875.
                         For the VV6300/VV6301, the default is FPS7_5.
                         For the OV6620, the default is FPS1_875.
                         For the OV7620, the default is FPS0_48375.
        Output:         NONE
        Semantics:      Set camera parameters


int CAMGet (int *para1, int *para2 ,int *para3);
        Note:           parameters have different meanings for different cameras
        Input:QuickCam  (para1) pointer for camera brightness
                        (para2) pointer for offset (b/w camera) / hue (color cam)
                        (para3) pointer for contrast (b/w cam) / sat. (color cam)
                        Valid values are: 0-255
                        --------------------------------------------------
              EyeCam    (para1) frame rate in frames per second
                        (para2) full image width
                        (para3) full image height
        Output:         NONE
        Semantics:      Get camera hardware parameters


int CAMMode (int mode);
        Input:          (mode) the camera mode you want
                        Valid values are: (NO)AUTOBRIGHTNESS
        Output:         NONE
        Semantics:      Set the display to the given mode
                        AUTOBRIGHTNESS: the brightness value of the
                          camera is automatically adjusted
                        NOAUTOBRIGHTNESS: the brightness value is not
                          automatically adjusted
```

B.5.5 System Functions

Miscellaneous system functions:

```
char *OSVersion (void);
Input:          NONE
Output:         OS version
Semantics:      Returns string containing running RoBIOS version.
Example:        "3.1b"

int OSError (char *msg,int number,BOOL dead);
Input:          (msg) pointer to message
                (number) int number
                (dead) switch to choose dead end or key wait
                Valid values are:      0 = no dead end
                                       1 = dead end
Output:         NONE
Semantics:      Print message and number to display then
                stop processor (dead end) or wait for key

int OSMachineType (void);
Input:          NONE
Output:         Type of used hardware
                Valid values are:
                VEHICLE, PLATFORM, WALKER
Semantics:      Inform the user in which environment the program runs.

int OSMachineSpeed (void);
Input:          NONE
Output:         actual clockrate of CPU in Hz
Semantics:      Inform the user how fast the processor runs.

char* OSMachineName (void);
Input:          NONE
Output:         Name of actual Eyebot
Semantics:      Inform the user with which name the Eyebot is
                titled (entered in HDT).

unsigned char OSMachineID (void);
Input:          NONE
Output:         ID of actual Eyebot
Semantics:      Inform the user with which ID the Eyebot is titled
                (entered in HDT).

void *HDTFindEntry(TypeID typeid,DeviceSemantics semantics);
       Input:   (typeid)   Type identifier tag of the category
                           (e.g. MOTOR, for a motor type)
                (semantics) Semantics itentifier tag (e.g. MOTOR_LEFT,
                           specifying which of several motors)
       Output:            Reference to matching HDT entry
       Semantics:         This function is used by device drivers to search for
                          first entry that matches the semantics and returns a
                          pointer to the corresponding data structure.
                          See HDT description in HDT.txt .
```

Interrupts:

```
int OSEnable (void);
Input:          NONE
Output:         NONE
Semantics:      Enable all cpu-interrupts

int OSDisable (void);
Input:          NONE
```

```
Output:         NONE
Semantics:      Disable all cpu-interrupts
```

Saving of variables in TPU-RAM (SAVEVAR1-3 occupied by RoBiOS):

```
int OSGetVar (int num);
Input:          (num) number of tpupram save location
Valid values:   SAVEVAR1-4 for word saving
                SAVEVAR1a-4a/1b-4b for byte saving

Output:         (returncode) the value saved
                  Valid values are:  0-65535 for word saving
                                     0-255 for byte saving
Semantics:      Get the value from the given save location

int OSPutVar (int num, int value);
Input:          (num) number of tpupram save location
                valid values are: SAVEVAR1-4 for word saving
                                  SAVEVAR1a-4a/1b-4b for byte saving
                (value) value to be stored
                Valid values are: 0-65535 for word saving
                                  0-255 for byte saving
Output:         NONE
Semantics:      Save the value to the given save location
```

B.5.6 Multitasking

RoBiOS implements both preemptive and cooperative multitasking. One of these modes needs to be selected when initializing multitasking operation.

```
int OSMTInit (BYTE mode);
Input:          (mode) operation mode
                 Valid values are: COOP=DEFAULT,PREEMPT
Output:         NONE
Semantics:      Initialize multithreading environment

tcb *OSSpawn (char *name,int code,int stksiz,int pri,int uid);
Input:          (name) pointer to thread name
                (code) thread start address
                (stksize) size of thread stack
                (pri) thread priority
                Valid values are: MINPRI-MAXPRI
                (uid) thread user id
Output:         (returncode) pointer to initialized thread
                control block
Semantics:      Initialize new thread, tcb is initialized and
                inserted in scheduler queue but not set to
                READY

int OSMTStatus (void);
Input:          NONE
Output:         PREEMPT, COOP, NOTASK
Semantics:      returns actual multitasking mode (preemptive,
                cooperative or sequential)

int OSReady (struct tcb *thread);
Input:          (thread) pointer to thread control block
Output:         NONE
Semantics:      Set status of given thread to READY

int OSSuspend (struct tcb *thread);
Input:          (thread) pointer to thread control block
```

```
Output:         NONE
Semantics:      Set status of given thread to SUSPEND

int OSReschedule (void);
Input:          NONE
Output:         NONE
Semantics:      Choose new current thread

int OSYield (void);
Input:          NONE
Output:         NONE
Semantics:      Suspend current thread and reschedule

int OSRun (struct tcb *thread);
Input:          (thread) pointer to thread control block
Output:         NONE
Semantics:      READY given thread and reschedule

int OSGetUID (thread);
Input:          (thread) pointer to thread control block
                (tcb *)0 for current thread
Output:         (returncode) UID of thread
Semantics:      Get the UID of the given thread

int OSKill (struct tcb *thread);
Input:          (thread) pointer to thread control block
Output:         NONE
Semantics:      Remove given thread and reschedule

int OSExit (int code);
Input:          (code) exit code
Output:         NONE
Semantics:      Kill current thread with given exit code and message

int OSPanic (char *msg);
Input:          (msg) pointer to message text
Output:         NONE
Semantics:      Dead end multithreading error, print message to display
                and stop processor

int OSSleep (int n)
Input:          (n) number of 1/100 secs to sleep
Output:         NONE
Semantics:      Let current thread sleep for at least n*1/100
                seconds.  In multithreaded mode, this will
                reschedule another thread.  Outside
                multi-threaded mode, it will call OSWait().

int OSForbid (void)
Input:          NONE
Output:         NONE
Semantics:      disable thread switching in preemptive mode

int OSPermit (void)
Input:          NONE
Output:         NONE
Semantics:      enable thread switching in preemptive mode
```

In the functions described above the parameter "thread" can always be a pointer to a tcb or 0 for current thread.

Semaphores:

```
int OSSemInit (struct sem *sem, int val);
Input:          (sem) pointer to a semaphore
```

```
                           (val) start value
Output:              NONE
Semantics:           Initialize semaphore with given start value

int OSSemP (struct sem *sem);
Input:               (sem) pointer to a semaphore
Output:              NONE
Semantics:           Do semaphore P (down) operation

int OSSemV (struct sem *sem);
Input:               (sem) pointer to a semaphore
Output:              NONE
Semantics:           Do semaphore V (up) operation
```

B.5.7 Timer

```
int OSSetTime (int hrs,int mins,int secs);
Input:               (hrs) value for hours
                     (mins) value for minutes
                     (secs) value for seconds
Output:              NONE
Semantics:           Set system clock to given time

int OSGetTime (int *hrs,int *mins,int *secs,int *ticks);
Input:               (hrs) pointer to int for hours
                     (mins) pointer to int for minutes
                     (secs) pointer to int for seconds
                     (ticks) pointer to int for ticks
Output:              (hrs) value of hours
                     (mins) value of minutes
                     (secs) value of seconds
                     (ticks) value of ticks
Semantics:           Get system time, one second has 100 ticks

int OSShowTime (void);
Input:               NONE
Output:              NONE
Semantics:           Print system time to display

int OSGetCount (void);
Input:               NONE
Output:              (returncode) number of 1/100 seconds since last reset
Semantics:           Get the number of 1/100 seconds since last reset.
                     Type int is 32 bits, so this value will wrap
                     around after ~248 days.

int OSWait (int n);
Input:               (n) time to wait
Output:              NONE
Semantics:           Busy loop for n*1/100 seconds.


Timer-IRQ:

TimerHandle OSAttachTimer (int scale, TimerFnc function);
Input:               (scale) prescale value for 100Hz Timer (1 to ...)
                     (TimerFnc) function to be called periodically
Output:              (TimerHandle) handle to reference the IRQ-slot
                     A value of 0 indicates an error due to a full list
                     (max. 16).
Semantics:           Attach an irq-routine (void function(void)) to the irq-list.
                     The scale parameter adjusts the call frequency (100/scale Hz)
```

of this routine to allow many different applications.

```
int OSDetachTimer (TimerHandle handle)
Input:          (handle) handle of a previous installed timer irq
Output:         0 = handle not valid
                1 = function successfully removed from timer irq list
Semantics:      Detach a previously installed irq-routine from the irq-list.
```

B.5.8 Serial Communication (RS232)

```
int OSDownload (char *name,int *bytes,int baud,int handshake,int interface);
Input:          (name) pointer to program name array
                (bytes) pointer to bytes transferred int
                (baud) baud rate selection
                Valid values are: SER4800, SER9600,SER19200,SER38400,
                SER57600, SER115200
                (handshake) handshake selection
                Valid values are: NONE,RTSCTS
                (interface): serial interface
                 Valid values are: SERIAL1-3
Output:         (returncode)
                 0 = no error, download incomplete - call again
                99 = download complete
                 1 = receive timeout error
                 2 = receive status error
                 3 = send timeout error
                 5 = srec checksum error
                 6 = user canceled error
                 7 = unknown srecord error
                 8 = illegal baud rate error
                 9 = illegal startadr. error
                10 = illegal interface

Semantics:      Load user program with the given serial setting
                and get name of program.  This function must
                be called in a loop until the returncode is
                !=0. In the loop the bytes that have been
                transferred already can be calculated from the
                bytes that have been transferred in this round.
                Note: do not use in application programs.

int OSInitRS232 (int baud,int handshake,int interface);
Input:          (baud) baud rate selection
                Valid values are:
                SER4800,SER9600,SER19200,SER38400,SER57600,SER115200
                (handshake) handshake selection
                Valid values are: NONE,RTSCTS
                (interface) serial interface
                Valid values are: SERIAL1-3
Output:         (returncode)
                 0 = ok
                 8 = illegal baud rate error
                10 = illegal interface
Semantics:      Initialize rs232 with given setting

int OSSendCharRS232 (char chr,int interface);
Input:          (chr) character to send
                (interface) serial interface
                Valid values are: SERIAL1-3
Output:         (returncode)
                 0 = good
                 3 = send timeout error
```

```
                    10 = illegal interface
Semantics:          Send a character over rs232

int OSSendRS232 (char *chr,int interface);
Input:              (chr) pointer to character to send
                    (interface) serial interface
                    Valid values are: SERIAL1-3
Output:             (returncode)
                    0 = good
                    3 = send timeout error
                    10 = illegal interface
Semantics:          Send a character over rs232.  Use OSSendCharRS232()
                    instead.  This function will be removed in the future.

int OSRecvRS232 (char *buf,int interface);
Input:              (buf) pointer to a character array
                    (interface) serial interface
                    Valid values are: SERIAL1-3
Output:             (returncode)
                    0 = good
                    1 = receive timeout error
                    2 = receive status error
                    10 = illegal interface
Semantics:          Receive a character over rs232

int OSFlushInRS232 (int interface);
Input:              (interface) serial interface
                    Valid values are: SERIAL1-3
Output:             (returncode)
                    0 = good
                    10 = illegal interface
Semantics:          resets status of receiver and flushes its
                    FIFO. Very useful in NOHANDSHAKE-mode to bring
                    the FIFO in a defined condition before
                    starting to receive

int OSFlushOutRS232 (int interface);
Input:              (interface) serial interface
                    Valid values are: SERIAL1-3
Output:             (returncode)
                    0 = good
                    10 = illegal interface
Semantics:          flushes the transmitter-FIFO.  very useful to abort
                    current transmission to host (E.g.: in the case
                    of a not responding host)

int OSCheckInRS232 (int interface);
Input:              (interface) serial interface
                    Valid values are: SERIAL1-3
Output:             (returncode) >0 : the number of chars currently
                                        available in FIFO
                    <0 : 0xffffff02 receive status error
                          (no chars available)
                    0xffffff0a illegal interface
Semantics:          useful to read out only packages of a certain size

int OSCheckOutRS232 (int interface);
Input:              (interface) serial interface
                    Valid values are: SERIAL1-3
Output:             (returncode) >0 : the number of chars currently
                    waiting in FIFO
                    <0 : 0xffffff0a illegal interface
Semantics:          useful to test if the host is receiving
                    properly or to time transmission of packages
                    in the speed the host can keep up with
```

```
int USRStart (void);
Input:          NONE
Output:         NONE
Semantics:      Start loaded user program.
                Note: do not use in application programs.


int USRResident (char *name, BOOL mode);
Input:          (name) pointer to name array
                (mode) mode
                Valid values are: SET,GET
Output:         NONE
Semantics:      Make loaded user program reset resistant
                SET   save startaddress and program name.
                GET   restore startaddress and program name.
                Note: do not use in application programs.
```

B.5.9 Audio

```
Audio files can be generated with a conversion program on a PC.
Sampleformat: WAV or AU/SND (8bit, pwm or mulaw)
Samplerate: 5461, 6553, 8192, 10922, 16384, 32768 (Hz)
Tonerange: 65 Hz to 21000 Hz
Tonelength: 1 msec to 65535 msecs


int AUPlaySample (char* sample);
Input:          (sample) pointer to sample data
Output:         (returncode) playfrequency for given sample
                0 if unsupported sampletype
Semantics:      Plays a given sample (nonblocking)
                supported formats are:
                WAV or AU/SND  (8bit, pwm or mulaw)
                5461, 6553, 8192, 10922, 16384, 32768 (Hz)


int AUCheckSample (void);
Input:          NONE
Output:         FALSE while sample is playing
Semantics:      nonblocking test for sampleend


int AUTone (int freq, int msec);
Input:          (freq) tone frequency
                (msecs) tone length
Output:         NONE
Semantics:      Plays tone with given frequency for the given
                time (nonblocking)
                supported formats are:
                freq = 65 Hz to 21000 Hz
                msecs = 1 msec to 65535 msecs


int AUCheckTone (void);
Input:          NONE
Output:         FALSE while tone is playing
Semantics:      nonblocking test for toneend


int AUBeep (void);
Input:          NONE
Output:         NONE
Semantics:      BEEP!


int AURecordSample (BYTE* buf, long len, long freq);
Input:          (buf) pointer to buffer
                (len) bytes to sample + 28 bytes header
```

```
                         (freq) desired samplefrequency
Output:                  (returncode) real samplefrequency
Semantics:               Samples from microphone into buffer with given
                         frequency (nonblocking)
                         Recordformat: AU/SND (pwm) with unsigned 8bit samples


int AUCheckRecord (void);
Input:                   NONE
Output:                  FALSE while recording
Semantics:               nonblocking test for recordend


int AUCaptureMic (void);
Input:                   NONE
Output:                  (returncode) microphone value (10bit)
Semantics:               Get microphone input value
```

B.5.10 Position Sensitive Devices (PSDs)

```
Position Sensitive Devices (PSDs) use infrared beams to measure
distance. The accuracy varies from sensor to sensor, and they need to
be calibrated in the HDT to get correct distance readings.

PSDHandle PSDInit (DeviceSemantics semantics);
Input:                   (semantics) unique definition for desired PSD (see hdt.h)
Output:                  (returncode) unique handle for all further operations
Semantics:               Initialize single PSD with given name (semantics)
                         Up to 8 PSDs can be initialized


int PSDRelease (void);
Input:                   NONE
Output:                  NONE
Semantics:               Stops all measurings and releases all initialized PSDs


int PSDStart (PSDHandle bitmask, BOOL cycle);
Input:                   (bitmask) sum of all handles to which parallel
                          measuring should be applied
                         (cycle)   TRUE  = continuous measuring
                                   FALSE = single measuring
Output:                  (returncode) status of start-request
                          -1 = error (false handle)
                           0 = ok
                           1 = busy (another measuring blocks driver)
Semantics:               Starts a single/continuous PSD-measuring.  Continuous
                         gives new measurement ca. every 60ms.


int PSDStop (void);
Input:                   NONE
Output:                  NONE
Semantics:               Stops actual continuous PSD-measuring after
                         completion of the current shot


BOOL PSDCheck (void);
Input:                   NONE
Output:                  (returncode) TRUE if a valid result is available
Semantics:               nonblocking test if a valid PSD-result is available


int PSDGet (PSDHandle handle);
Input:                   (handle) handle of the desired PSD
                          0 for timestamp of actual measure-cycle
Output:                  (returncode) actual distance in mm (converted through
                         internal table)
Semantics:               Delivers actual timestamp or distance measured by
```

the selected PSD. If the raw reading is out of
range for the given sensor, PSD_OUT_OF_RANGE(=9999)
is returned.

```
int PSDGetRaw (PSDHandle handle);
Input:          (handle) handle of the desired PSD
                0 for timestamp of actual measure-cycle
Output:         (returncode) actual raw-data (not converted)
Semantics:      Delivers actual timestamp or raw-data measured by
                the selected PSD
```

B.5.11 Servos and Motors

```
ServoHandle SERVOInit (DeviceSemantics semantics);
Input:          (semantics) semantic (see hdt.h)
Output:         (returncode) ServoHandle
Semantics:      Initialize given servo
```

```
int SERVORelease (ServoHandle handle)
Input:          (handle) sum of all ServoHandles which should be released
Output:         (returncode)
                0 = ok
                errors (nothing is released):
                0x11110000 = totally wrong handle
                0x0000xxxx = the handle parameter in which only those bits
                             remain set that are connected to a releasable
                             TPU-channel
Semantics:      Release given servos
```

```
int SERVOSet (ServoHandle handle,int angle);
Input:          (handle) sum of all ServoHandles which should be set parallel
                (angle) servo angle
                Valid values: 0-255
Output:         (returncode)
                0 = ok
                -1 = error wrong handle
Semantics:      Set the given servos to the same given angle
```

```
MotorHandle MOTORInit (DeviceSemantics semantics);
Input:          (semantics) semantic (see hdt.h)
Output:         (returncode) MotorHandle
Semantics:      Initialize given motor
```

```
int MOTORRelease (MotorHandle handle)
Input:          (handle) sum of all MotorHandles which should be released
Output:         (returncode)
                0 = ok
                errors (nothing is released):
                0x11110000 = totally wrong handle
                0x0000xxxx = the handle parameter in which only those bits
                             remain set that are connected to a releasable
                             TPU-channel
Semantics:      Release given motor
```

```
int MOTORDrive (MotorHandle handle,int speed);
Input:          (handle) sum of all MotorHandles which should be driven
                (speed) motor speed in percent
                Valid values: -100 - 100 (full backward to full forward)
                              0 for full stop
Output:         (returncode)
                0 = ok
                -1 = error wrong handle
```

```
Semantics:       Set the given motors to the same given speed


QuadHandle QuadInit (DeviceSemantics semantics);
Input:           (semantics) semantic
Output:          (returncode) QuadHandle or 0 for error
Semantics:       Initialize given Quadrature-Decoder (up to 8 decoders are
                 possible)


int QuadRelease (QuadHandle handle);
Input:           (handle) sum of decoder-handles to be released
Output:           0 = ok
                 -1 = error wrong handle
Semantics:       Release one or more Quadrature-Decoder


int QuadReset (QuadHandle handle);
Input:           (handle) sum of decoder-handles to be reset
Output:           0 = ok
                 -1 = error wrong handle
Semantics:       Reset one or more Quadrature-Decoder


int QuadRead (QuadHandle handle);
Input:           (handle) ONE decoder-handle
Output:          32bit counter-value (0 to 2^32-1)
                 a wrong handle will ALSO result in an 0 counter-value!!
Semantics:       Read actual Quadrature-Decoder counter


DeviceSemantics QUADGetMotor (DeviceSemantics semantics);
Input:           (handle) ONE decoder-handle
Output:          semantic of the corresponding motor
                  0 = wrong handle
Semantics:       Get the semantic of the corresponding motor


float QUADODORead (QuadHandle handle);
Input:           (handle) ONE decoder-handle
Output:          meters since last odometer-reset
Semantics:       Get the distance from the last resetpoint of a single motor!
                 It is not the overall meters driven since the last reset!
                 It is just the nr of meters left to go back to the startpoint.
                 Useful to implement a PID-control


int QUADODOReset (QuadHandle handle);
Input:           (handle) sum of decoder-handles to be reset
Output:           0 = ok
                 -1 = error wrong handle
Semantics:       Resets the simple odometer(s) to define the startpoint
```

B.5.12 Driving Interface vω

```
This is a high level wheel control API using the motor and quad primitives to
drive the robot.

Data Types:
     typedef float meterPerSec;
     typedef float radPerSec;
     typedef float meter;
     typedef float radians;

     typedef struct
     { meter x;
       meter y;
```

```
            radians phi;
        } PositionType;

        typedef struct
        { meterPerSec v;
          radPerSec w;
        } SpeedType;
```

VWHandle **VWInit** (DeviceSemantics semantics, int Timescale);
Input: (semantics) semantic
                    (Timescale) prescale value for 100Hz IRQ (1 to ...)
Output: (returncode) VWHandle or 0 for error
Semantics: Initialize given VW-Driver (only 1 can be initialized!)
                    The motors and encoders are automatically reserved!!
                    The Timescale allows to adjust the tradeoff between
                    accuracy (scale=1, update at 100Hz) and speed(scale>1,
                    update at 100/scale Hz).

int **VWRelease** (VWHandle handle);
Input: (handle) VWHandle to be released
Output: 0 = ok
                    -1 = error wrong handle
Semantics: Release VW-Driver, stop motors

int **VWSetSpeed** (VWHandle handle, meterPerSec v, radPerSec w);
Input: (handle) ONE VWHandle
                    (v) new linear speed
                    (w) new rotation speed
Output: 0 = ok
                    -1 = error wrong handle
Semantics: Set the new speed: v(m/s) and w(rad/s not degree/s)

int **VWGetSpeed** (VWHandle handle, SpeedType* vw);
Input: (handle) ONE VWHandle
                    (vw) pointer to record to store actual v, w values
Output: 0 = ok
                    -1 = error wrong handle
Semantics: Get the actual speed: v(m/s) and w(rad/s not degree/s)

int **VWSetPosition** (VWHandle handle, meter x, meter y, radians phi);
Input: (handle) ONE VWHandle
                    (x) new x-position
                    (y) new y-position
                    (phi) new heading
Output: 0 = ok
                    -1 = error wrong handle
Semantics: Set the new position: x(m), y(m) phi(rad not degree)

int **VWGetPosition** (VWHandle handle, PositionType* pos);
Input: (handle) ONE VWHandle
                    (pos) pointer to record to store actual position (x,y,phi)
Output: 0 = ok
                    -1 = error wrong handle
Semantics: Get the actual position: x(m), y(m) phi(rad not degree)

int **VWStartControl** (VWHandle handle, float Vv, float Tv, float Vw, float Tw);
Input: (handle) ONE VWHandle
                    (Vv) the parameter for the proportional component of the
                    v-controller
                    (Tv) the parameter for the integrating component of the
                    v-controller
                    (Vw) the parameter for the proportional component of the
                    w-controller
                    (Tv) the parameter for the integrating component of the
                    w-controller

```
Output:              0 = ok
                     -1 = error wrong handle
Semantics:           Enable the PI-controller for the vw-interface and set
                     the parameters.
                     As default the PI-controller is deactivated when the
                     vw-interface is initialized. The controller tries to keep the
                     desired speed (set with VWSetSpeed) stable by adapting the
                     energy of the involved motors.
                     The parameters for the controller have to be choosen carefully!
                     The formula for the controller is:
```

$$new(t) = V*(diff(t) + 1/T * \int_0^t diff(t)dt)$$

```
                     V: a value usually around 1.0
                     T: a value usually between 0 and 1.0
                     After enabling the controller the last set speed (VWSetSpeed)
                     is taken as the speed to be held stable.


int VWStopControl (VWHandle handle);
Input:               (handle) ONE VWHandle
Output:              0 = ok
                     -1 = error wrong handle
Semantics:           Disable the controller immediately. The vw-interface continues
                     normally with the last valid speed of the controller.


int VWDriveStraight (VWHandle handle, meter delta, meterpersec v)
Input:               (handle) ONE VWHandle
                     (delta)  distance to drive in m (pos. -> forward)
                                                      (neg. -> backward)
                     (v)      speed to drive with (always positive!)
Output:              0 = ok
                     -1 = error wrong handle
Semantics:           Drives distance "delta" with speed v straight ahead
                     (forward or backward).
                     Any subsequent call of VWDriveStraight, -Turn, -Curve or
                     VWSetSpeed, while this one is still being executed, results in
                     an immediate interruption of this command


int VWDriveTurn (VWHandle handle, radians delta, radPerSec w)
Input:               (handle) ONE VWHandle
                     (delta)  degree to turn in radians (pos. -> counter-clockwise)
                                                         (neg. -> clockwise)
                     (w)      speed to turn with (always positive!)
Output:              0 = ok
                     -1 = error wrong handle
Semantics:           turns about "delta" with speed w on the spot (clockwise
                     or counter-clockwise)
                     any subsequent call of VWDriveStraight, -Turn, -Curve or
                     VWSetSpeed, while this one is still being executed, results in
                     an immediate interruption
                     of this command


int VWDriveCurve (VWHandle handle, meter delta_l, radians delta_phi,
                  meterpersec v)
Input:               (handle)    ONE VWHandle
                     (delta_l)   length of curve_segment to drive in m
                     (pos. -> forward)
                     (neg. -> backward)
                     (delta_phi) degree to turn in radians
                     (pos. -> counter-clockwise)
                     (neg. -> clockwise)
                     (v)            speed to drive with (always positive!)
Output:              0 = ok
                     -1 = error wrong handle
Semantics:           drives a curve segment of length "delta_l" with overall vehicle
```

403

turn of "delta_phi"
with speed v (forw. or backw. / clockw. or counter-clockw.).
any subsequent call of VWDriveStraight, -Turn, -Curve or
VWSetSpeed, while this one is still being executed,
results in an immediate interruption of this command

float **VWDriveRemain** (VWHandle handle)
Input: (handle) ONE VWHandle
Output: 0.0 = previous VWDriveX command has been completed
                any other value = remaining distance to goal
Semantics: remaining distance to goal set by VWDriveStraight, -Turn
                (for -Curve only the remaining part of delta_l is reported)

int **VWDriveDone** (VWHandle handle)
Input: (handle) ONE VWHandle
Output: -1 = error wrong handle
                 0 = vehicle is still in motion
                 1 = previous VWDriveX command has been completed
Semantics: checks if previous VWDriveX() command has been completed

int **VWDriveWait** (VWHandle handle)
Input: (handle) ONE VWHandle
Output: -1 = error wrong handle
                 0 = previous VWDriveX command has been completed
Semantics: blocks the calling process until the previous VWDriveX()
                command has been completed

int **VWStalled** (VWHandle handle)
Input: (handle) ONE VWHandle
Output: -1 = error wrong handle
                 0 = vehicle is still in motion or
                     no motion command is active
                 1 = at least one vehicle motor is stalled during
                     VW driving command
Semantics: checks if at least one of the vehicle's motors is stalled
                right now

B.5.13 Bumper and Infrared Sensors

Tactile bumpers and infrared proximity sensors have been used in some previous
robot models. They are currently not used for the SoccerBots, but may be used,
e.g. for integrating additional sensors.

BumpHandle **BUMPInit** (DeviceSemantics semantics);
Input: (semantics) semantic
Output: (returncode) BumpHandle or 0 for error
Semantics: Initialize given bumper (up to 16 bumpers are possible)

int **BUMPRelease** (BumpHandle handle);
Input: (handle) sum of bumper-handles to be released
Output: (returncode)
                0 = ok
                errors (nothing is released):
                0x11110000 = totally wrong handle
                0x0000xxxx = the handle parameter in which only those
                            bits remained set that are connected to a releasable
                            TPU-channel
Semantics: Release one or more bumper

int **BUMPCheck** (BumpHandle handle, int* timestamp);
Input: (handle) ONE bumper-handle
                (timestamp) pointer to an int where the timestamp is placed

```
Output:             (returncode)
                    0 = bump occurred, in *timestamp is now a valid stamp
                    -1 = no bump occurred or wrong handle, *timestamp is cleared
Semantics:          Check occurrence of a single bump and return the
                    timestamp(TPU).
The first bump is recorded and held until BUMPCheck is called.

IRHandle IRInit (DeviceSemantics semantics);
Input:              (semantics) semantic
Output:             (returncode) IRHandle or 0 for error
Semantics:          Initialize given IR-sensor (up to 16 sensors are possible)

int IRRelease (IRHandle handle);
Input:              (handle) sum of IR-handles to be released
Output:             (returncode)
                    0 = ok
                    errors (nothing is released):
                    0x11110000 = totally wrong handle
                    0x0000xxxx = the handle parameter in which only those bits
                    remain set that are connected to a releasable TPU-channel
Semantics:          Release one or more IR-sensors

int IRRead (IRHandle handle);
Input:              (handle) ONE IR-handle
Output:             (returncode)
                    0/1 = actual pinstate of the TPU-channel
                    -1 = wrong handle
Semantics:          Read actual state of the IR-sensor
```

B.5.14 Latches

```
Latches are low-level IO buffers.

BYTE OSReadInLatch (int latchnr);
Input:              (latchnr) number of desired Inlatch (range: 0..3)
Output:             actual state of this inlatch
Semantics:          reads contents of selected inlatch

BYTE OSWriteOutLatch (int latchnr, BYTE mask, BYTE value);
Input:              (latchnr) number of desired Outlatch (range: 0..3)
                    (mask)    and-bitmask of pins which should be cleared
                    (inverse!)
                    (value)   or-bitmask of pins which should be set
Output:             previous state of this outlatch
Semantics:          modifies an outlatch and keeps global state consistent
                    example: OSWriteOutLatch(0, 0xF7, 0x08); sets bit4
                    example: OSWriteOutLatch(0, 0xF7, 0x00); clears bit4

BYTE OSReadOutLatch (int latchnr);
Input:              (latchnr) number of desired Outlatch (range: 0..3)
Output:             actual state of this outlatch
Semantics:          reads global copy of outlatch
```

B.5.15 Parallel Port

```
BYTE OSReadParData (void);
Input:              NONE
Output:             actual state of the 8bit dataport
Semantics:          reads contents of parallelport (active high)
```

```
void OSWriteParData (BYTE value);
Input:          (value) new output-data
Output:         NONE
Semantics:      writes out new data to parallelport (active high)

BYTE OSReadParSR (void);
Input:          NONE
Output:         actual state of the 5 statuspins
Semantics:      reads state of the 5 statuspins active-high!
                (BUSY(4), ACK(3), PE(2), SLCT(1), ERROR(0)):

void OSWriteParCTRL (BYTE value);
Input:          (value) new ctrl-pin-output (4bits)
Output:         NONE
Semantics:      writes out new ctrl-pin-states active high!
                (SLCTIN(3), INT(2), AUTOFDXT(1), STROBE(0))

BYTE OSReadParCTRL (void);
Input:          NONE
Output:         actual state of the 4 ctrl-pins
Semantics:      reads state of the 4 ctrl-pins active-high!
                (SLCTIN(3), INT(2), AUTOFDXT(1), STROBE(0))
```

B.5.16 Analog-Digital Converter

```
int OSGetAD (int channel);
Input:          (channel) desired AD-channel range: 0..15
Output:         (returncode) 10 bit sampled value
Semantics:      Captures one single 10bit value from specified
                AD-channel

int OSOffAD (int mode);
Input:          (mode) 0 = full powerdown
                       1 = fast powerdown
Output:         none
Semantics:      Powers down the 2 AD-converters (saves energy)
                A call of OSGetAD awakens the AD-converter again
```

B.5.17 Radio Communication

Note: Additional hardware and software (*Radio-Key*) are required to use these library routines.
"EyeNet" network among arbitrary number of EyeBots and optional workstation host. Network operates as virtual token ring and has fault tolerant aspects. A net Master is negotiated autonomously, new EyeBots will automatically be integrated into the net by "wildcard" messages, and dropped out EyeBots will be eliminated from the network. This network uses a RS232 interface and can be run over cable or wireless.

The communication is 8-bit clean and all packets are sent with checksums to detect transmission errors. The communication is unreliable, meaning there is no retransmit on error and delivery of packets are not guaranteed.

```
int RADIOInit (void);
Input:          none
Output:         returns 0 if OK
Semantics:      Initializes and starts the radio communication.
```

```
int RADIOTerm (void);
Input:          none
Output:         returns 0 if OK
Semantics:      Terminate network operation.

int RADIOSend (BYTE id, int byteCount, BYTE* buffer);
Input:          (id) the EyeBot ID number of the message destination
                (byteCount) message length
                (buffer)    message contents
Output:         returns 0 if OK
                returns 1 if send buffer is full or message is too long.
Semantics:      Send message to another EyeBot. Send is buffered,
                so the sending process can continue while the
                message is sent in the background.  Message
                length must be below or equal to MAXMSGLEN.
                Messages are broadcasted by sending them to
                the special id BROADCAST.

int RADIOCheck (void);
Input:          none
Output:         returns the number of user messages in the buffer
Semantics:      Function returns the number of buffered messages.
                This function should be called before
                receiving, if blocking is to be avoided.

int RADIORecv (BYTE* id, int* bytesReceived, BYTE* buffer);
Input:          none
Output:         (id) EyeBot ID number of the message source
                (bytesReceived) message length
                (buffer) message contents
Semantics:      Returns the next message buffered. Messages are
                returned in the order they are
                received. Receive will block the calling
                process if no message has been received until
                the next one comes in.  The buffer must have
                room for MAXMSGLEN bytes.

Data Type:
  struct RadioIOParameters_s{
    int interface;    /* SERIAL1, SERIAL2 or SERIAL3 */
    int speed;        /* SER4800,SER9600,SER19200,SER38400,SER57600,SER115200*/
    int id;           /* machine id */
    int remoteOn;     /* non-zero if remote control is active */
    int imageTransfer; /* if remote on: 0 off, 2 full, 1 reduced */
    int debug;        /* 0 off, 1..100 level of debugging spew */
  };


void RADIOGetIoctl (RadioIOParameters* radioParams);
Input:          none
Output:         (radioParams) current radio parameter settings
Semantics:      Reads out current radio parameter settings.

void RADIOSetIoctl (RadioIOParameters* radioParams);
Input:          (radioParams) new radio parameter settings
Output:         none
Semantics:      Changes radio parameter settings.  This should
                be done before calling RADIOInit().

int RADIOGetStatus(RadioStatus *status);
        Input:          NONE
        Output:         (status) current radio communication status.
        Semantics:      Return current status info from RADIO communication.
```

B.5.18 Compass

These routines provide an interface to a digital compass.

Sample HDT Setting:
```
compass_type compass = {0,13,(void*)OutBase, 5,(void*)OutBase, 6,
(BYTE*)InBase, 5};

HDT_entry_type HDT[] =
{ ...
  {COMPASS,COMPASS,"COMPAS",(void *)&compass},
  ...
};
```

```
int COMPASSInit(DeviceSemantics semantics);
```
        Input: Unique definition for desired COMPASS (see hdt.h)
        Output: (return code) 0 = OK
                                      1 = error
        Semantics: Initialize digital compass device

```
int COMPASSStart(BOOL cycle);
```
        Input: (cycle) 1 for cyclic mode
                                0 for single measurement
        Output: (return code) 1 = module has already been started
                                      0 = OK
        Semantics: This function starts the measurement of the actual
                        heading. The cycle parameter chooses the operation mode
                        of the compass-module.
                        In cyclic mode (1), the compass delivers as fast as
                        possible the actual heading without pause. In normal mode
                        (0) a single measurement is requested and allows the
                        module to go back to sleep mode afterwards.

```
int COMPASSCheck();
```
        Input: NONE
        Output: (return code) 1 = result is ready
                                      0 = result is not yet ready
        Semantics: If a single shot was requested this function allows to
                        check if the result is already available. In the cyclic
                        mode this function is useless because it always indicates
                        'busy'. Usually a user uses a loop to wait for a result:
                            int heading;
                            COMPASSStart(FALSE);
                            while(!COMPASSCheck());
                              //In single tasking! Otherwise yield to other tasks
                            heading = COMPASSGet();

```
int COMPASSStop();
```
        Input: NONE
        Output: (return code) 0 = OK
                                      1 = error
        Semantics: To stop the initiated cyclic measurement this function
                        WAITS for the current measurement to be finished and
                        stops the module. This function therefore will
                        return after 100msec at latest or will deadlock if no
                        compass module is connected to the EyeBot!

```
int COMPASSRelease();
```
        Input: NONE
        Output: (return code) 0 = OK
                                      1 = error
        Semantics: This function shuts down the driver and aborts any
                        ongoing measurement directly.

```
int COMPASSGet();
        Input:          NONE
        Output:         (return code) Compass heading data: [0..359]
                                     -1 = no heading has been calculated yet
                                          (wait after initializing).
        Semantics:      This function delivers the actual compass heading.

int COMPASSCalibrate(int mode);
        Input:          (mode) 0 to reset calibration data of compass module
                                 (requires about 0.8s)
                               1 to perform normal calibration.
        Output:         (return code) 0 = OK
                                      1 = error
        Semantics:      This function has two tasks. With mode=0 it resets the
                        calibration data of the compass module. With mode=1 the
                        normal calibration is performed. It has to be called
                        twice (first at any position, second at 180degree to the
                        first position).
                        Normally you will perform the following steps:
                          COMPASSCalibrate(1);
                          VWDriveTurn(VWHandle handle, M_PI, speed);
                              // turn EyeBot 180deg in place
                          COMPASSCalibrate(1);
```

B.5.19 IR Remote Control

These commands allow sending commands to an EyeBot via a standard TV remote.

```
Include:
#include "irtv.h"     /* only required for HDT files */
#include "IRu170.h"; /* depending on remote control, e.g. also "IRnokia.h" */


Sample HDT Setting:
/* infrared remote control on Servo S10 (TPU11)*/
/* SupportPlus 170 */
irtv_type irtv =     {1, 13, TPU_HIGH_PRIO, REMOTE_ON,
                      MANCHESTER_CODE, 14, 0x0800, 0x0000, DEFAULT_MODE, 4,300,
                      RC_RED, RC_YELLOW, RC_BLUE, 0x303C};

/* NOKIA */
irtv_type irtv =     {1, 13, TPU_HIGH_PRIO, REMOTE_ON,
                      SPACE_CODE,      15, 0x0000, 0x03FF, DEFAULT_MODE, 1,  -1,
                      RC_RED, RC_GREEN, RC_YELLOW, RC_BLUE};


HDT_entry_type HDT[] =
{ ...
  {IRTV,IRTV,"IRTV",(void *)&irtv},
  ...
};


int IRTVInitHDT(DeviceSemantics semantics);
        Input:          (semantics) unique def. for desired IRTV (see hdt.h)
        Output:         (return code) 0 = ok
                                      1 = illegal type or mode (in HDT IRTV entry)
                                      2 = invalid or missing "IRTV" HDT entry
                                             for this semantics
        Semantics:      Initializes the IR remote control decoder by calling
                        IRTVInit() with the parameters found in the correspond.
```

HDT entry. Using this function applications are indep.
of the used remote control since the defining param.
are located in the HDT.

```
int IRTVInit(int type, int length, int tog_mask, int inv_mask, int mode,
         int bufsize, int delay);
```
Input: (type)   the used code type
                                 Valid values are:
                                 SPACE_CODE, PULSE_CODE, MANCHESTER_CODE,
                                 RAW_CODE
                    (length)   code length (number of bits)
                    (tog_mask) bitmask that selects "toggle bits" in a code
                                 (bits that change when the same key is pressed
                                     repeatedly)
                    (inv_mask) bitmask that selects inverted bits in a code
                                 (for remote controls with alternating codes)
                    (mode)   operation mode
                                 Valid values are: DEFAULT_MODE, SLOPPY_MODE,
                                 REPCODE_MODE
                    (bufsize)   size of the internal code buffer
                    Valid values are: 1-4
                    (delay)   key repetition delay
                                 >0: number of 1/100 sec (should be >20)
                                 -1: no repetition
Output: (return code) 0 = ok
                                 1 = illegal type or mode
                                 2 = invalid or missing "IRTV" HDT entry
Semantics: Initializes the IR remote control decoder.
                    To find out the correct values for the "type", "length",
                    "tog_mask", "inv_mask" and "mode" parameters, use the IR
                    remote control analyzer program (IRCA).
                    SLOPPY_MODE can be used as alternative to DEFAULT_MODE.
                    In default mode, at least two consecutive identical code
                    sequences must be received before the code becomes
                    valid. When using sloppy mode, no error check is
                    performed, and every code becomes valid immediately.
                    This reduces the delay between pressing the key and
                    the reaction.
                    With remote controls that use a special repetition
                    coding, REPCODE_MODE must be used (as suggested by the
                    analyzer).

```
Typical param. | Nokia (VCN 620)    | RC5 (Philips)
---------------+--------------------+--------------
type           | SPACE_CODE         | MANCHESTER_CODE
length         | 15                 | 14
tog_mask       | 0                  | 0x800
inv_mask       | 0x3FF              | 0
mode           | DEFAULT_MODE /     | DEFAULT_MODE /
               | SLOPPY_MODE        | SLOPPY_MODE
```

                    The type setting RAW_CODE is intended for code analysis
                    only. If RAW_CODE is specified, all of the other
                    parameters should be set to 0. Raw codes must be handled
                    by using the IRTVGetRaw and IRTVDecodeRaw functions.

```
void IRTVTerm(void);
```
    Input: NONE
    Output: NONE
    Semantics: Terminates the remote control decoder and releases the
                    occupied TPU channel.

```
int IRTVPressed(void);
```
    Input: NONE
    Output: (return code) Code of the remote key that is currently

```
                              being pressed
                                      0 = no key
             Semantics:      Directly reads the current remote key code. Does not
                             touch the code buffer. Does not wait.


int IRTVRead(void);
             Input:          NONE
             Output:         (return code) Next code from the buffer
                                      0 = no key
             Semantics:      Reads and removes the next key code from code buffer.
                             Does not wait.


int IRTVGet(void);
             Input:          NONE
             Output:         (return code) Next code from the buffer (!=0)
             Semantics:      Reads and removes the next key code from  code buffer.
                           If the buffer is empty, the function waits until a remote
                             key is pressed.


void IRTVFlush(void);
             Input:          NONE
             Output:         NONE
             Semantics:      The code buffer is emptied.


void IRTVGetRaw(int bits[2], int *count, int *duration, int *id, int *clock);
             Input:          NONE
             Output:         (bits)        contains the raw code
                                           bit #0 in bits[0] represents the 1st pulse
                                                         in code sequence
                                           bit #0 in bits[1] represents the 1st space
                                           bit #1 in bits[0] represents the 2nd pulse
                                           bit #1 in bits[1] represents the 2nd space
                                           ...
                                           A cleared bit stands for a short signal,
                                           a set bit for a long signal.
                             (count)    number of signals (= pulses + spaces) received
                             (duration) the logical duration of the code sequence
                                           duration = (number of short signals) +
                                                             2*(num. of long signals)
                             (id)          a unique ID for the current code
                                           (incremented by 1 each time)
                             (clock)      the time when the code was received
             Semantics:      Returns information about the last received raw code.
                             Works only if type setting == RAW_CODE.


int IRTVDecodeRaw(const int bits[2], int count, int type);
             Input:          (bits)   raw code to be decoded (see IRTVGetRaw)
                             (count) number of signals (= pulses + spaces) in raw code
                             (type)   the decoding method
                                           Valid values are: SPACE_CODE, PULSE_CODE,
                                                            MANCHESTER_CODE
             Output:         (return code) The decoded value (0 on an illegal
                             Manchester code)
             Semantics:      Decodes the raw code using the given method.
```

Thomas Bräunl, Klaus Schmitt, Michael Kasper 1996-2006

C HARDWARE DESCRIPTION TABLE

C.1 HDT Overview

The Hardware Description Table (HDT) is the link between the RoBIOS operating system and the actual hardware configuration of a robot. This table allows us to run the same operating system on greatly varying robot structures with different mechanics and sensor/actuator equipment. Every sensor, every actuator, and all auxiliary equipment that is connected to the controller are listed in the HDT with its exact I/O pin and timing configurations. This allows us to change, for example, motor and sensor ports transparent to the user program – there is no need to even re-compile it. The HDT comprises:

- HDT access procedures
- HDT data structures

The HDT resides in the EyeCon's flash-ROM and can be updated by uploading a new HDT hex-file. Compilation of HDT files is done with the script `gcchdt` instead of the standard script `gcc68` for user programs.

The following procedures are part of RoBiOS and are used by hardware drivers to determine where and if a hardware component exists. These procedures cannot be called from a user program.

```
int HDT_Validate(void);
/* used by RoBiOS to check and initialize the HDT data structure. */

void *HDT_FindEntry(TypeID typeid, DeviceSemantics semantics);
/* used by device drivers to search for first entry that matches semantics and
returns pointer to the corresponding data structure. */

DeviceSemantics HDT_FindSemantics(TypeID typeid, int x);
/* look for xth entry of given Typ and return its semantics */

int HDT_TypeCount(TypeID typeid);
/* count entries of given Type */
```

```
char *HDT_GetString(TypeID typeid,DeviceSemantics semantics)
/* get semantic string */
```

The HDT data structure is a separate data file (sample sources in directory `hdtdata`). Each controller is required to have a compiled HDT file in ROM in order to operate.

Each HDT data file contains complete information about the connection and control details of all hardware components used in a specific system configuration. Each source file usually contains descriptions of all required data structures of HDT components, plus (at the end of the source file) the actual list of components, utilizing the previous definitions.

Example HDT data entry for a DC motor (see include file `hdt.h` for specific type and constant definitions):

```
motor_type motor0 = {2,   0, TIMER1, 8196, (void*)(OutBase+2), 6, 7,
                        (BYTE*)&motconv0};
2                 : the maximum driver version for which this entry is sufficient
0                 : the tpu channel the motor is attached to
TIMER2            : the tpu timer that has to be used
8196              : pwm period in Hz
OutBase+2         : the I/O Port address the driver has to use
6                 : the portbit for forward drive
7                 : the portbit for backward drive
motconv0          : the pointer to a conversion table to adjust different motors
```

The following example HDT list contains all hardware components used for a specific system configuration (entries `INFO` and `END_OF_HDT` are mandatory for all HDTs):

```
HDT_entry_type HDT[] =
{
    MOTOR,MOTOR_RIGHT,"RIGHT",(void *)&motor0,
    MOTOR,MOTOR_LEFT,"LEFT",(void *)&motor1,
    PSD,PSD_FRONT,"FRONT",(void *)&psd1,
    INFO,INFO,"INFO",(void *)&roboinfo,
    END_OF_HDT,UNKNOWN_SEMANTICS,"END",(void *)0
};
```

Explanations for first HDT entry:

```
MOTOR             : it is a motor
MOTOR_LEFT        : its semantics
"LEFT"            : a readable string for testroutines
&motor0           : a pointer to the motor0 data structure
```

From the user program point of view, the following describes how to make use of HDT entries, using the motor entry as an example. Firstly, a handle to the device has to be defined:

```
MotorHandle       leftmotor;
```

Next, the handle needs to be initialized by calling `MOTORInit` with the *semantics* (HDT name) of the motor. `MOTORInit` now searches the HDT for a motor with the given semantics and if found calls the motor driver to initialize the motor.

```
leftmotor = MOTORInit(LEFTMOTOR);
```

Now the motor can be used by the access routines provided, e.g. setting a certain speed. The following function calls the motor driver and sets the speed on a previously initialized motor:

```
MOTORDrive (leftmotor,50);
```

After finishing using a device (here: the motor), it is required to release it, so it can be used by other applications:

```
MOTORRelease (leftmotor);
```

Using the HDT entries for all other hardware components works in a similar way. See the following description of HDT information structures as well as the RoBIOS details in Appendix B.5.

C.2 Battery Entry

```
typedef struct
{
  int      version;
  short    low_limit;
  short    high_limit;
}battery_type;

e.g.
battery_type battery = {0,550,850};

int version:
The maximum driver version for which this entry is compatible.
Because newer drivers will surely need more information, this tag prevents this
driver from reading more information than actually available.

short low_limit:
The value the AD-converter channel 1 measures shortly before the batteries are
empty. This defines the lower limit of the tracked battery voltage.

short   high_limit:
The value the AD-converter channel 1 measures with fully loaded batteries.
This defines the upper limit of the tracked battery voltage.
```

C.3 Bumper Entry

```
typedef struct
{
    int      driver_version;
    int      tpu_channel;
    int      tpu_timer;
    short    transition;
}bump_type;

e.g.
bump_type bumper0 = {0, 6, TIMER2, EITHER};
```

```
int driver_version:
```
The maximum driver version for which this entry is compatible.
Because newer drivers will surely need more information, this tag prevents this
driver from reading more information than actually available.

```
int tpu_channel:
```
The tpu channel the bumper is attached to. Valid values are 0..15
Each bumper needs a tpu channel to signal a 'bump'-occurrence.

```
int tpu_timer:
```
The tpu timer that has to be used. Valid values are TIMER1, TIMER2
If a 'bump' is detected the corresponding timer-value is stored for later cal-
culations.
TIMER1 runs at a speed of 4MHz-8MHz (depending on CPUclock)
TIMER2 runs at a speed of 512kHz-1MHz (depending on CPUclock)

```
short transition:
```
React on a certain transition. Valid values are RISING, FALLING, EITHER
To alter the behaviour of the bumper, the type of transition the TPU reacts on
can be choosen.

C.4 Compass Entry

```
typedef struct
{
  short    version;
  short    channel;
  void*    pc_port;
  short    pc_pin;
  void*    cal_port;
  short    cal_pin;
  void*    sdo_port;
  short    sdo_pin;
}compass_type;
```

e.g.
```
compass_type compass = {0,13,(void*)IOBase, 2,(void*)IOBase, 4, (BYTE*)IOBase,
0};
```

```
short version:
```
The maximum driver version for which this entry is compatible.
Because newer drivers will surely need more information, this tag prevents this
driver from reading more information than actually available.

```
short channel:
```
TPU channel that is connected to the compass for clocking the data transfer.
Valid values are 0..15

```
void* pc_port:
```
Pointer to an 8Bit register/latch (out). PC is the start signal for the compass

```
short pc_pin:
```
This is the bit number in the register/latch addressed by pc_port. Valid values
are 0..7

```
void* cal_port:
```
Pointer to an 8Bit register/latch (out). CAL is the calibration start signal
for the compass.
It can be set to NULL if no calibration is needed (In this case never call the
calibration function).

```
short cal_pin:
```

This is the bitnumber in the register/latch addressed by cal_port. Valid values are 0..7

void* sdo_port:
Pointer to an 8Bit register/latch (in). SDO is the serial data output connection of the compass. The driver will read out the serial data timed by the TPU channel.

short sdo_pin:
This is the bitnumber in the register/latch addressed by sdo_port. Valid values are 0..7

C.5 Information Entry

```
typedef struct
{
    int     version;
    int     id;
    int     serspeed;
    int     handshake;
    int     interface;
    int     auto_download;
    int     res1;
    int     cammode;
    int     battery_display;
    int     CPUclock;
    float   user_version;
    String10 name;
    unsigned char res2;
}info_type;
```

```
e.g.
info_type roboinfo0  = {0,VEHICLE,SER115200,RTSCTS,SERIAL2,AUTOLOAD,0,
                        AUTOBRIGHTNESS,BATTERY_ON,16,VERSION,NAME,0};
```

int version:
The maximum driver version for which this entry is compatible.
Because newer drivers will surely need more information, this tag prevents this driver from reading more information than actually available.

int id:
The current environment on which RoBiOS is running. Valid values are PLATFORM, VEHICLE, WALKER
It is accessible via OSMachineType().

int serspeed:
The default baudrate for the default serial interface.
Valid values are SER9600, SER19200, SER38400, SER57600 SER115200

int handshake:
The default handshake mode for the default serial interface.
Valid values are NONE, RTSCTS

int interface:
The default serial interface for the transfer of userprograms.
Valid values are SERIAL1, SERIAL2, SERIAL3

int auto_download;
The download mode during the main menu of RoBIOS. After startup of RoBIOS it can permanently scan the default serial port for a file-download. If it detects a file it automatically downloads it (set to AUTOLOAD).

If it should automatically run this file too set the value to (AUTOLOADSTART).
If it is set to NO_AUTOLOAD no scanning is performed.

int res1:
this is a reserved value (formerly it was used for the state of the radio remote
control which has now its own HDT entry. So always set it to 0)

int cammode:
The default camera mode. Valid values are AUTOBRIGHTNESS, NOAUTOBRIGHTNESS

int battery_display:
Switch the battery status display on or off. Valid values are BATTERY_ON,
BATTERY_OFF

int CPUclock:
The clock rate(MHz) the MC68332 microprocessor should run with.
It is accessible via OSMachineSpeed().

float user_version:
The user defined version number of the actual HDT. This nr is just for information and will be displayed in the HRD-menue of the RoBiOS!

String10 name;
The user defined unique name for this Eyebot. This name is just for information
and will be displayed in the main menu of the RoBiOS! It is accessible via
OSMachineName().

unsigned char robi_id;
The user defined unique id for this Eyebot. This id is just for information and
will be displayed in the main-menu of the RoBiOS! Is is accessible via OSMachi-
neID(). It can temporarily be changed in Hrd/Set/Rmt

unsigned char res2:
this is a reserved value (formerly it was used for the robot-ID of the radio
remote control which has now its own HDT entry. So always set it to 0)

C.6 Infrared Sensor Entry

```
typedef struct
{
    int     driver_version;
    int     tpu_channel;
}ir_type;

e.g.
ir_type   ir0 = {0, 8};
```

int driver_version:
The maximum driver version for which this entry is compatible.
Because newer drivers will surely need more information this tag prevents this
driver from reading more information than actually available.

int tpu_channel:
The tpu channel the ir-sensor is attached to. Valid values are 0..15
Each ir-sensor needs a tpu channel to signal the recognition of an obstacle.

C.7 Infrared TV Remote Entry

```
typedef struct
{
  short    version;
  short    channel;
  short    priority;
  /* new in version 1: */
  short    use_in_robios;
  int      type;
  int      length;
  int      tog_mask;
  int      inv_mask;
  int      mode;
  int      bufsize;
  int      delay;
  int      code_key1;
  int      code_key2;
  int      code_key3;
  int      code_key4;
} irtv_type;
```

This is the new extended IRTV struct. RoBIOS can still handle the old version
0-format which will cause RoBIOS to use the settings for the standard Nokia VCN
620. But only with the new version 1 is it possible to use the IRTV to control
the 4 keys in RoBIOS.

old settings (version 0):
e.g. for a SoccerBot:
irtv_type irtv = {0, 11, TPU_HIGH_PRIO}; /* Sensor connected to TPU 11 (=S10)*/

e.g. for an EyeWalker:
irtv_type irtv = {0, 0, TPU_HIGH_PRIO};   /* Sensor connected to TPU 0 */

new settings (version 1 for Nokia VCN620 and activated RoBIOS control):
irtv_type irtv = {1, 11, TPU_HIGH_PRIO, REMOTE_ON, SPACE_CODE, 15, 0x0000,
0x03FF, DEFAULT_MODE, 1, -1, RC_RED, RC_GREEN, RC_YELLOW, RC_BLUE};

short version:
The maximum driver version for which this entry is compatible.
Because newer drivers will surely need more information, this tag
prevents this driver from reading more information than actually available.

short channel:
The TPU channel the IRTV-sensor is attached to. Valid values are 0..15.
Normally, the sensor is connected to a free servo port. However on the
EyeWalker there is no free servo connector so the sensor should be
connected to a motor connector (a small modification is needed for this
- see manual).

short priority:
The IRQ-priority of the assigned TPU channel. This should be set to
TPU_HIGH_PRIO to make sure that no remote commands are missed.

short use_in_robios:
If set to REMOTE_ON, the remote control can be used to control the 4 EyeCon keys
in RoBIOS. Use REMOTE_OFF to disable this feature.

int type:
int length:
int tog_mask:
int inv_mask:
int mode:
int bufsize:

int delay:
These are the settings to configure a certain remote control. They are exactly the same as the parameters for
the IRTVInit() system call. Above is an example for the default Nokia VCN620 control. The settings can be found by using the irca-program.

int code_key1:
int code_key2:
int code_key3:
int code_key4:
These are the codes of the 4 buttons of the remote control that should match the 4 EyeCon keys. For the Nokia remote control all codes can be found in the header file 'IRnokia.h'.

C.8 Latch Entry

With this entry RoBIOS is told where to find the In/Out-Latches and how many of them are installed.

```
typedef struct
{
  short    version;
  BYTE*    out_latch_address;
  short    nr_out;
  BYTE*    in_latch_address;
  short    nr_in;
} latch_type;
```

e.g.
latch_type latch = {0, (BYTE*)IOBase, 1 , (BYTE*)IOBase, 1};

int version:
The maximum driver version for which this entry is compatible.
Because newer drivers will surely need more information, this tag prevents this driver from reading more information than actually available.

BYTE*   out_latch_address:
Start address of the out-latches.

short   nr_out:
Amount of 8Bit out-latches

BYTE*   in_latch_address;
Start address of the in-latches.

short   nr_in;
Amount of 8Bit in-latches

C.9 Motor Entry

```
typedef struct
{
  int      driver_version;
  int      tpu_channel;
  int      tpu_timer;
  int      pwm_period;
  BYTE*    out_pin_address;
  short    out_pin_fbit;
  short    out_pin_bbit;
```

```
    BYTE*    conv_table;       /* NULL if no conversion needed */
    short    invert_direction; /* only in driver_version > 2 */
}motor_type;
```

e.g.
```
motor_type motor0 = {3,  0, TIMER1, 8196, (void*)(OutBase+2), 6, 6,
(BYTE*)&motconv0), 0};
```

int driver_version:
The maximum driver version for which this entry is compatible.
Because newer drivers will surely need more information this tag prevents this
driver from reading more information than actually available.
Use driver_version = 2 for hardware versions < MK5 to utilize the two bits for
the motor direction setting.
Use driver_version = 3 for hardware version >= MK5 to utilize only one bit
(_fbit) for the direction setting.

int tpu_channel:
The tpu channel the motor is attached to. Valid values are 0..15
Each motor needs a pwm (pulse width modulated) signal to drive with different
speeds.
The internal TPU of the MC68332 is capable of generating this signal on up to 16
channels. The value to be entered here is given through the actual hardware
design.

int tpu_timer:
The tpu timer that has to be used. Valid values are TIMER1, TIMER2
The tpu generates the pwm signal on an internal timer basis. There are two different timers that can be used to determine the actual period for the pwm signal.
TIMER1 runs at a speed of 4MHz up to 8MHz depending on the actual CPU-clock
which allows periods between 128Hz and 4MHz (with 4MHz basefrq) up to 256Hz -
8MHz (with 8MHz)
TIMER2 runs at a speed of 512kHz up to 1MHz depending on the actual CPU-clock
which allows periods between 16Hz and 512kHz (512kHz base) up to 32Hz - 1MHz
(1MHz base)
To determine the actual TIMERx speed use the following equation:
$TIMER1[MHz] = 4MHZ * (16MHz + (CPUclock[MHz] \% 16))/16$
$TIMER2[MHz] = 512kHZ * (16MHz + (CPUclock[MHz] \% 16))/16$

int pwm_period:
This value sets the length of one pwm period in Hz according to the selected
timer.
The values are independent (in a certain interval) of the actual CPU-clock.
The maximal frequency is the actual TPU-frequency divided by 100 in order
to guarantee 100 different energy levels for the motor. This implies a maximum
period of
40-80kHz with TIMER1 and 5-10kHz with TIMER2 (depending on the cpuclock).
The minimal frequency is therefore the Timerclock divided by 32768 which
implies 128-256Hz (Timer1) and 16-32Hz (Timer2) as longest periods (depending
on CPUclock).
To be independent of the actual CPUclock a safe interval is given by 256Hz -
40kHz (Timer1) and 32Hz - 5kHz (Timer2).
To avoid a 'stuttering' of the motor, the period should not be set too slow. But
on the other hand setting the period too fast, will decreases the remaining
calculation time of the TPU.

BYTE* out_pin_address:
The I/O Port address the driver has to use. Valid value is a 32bit address.
To control the direction a motor is spinning a H-bridge is used. This type of
hardware is normally connected via two pins to a latched output. The out-
latches of the EyeCon controller are for example located at IOBASE and the suc-
ceeding addresses.
One of these two pins is set for forward movement and the other for backward
movement.

```
short out_pin_fbit:
```
The portbit for forward drive. Valid values are 0..7
This is the bitnumber in the latch addressed by out_pin_address.

```
short out_pin_bbit:
```
The portbit for backward drive. Valid values are 0..7
This is the bitnumber in the latch addressed by out_pin_address.
If driver_version is set to 3 this bit is not used and should be set to the same value as the fbit.

```
BYTE* conv_table:
```
The pointer to a conversion table to adjust differently motors.
Valid values are NULL or a pointer to a table containing 101 bytes.
Usually two motors behave slightly different when they get exactly the same amount of energy. This will for example show up in a differential drive, when a vehicle should drive in a straight line but moves in a curve. To adjust one motor to another a conversion table is needed. For each possible speed (0..100%) an appropriate value has to be entered in the table to obtain the same speed for both motors. It is wise to adapt the faster motor because at high speeds the slower one can't keep up, you would need speeds of more than 100% !
Note: The table can be generated by software using the connected encoders.

```
short invert_direction:
```
This flag is only used if driver_version is set to 3. This flag indicates to the driver to invert the spinning direction.
If driver_version is set to 2, the inversion will be achieved by swapping the bit numbers of fbit and bbit and this flag will not be regarded.

C.10 Position Sensitive Device (PSD) Entry

```
typedef struct
{
  short    driver_version;
  short    tpu_channel;
  BYTE*    in_pin_address;
  short    in_pin_bit;
  short    in_logic;
  BYTE*    out_pin_address;
  short    out_pin_bit;
  short    out_logic;
  short*   dist_table;
}psd_type;

e.g.
psd_type psd0 = {0, 14, (BYTE*)(Ser1Base+6), 5, AL, (BYTE*)(Ser1Base+4), 0, AL,
                (short*)&dist0};
psd_type psd1 = {0, 14, (BYTE*)IOBase, 2, AH, (BYTE*)IOBase, 0, AH,
                (short*)&dist1};
```

```
int driver_version:
```
The maximum driver version for which this entry is compatible.
Because newer drivers will surely need more information, this tag prevents this driver from reading more information than actually available.

```
short tpu_channel:
```
The master TPU channel for serial timing of the PSD communication. Valid values are 0..15
This TPU channel is not used as an input or output. It is just used as a high resolution timer needed to generate exact communication timing. If there are more than 1 PSD connected to the hardware each PSD has to use the same TPU channel. The complete group or just a selected subset of PSDs can 'fire' simultane-

ously. Depending on the position of the PSDs it is preferable to avoid measure
cycles of adjacent sensors to get correct distance values.

BYTE* in_pin_address:
Pointer to an 8Bit register/latch to receive the PSD measuring result.

short in_pin_bit:
The portbit for the receiver. Valid values are 0..7
This is the bitnumber in the register/latch addressed by in_pin_address.

short in_logic:
Type of the received data. Valid values are AH, AL
Some registers negate the incoming data. To compensate this, active low(AL) has
to be selected.

BYTE* out_pin_address:
Pointer to an 8Bit register/latch to transmit the PSD control signal.
If two or more PSDs are always intended to measure simultaneously the same out-
pin can be connected to all of these PSDs. This saves valuable register bits.

short out_pin_bit:
The portbit for the transmitter. Valid values are 0..7
This is the bitnumber in the register/latch addressed by out_pin_address.

short out_logic:
Type of the transmitted data. Valid values are AH, AL
Some registers negate the outgoing data. To compensate this, active low(AL) has
to be selected.

short* dist_table:
The pointer to a distance conversion table.
A PSD delivers an 8bit measure result which is just a number. Due to inaccuracy
of the result only the upper 7 bits are used (div 2). To obtain the corresponding distance in mm, a lookup table with 128 entries is needed. Since every PSD
slightly deviates in its measured distance from each other, each PSD needs its
own conversion table to guarantee correct distances. The tables have to be generated 'by hand'. The testprogram included in RoBiOS shows the raw 8bit PSD
value for the actual measured distance. By slowly moving a plane object away
from the sensor the raw values change accordingly. Now take every second raw
value and write down the corresponding distance in mm.

C.11 Quadrature Encoder Entry

```
typedef struct
{
  int     driver_version;
  int     master_tpu_channel;
  int     slave_tpu_channel;
  DeviceSemantics motor;
  unsigned int clicksPerMeter;
  float   maxspeed;       /* (in m/s) only needed for VW-Interface */
}quad_type;
```

e.g.
quad_type decoder0 = {0, 3, 2, MOTOR_LEFT, 1234, 2.34};

int driver_version:
The maximum driver version for which this entry is compatible.
Because newer drivers will surely need more information, this tag prevents this
driver from reading more information than actually available.

int master_tpu_channel:

The first TPU channel used for quadrature decoding. Valid values are 0..15
To perform decoding of the motor encoder signals the TPU occupies two adjacent
channels. By changing the order of the two channels the direction of counting
can be inverted.

int slave_tpu_channel:
The second TPU channel used for quadrature decoding. Valid values are
master_tpu_channel +|- 1

DeviceSemantics motor:
The semantics of the attached motor.
To test a specific encoder via the internal RoBiOS function the semantics of
the coupled motor is needed.

unsigned int clicksPerMeter:
This parameter is used only if the the connected motor powers a driving wheel.
It is the number of clicks that are delivered by the encoder covering the dis-
tance of 1 meter.

float maxspeed:
This parameter is used only if the connected motor powers a driving wheel.
It is the maximum speed of this wheel in m/s.

C.12 Remote Control Entry

With this entry the default behavior of the (wireless) remote control can be
specified.

```
typedef struct
{
    int version;
    short robi_id;
    short remote_control;
    short interface;
    short serspeed;
    short imagemode;
    short protocol;
} remote_type;
```

e.g.
remote_type remote = {1, ID, REMOTE_ON, SERIAL2, SER115200, IMAGE_FULL,
                      RADIO_BLUETOOTH};

int version:
The maximum driver version for which this entry is compatible.
Because newer drivers will surely need more information this tag prevents this
driver from reading more information than actually available.

short robi_id;
The user defined unique id (0-255) for this EyeCon. This id is just for information and will
be displayed in the main menu of the RoBiOS! Is is accessible via OSMachi-
neID(). It can temporarily be changed in Hrd/Set/Rmt

short remote_control:
The default control mode for the EyeCon. Valid values are:
REMOTE_ON (the display is forwarded to and the keys are sent from a remote PC),
REMOTE_OFF (normal mode),
REMOTE_PC (only the PC sends data i.e. button press is activated only)
REMOTE_EYE (only the EyeCon sends data i.e. display information only)

short interface:

```
The default serial interface for the radio transfer
Valid values are SERIAL1, SERIAL2, SERIAL3

short serspeed:
The default baudrate for the selected serial interface.
Valid values are SER9600, SER19200, SER38400, SER57600, SER115200

short imagemode:
The mode in which the images of the camera should be transferred to the PC.
Valid values are IMAGE_OFF (no image), IMAGE_REDUCED (reduced quality),
IMAGE_FULL (original frame)

short protocol:
This specifies the module type connected to the serial port.
Valid values are RADIO_METRIX (message length 50 Bytes), RADIO_BLUETOOTH
(mes.len. 64KB), RADIO_WLAN (message lenngth 64KB)
```

C.13 Servo Entry

```
typedef struct
{
  int     driver_version;
  int     tpu_channel;
  int     tpu_timer;
  int     pwm_period;
  int     pwm_start;
  int     pwm_stop;
}servo_type;

e.g.
servo_type servo0 = {1,  0, TIMER2, 20000, 700, 1700};

int driver_version:
The maximum driver version for which this entry is compatible.
Because newer drivers will surely need more information, this tag prevents this
driver from reading more information than actually available.

int tpu_channel:
The tpu channel the servo is attached to. Valid values are 0..15
Each servo needs a pwm (pulse width modulated) signal to turn into different
positions.
The internal TPU of the MC68332 is capable of generating this signal on up to 16
channels. The value to be entered here is given through the actual hardware
design.

int tpu_timer:
The tpu timer that has to be used. Valid values are TIMER1, TIMER2
The tpu generates the pwm signal on an internal timer basis. There are two dif-
ferent timers that can be used to determine the actual period for the pwm sig-
nal.
TIMER1 runs at a speed of 4MHz up to 8MHz depending on the actual CPU-clock
which allows periods between 128Hz and 4MHz (with 4MHz basefrq) up to 256Hz -
8MHz (with 8MHz)
TIMER2 runs at a speed of 512kHz up to 1MHz depending on the actual CPU-clock
which allows periods between 16Hz and 512kHz (512kHz base) up to 32Hz - 1MHz
(1MHz base)
To determine the actual TIMERx speed use the following equation:
TIMER1[MHz] = 4MHZ * (16MHz + (CPUclock[MHz] % 16))/16
TIMER2[MHz] = 512kHZ * (16MHz + (CPUclock[MHz] % 16))/16

int pwm_period:
This value sets the length of one pwm period in microseconds (us).
```

A normal servo needs a pwm_period of 20ms which equals 20000us. For any exotic servo this value can be changed accordingly. It is always preferable to take TIMER2 because only here are enough discrete steps available to position the servo accurately. The values are in a certain interval (see motor), independent of the CPUclock.

int pwm_start:
This is the minimal hightime of the pwm period in us. Valid values are
0..pwm_period
To position a servo the two extreme positions for it have to be defined. In the normal case a servo needs to have a minimal hightime of 0.7ms (700us) at the beginning of each pwm period. This is also one of the two extreme positions a servo can take.

int pwm_stop:
This is the maximum hightime of the pwm period. Valid values are 0..pwm_period. Depending on the rotation direction of a servo, one may choose pwm_stop less than or greater than pwm_start.
To position a servo the two extreme positions for it have to be defined. In the normal case a servo needs to have a maximum hightime of 1.7ms (1700us) at the beginning of each pwm period. This is also one of the two extreme positions a servo can take.
All other positions of the servo are linear interpolated in 256 steps between these two extremes.
Hint: If you don't need the full range the servo offers you can adjust the start and stop parameters to a smaller 'window' like 1ms to 1.5ms and gain a higher resolution in these bounds. Or the other way around, you can enlarge the 'window' to adjust the values to the real degrees the servo changes its position: Take for example a servo that covers a range of 210 degrees. Simply adjust the stop value to 1.9ms. If you now set values between 0 and 210 you will reach the two extremes in steps corresponding to the real angles. Values higher than 210 would not differ from the result gained by the value of 210.

C.14 Startimage Entry

```
typedef BYTE image_type[16*64];
```

```
e.g.
image_type startimage = {0xB7,0x70,0x1C,...0x00};
```

Here a user-defined startup image can be entered as a byte array
(16*64 = 1024Bytes).
This is a 128x64 Pixel B/W picture where each pixel is represented by a bit.

C.15 Startmelody Entry

```
no typedef
```

```
e.g.
int startmelody[] = {1114,200, 2173,200, 1114,200, 1487,200, 1669,320, 0};
```

Here you can enter your own melody that will be played at startup. It is a list of integer pairs. The first value indicates the frequency, the second the duration in 1/100s of the tone. As last value there must be single 0 in the list.

C.16 VW Drive Entry

```
typedef struct
{
  int     version;
  int     drive_type;
  drvspec drive_spec; /* -> diff_data */
}vw_type;

typedef struct
{
  DeviceSemantics quad_left;
  DeviceSemantics quad_right;
  float           wheel_dist; /* meters */
}diff_data;
```

```
e.g.
vw_type drive = {0, DIFFERENTIAL_DRIVE, {QUAD_LEFT, QUAD_RIGHT, 0.21}};
```

int driver_version:
The maximum driver version for which this entry is compatible.
Because newer drivers will surely need more information, this tag prevents this
driver from reading more information than actually available.

int drive_type:
Define the type of the actual used drive.
Valid values are DIFFERENTIAL_DRIVE (ACKERMAN_DRIVE, SYNCHRO_DRIVE,
TRICYCLE_DRIVE)
The following parameters depend on the selected drive type.

DIFFERENTIAL_DRIVE:
The differential drive is made up of two parallel independent wheels with the
kinematic center right between them. Obviously two encoders with the connected
motors are needed.

    DeviceSemantics quad_left:
    The semantics of the encoder used for the left wheel.

    DeviceSemantics quad_right:
    The semantics of the encoder used for the right wheel.

    float wheel_dist:
    The distance (meters) between the two wheels to determine the kinematic
    center.

C.17 Waitstates Entry

```
typedef struct
{
  short     version;
  short     rom_ws;
  short     ram_ws;
  short     lcd_ws;
  short     io_ws;
  short     serpar_ws;
}waitstate_type;
```

```
e.g.
waitstate_type waitstates = {0,3,0,1,0,2};
```

int version:

The maximum driver version for which this entry is compatible.
Because newer drivers will surely need more information, this tag prevents this
driver from reading more information than actually available.

short rom_ws:
Waitstates for the ROM access
Valid values (for all waitstates):
waitstates = 0..13, Fast Termination = 14, External = 15

short ram_ws:
Waitstates for the RAM access

short lcd_ws:
Waitstates for the LCD access

short io_ws:
Waitstates for the Input/Output latches access

short serpar_ws:
Waitstates for the 16c552 Serial/Parallel Port Interface access

---

Thomas Bräunl, Klaus Schmitt, Michael Kasper 1996-2006

HARDWARE SPECIFICATION

The following tables speficy details of the EyeCon controller hardware.

Version	Features
Mark 1	First prototypes, two boards, double-sided, rectangular push button, no speaker
Mark 2	Major change: two boards, double-sided, speaker and microphone on board, changed audio circuit
Mark 2.1	Minor change: connect digital and analog ground
Mark 3.0	Completely new design: single board design, four layers, direct-plug-in connectors for sensors and motors, motor controllers on board, BDM on board, wireless module and antenna on board
Mark 3.11	Minor change: miniature camera port added
Mark 3.12	Minor change: replaced fuse by reconstituting polyswitch
Mark 4.02	Major change: extension to 2MB RAM, adding fast camera framebuffer, additional connector for third serial port, redesign of digital I/O
Mark 5	Major redesign: camera plugs in directly into controller, new motor connectors, video out, additional servo connectors

Table D.1: Hardware versions

Chip Select	Function
CSBOOT	Flash-ROM
CS 0+1	RAM (1MB)
CS 2	LCD
CS 3+7	RAM (additional 1MB)
CS 4	Input/Output latch (IOBase)
CS 5	FIFO camera buffer
CS 6	Address A19
CS 7	Autovector acknowledge generation
CS 8	Parallel port of 16C552
CS 9	Serial port 1 of 16C552
CS 10	Serial port 2 of 16C552

Table D.2: Chip-select lines

Address	Memory Usage	Chip Selects
0x00000000	RoBIOS RAM (128KB)	CS0,1,3,7
0x00020000	User RAM (max. 2MB-128KB)	CS0,1,3,7
0x00200000	End of RAM	
. . .	*unused addresses*	
0x00a00000	TpuBase (2KB)	
0x00a00800	End of TpuBase	
. . .	*unused addresses*	
0x00c00000	Flash-ROM (512KB)	CS2
0x00c80000	End of Flash-ROM	
. . .	*unused addresses*	
0x00e00800	Latches	CS4
0x00e01000	FIFO or Latches	CS5
0x00e01800	Parallel Port/Camera	CS8

Table D.3: Memory map (continued)

Address	Memory Usage	Chip Selects
0x00e02000	Serial Port2	CS9
0x00e02800	Serial Port3	CS10
...	*unused addresses*	
0x00fff000	MCU68332 internal registers (4KB)	
0x01000000	End of registers and addressable RAM	

Table D.3: Memory map (continued)

IRQ	Function
1	FIFO half-full flag *(hardwired)*
2	INT-SIM (100Hz Timer, arbitration 15)
3	INT serial 1 (neg.)/serial 2 (neg.) of 16C552 *(hardwired)*
4	INT QSPI and SCI of the QSM (arbitration 13)
5	INT parallel port (neg.) of 16C552 *(hardwired)*
6	INT-TPU (arbitration 14)
7	*free*
Note	INT 1,3,5 are hardwired to FIFO or 16C552, respectively, all other INTs are set via software

Table D.4: Interrupt request lines

Port F	Key Function
PF0	KEY4
PF2	KEY3
PF4	KEY2
PF6	KEY1

Table D.5: Push buttons

Description	Value
Voltage	Required: between 6V and 12V DC, normally: 7.2V
Power consumption	EyeCon controller only: 235mA EyeCon controller with EyeCam CMOS camera: 270mA
Run-time	With 1,350mAh, 7.2V Li-ion rechargeable battery (approx.): 4 – 5 hours EyeCon controller only 1 – 2 hours EyeCon controller with SoccerBot robot and camera, constantly driving and sensing, depending on program and speed
Power limitation	Total power limit is 3A 3A polyswitch prohibits damage through higher current or wrong polarity Can drive DC motors with up to 1A each

Table D.6: Electrical characteristics

Description	Value
Size	Controller: 10.6cm × 10.0cm × 2.8cm (width × height × depth) EyeCam 3.0cm × 3.4cm × 3.2cm
Weight	Controller: 190g EyeCam: 25g

Table D.7: Physical characteristics

Port	Pins
Serial 1	**Download** (9 pin), standard RS232 serial port, 12V, **female** 1 - 2 Tx 3 Rx 4 - 5 GND 6 - 7 CTS 8 RTS 9 -
Serial 2	**Upload** (9 pin), standard RS232 serial port, 12V, **male** 1 - 2 Rx 3 Tx 4 - 5 GND 6 - 7 RTS 8 CTS 9 5V regulated
Serial 3	RS232 at TTL level (5V) 1 CD' 2 DTR' 3 Tx 4 CTS' 5 Rx 6 RTS' 7 DSR' 8 RI' 9 GND 10 Vcc (5V)

Table D.8: Pinouts EyeCon Mark 5 (continued)

Port	Pins

Digital camera	16 pin connector requires 1:1 connection (cable with female:female) to EyeCam digital color camera Note: The little pin on the EyeCon side of the cable has to point up:
	 `|--^--|`
	 `|-----|`
	 1 STB
	 2-9 Data 0-7
	 10 ACK
	 11 INT
	 12 BSY
	 13 KEY
	 14 SLC
	 15 Vcc (5V)
	 16 GND

Parallel	Standard parallel port 1 Strobe' 2 PD0 3 PD1 4 PD2 5 PD3 6 PD4 7 PD5 8 PD6 9 PD7 10 ACK 11 Busy' 12 PE 13 SLCT 14 Autofxdt' 15 Error 16 Init 17 Slctin' 18..25 GND
BDM	Motorola Background Debugger (10 pin), connects to PC parallel port

Table D.8: Pinouts EyeCon Mark 5 (continued)

Port	Pins
Motors	DC motor and encoder connectors (2 times 10 pin) Motors are mapped to TPU channels 0..1 Encoders are mapped to TPU channels 2..5 Note: Pins are labeled in the following way: | 1 | 3 | 5 | 7 | 9 | \- | 2 | 4 | 6 | 8 | 10| 1 Motor + 2 Vcc (unregulated) 3 Encoder channel A 4 Encoder channel B 5 GND 6 Motor − 7 -- 8 -- 9 -- 10 --
Servos	Servo connectors (12 times 3 pin) Servo signals are mapped to TPU channels 2..13 Note: If both DC motors are used, TPU 0..5 are already in use, so Servo connectors Servo1 (TPU2) .. Servo4 (TPU5) **cannot** be used. 1 Signal 2 Vcc (unregulated) 3 GND

Table D.8: Pinouts EyeCon Mark 5 (continued)

Port	Pins
Infrared	Infrared connectors (6 times 4 pin) Sensor outputs are mapped to digital input 0..3 1 GND 2 Vin (pulse) 3 Vcc (5V regulated) 4 Sensor output (digital)
Analog	Analog input connector (10 pin) Microphone, mapped to analog input 0 Battery-level gauge, mapped to analog input 1 1 Vcc (5V regulated) 2 Vcc (5V regulated) 3 analog input 2 4 analog input 3 5 analog input 4 6 analog input 5 7 analog input 6 8 analog input 7 9 analog GND 10 analog GND
Digital	Digital input/output connector (16 pin) [Infrared PSDs use digital output 0 and digital input 0..3] 1- 8 digital output 0..7 9-12 digital input 4..7 13-14 Vcc (5V) 15-16 GND

Table D.8: Pinouts EyeCon Mark 5 (continued)

LABORATORIES

· ·

Lab 1 Controller

The first lab uses the controller only and not the robot

EXPERIMENT 1 Etch-a-Sketch

Write a program that implements the "Etch-a-Sketch" children's game.

Use the four buttons in a consistent way for moving the drawing pen left/right and up/down. Do not erase previous dots, so pressing the buttons will leave a visible trail on the screen.

EXPERIMENT 2 Reaction Test Game

Write a program that implements the reaction game as given by the flow diagram.

To compute a random wait-time value, isolate the last digit of the current time using `OSGetCount()` and transform it into a value for `OSWait()` to wait between 1 and 8 seconds.

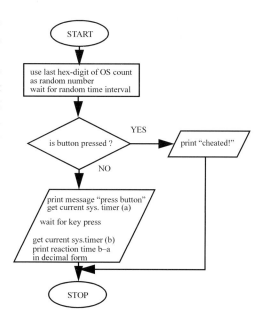

EXPERIMENT 3 Analog Input and Graphics Output

Write a program to plot the amplitude of an analog signal. For this experiment, the analog source will be the microphone. For input, use the following function:

```
AUCaptureMic(0)
```

It returns the current microphone intensity value as an integer between 0 and 1,023.

Plot the analog signal versus time on the graphics LCD. The dimension of the LCD is 64 rows by 128 columns. For plotting use the functions:

```
LCDSetPixel(row,col,1)
```

Maintain an array of the most recent 128 data values and start plotting data values from the leftmost column (0). When the rightmost column is reached (127), continue at the leftmost column (0) – but be sure to remove the column's old pixel before you plot the new value. This will result in an oscilloscope-like output.

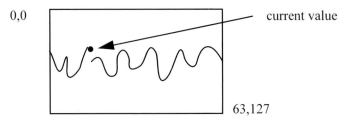

0,0   current value

      63,127

Lab 2 Simple Driving

Driving a robot using motors and shaft encoders

EXPERIMENT 4 Drive a Fixed Distance and Return

Write a robot program using VWDriveStraight and VWDriveTurn to let the robot drive 40cm straight, then turn 180°, drive back and turn again, so it is back in its starting position and orientation.

EXPERIMENT 5 Drive in a Square

Similar to experiment 4.

EXPERIMENT 6 Drive in a Circle

Use routine VWDriveCurve to drive in a circle.

Lab 3 Driving Using Infrared Sensors

Combining
sensor reading
with driving
routines **EXPERIMENT 7 Drive Straight toward an Obstacle and Return**

This is a variation of an experiment from the previous lab. This time the task is to drive until the infrared sensors detect an obstacle, then turn around and drive back the same distance.

Lab 4 Using the Camera

Using camera
and controller
without the
vehicle **EXPERIMENT 8 Motion Detection with Camera**

By subtracting the pixel value of two subsequent grayscale images, motion can be detected. Use an algorithm to add up grayscale differences in three different image sections (left, middle, right). Then output the result by printing the word "left", "middle", or "right".

Variation (a): Mark the detected motion spot graphically on the LCD.

Variation (b): Record audio files for speaking "left", "middle", "right" and have the EyeBot speak the result instead of print it.

EXPERIMENT 9 Motion Tracking

Detect motion like before. Then move the camera servo (and with it the camera) in the direction of movement. Make sure that you do not mistake the auto-motion of the camera for object motion.

Lab 5 Controlled Motion

Drive of the robot
using motors and
shaft encoders
only Due to manufacturing tolerances in the motors, the wheels of a the mobile robots will usually not turn at the same speed, when applying the same voltage. Therefore, a naive program for driving straight may lead in fact to a curve. In order to remedy this situation, the wheel encoders have to be read periodically and the wheel speeds have to be amended.

For the following experiments, use only the low-level routines MOTORDrive and QUADRead. Do not use any of the vω routines, which contain a PID controller as part their implementation.

EXPERIMENT 10 PID Controller for Velocity Control of a Single Wheel

Start by implementing a P controller, then add I and D components. The wheel should rotate at a specified rotational velocity. Increasing the load on the wheel (e.g. by manually slowing it down) should result in an increased motor output to counterbalance the higher load.

EXPERIMENT 11 PID Controller for Position Control of a Single Wheel

The previous experiment was only concerned with maintaining a certain rotational velocity of a single wheel. Now we want this wheel to start from rest, accelerate to the specified velocity, and finally brake to come to a standstill *exactly at a specified distance* (e.g. exactly 10 revolutions).

This experiment requires you to implement speed ramps. These are achieved by starting with a constant acceleration phase, then changing to a phase with (controlled) constant velocity, and finally changing to a phase with constant deceleration. The time points of change and the acceleration values have to be calculated and monitored during execution, to make sure the wheel stops at the correct position.

EXPERIMENT 12 Velocity Control of a Two-Wheeled Robot

Extend the previous PID controller for a single wheel to a PID controller for two wheels. There are two major objectives:

    a.   The robot should drive along a straight path.

    b.   The robot should maintain a constant speed.

You can try different approaches and decide which one is the best solution:

    a.   Implement two PID controllers, one for each wheel.

    b.   Implement one PID controller for forward velocity and one PID controller for rotational velocity (here: desired value is zero).

    c.   Implement only a single PID controller and use offset correction values for both wheels.

Compare the driving performance of your program with the built-in vω routines.

EXPERIMENT 13 PID Controller for Driving in Curves

Extend the PID controller from the previous experiment to allow driving in general curves as well as straight lines.

Compare the driving performance of your program with the built-in vω routines.

EXPERIMENT 14 Position Control of a Two-Wheeled Robot

Extend the PID controller from the previous experiment to enable position control as well as velocity control. Now it should be possible to specify a path (e.g. straight line or curve) plus a desired distance or angle and the robot should come to a standstill at the desired location after completing its path.

Compare the driving performance of your program with the built-in vω routines.

Lab 6 Wall-Following

This will be a useful subroutine for subsequent experiments **EXPERIMENT 15 Driving Along a Wall**

Let the robot drive forward until it detects a wall to its left, right, or front. If the closest wall is to its left, it should drive along the wall facing its left-hand side and vice versa for right. If the nearest wall is in front, the robot can turn to either side and follow the wall.

The robot should drive in a constant distance of 15cm from the wall. That is, if the wall is straight, the robot would drive in a straight line at constant distance to the wall. If the wall is curved, the robot would drive in the same curve at the fixed distance to the wall.

Lab 7 Maze Navigation

Have a look at the Micro Mouse Contest. This is an international competition for robots navigating mazes. **EXPERIMENT 16 Exploring a Maze and Finding the Shortest Path**

The robot has to explore and analyze an unknown maze consisting of squares of a known fixed size. An important sub-goal is to keep track of the robot's position, measured in squares in the x- and y-direction from the starting position.

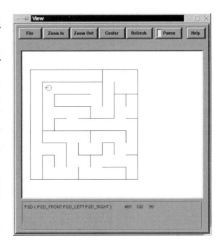

After searching the complete maze the robot is to return to its starting position. The user may now enter any square position in the maze and the robot has to drive to this location and back along the shortest possible path.

Lab 8 Navigation

Two of the classic and most challenging tasks for mobile robots **EXPERIMENT 17 Navigating a Known Environment**

The previous lab dealt with a rather simple environment. All wall segments were straight, had the same length, and all angles were 90°. Now imagine the task of navigating a somewhat more general environment, e.g. the floor of a building.

Specify a map of the floor plan, e.g. in "world format" (see EyeSim simulator), and specify a desired path for the robot to drive in map coordinates. The robot has to use its on-board sensors to carry out self-localization and navigate through the environment using the provided map.

EXPERIMENT 18 Mapping an Unknown Environment

One of the classic robot tasks is to explore an unknown environment and auto-matically generate a map. So, the robot is positioned at any point in its envi-ronment and starts exploration by driving around and mapping walls, obsta-cles, etc.

This is a very challenging task and greatly depends on the quality and com-plexity of the robot's on-board sensors. Almost all commercial robots today use laser scanners, which return a near-perfect 2D distance scan from the robot's location. Unfortunately, laser scanners are still several times larger, heavier, and more expensive than our robots, so we have to make do without them for now.

Our robots should make use of their wheel encoders and infrared PSD sen-sors for positioning and distance measurements. This can be augmented by image processing, especially for finding out when the robot has returned to its start position and has completed the mapping.

The derived map should be displayed on the robot's LCD and also be pro-vided as an upload to a PC.

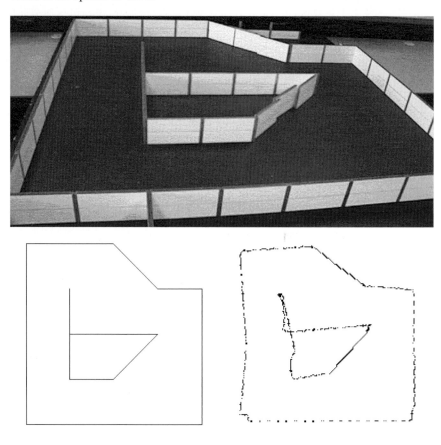

Lab 9 Vision

EXPERIMENT 19 Follow the Light

Assume the robot driving area is enclosed by a boundary wall. The robot's task is to find the brightest spot within a rectangular area, surrounded by walls. The robot should use its camera to search for the brightest spot and use its infrared sensors to avoid collisions with walls or obstacles.

Idea 1: Follow the wall at a fixed distance, then at the brightest spot turn and drive inside the area.

Idea 2: Let the robot turn a full circle (360°) and record the brightness levels for each angle. Then drive in the direction of the brightest spot.

EXPERIMENT 20 Line-Following

Mark a bright white line on a dark table, e.g. using masking tape. The robot's task is to follow the line.

This experiment is somewhat more difficult than the previous one, since not just the general direction of brightness has to be determined, but the position (and maybe even curvature) of a bright line on a dark background has to be found. Furthermore, the driving commands have to be chosen according to the line's curvature, in order to prevent the robot "losing the line", i.e. the line drifting out of the robot's field of view.

Special routines may be programmed for dealing with a "lost line" or for learning the maximum speed a robot can drive at for a given line curvature without losing the line.

Lab 10 Object Detection

EXPERIMENT 21 Object Detection by Shape

An object can be detected by its:

a. Shape

b. Color

c. Combination of shape and color

To make things easy at the beginning, we use objects of an easy-to-detect shape and color, e.g. a bright yellow tennis ball. A ball creates a simple circular image from all viewpoints, which makes it easy to detect its shape. Of course it is not that easy for more general objects: just imagine looking from different viewpoints at a coffee mug, a book, or a car.

443

There are textbooks full of image processing and detection tasks. This is a very broad and active research area, so we are only getting an idea of what is possible.

An easy way of detecting shapes, e.g. distinguishing squares, rectangles, and circles in an image, is to calculate "moments". First of all, you have to identify a continuous object from a pixel pattern in a binary (black and white) image. Then, you compute the object's area and circumference. From the relationship between these two values you can distinguish several object categories such as circle, square, rectangle.

EXPERIMENT 22 Object Detection by Color

Another method for object detection is color recognition, as mentioned above. Here, the task is to detect a colored object from a background and possibly other objects (with different colors).

Color detection is simpler than shape detection in most cases, but it is not as straightforward as it seems. The bright yellow color of a tennis ball varies quite a bit over its circular image, because the reflection depends on the angle of the ball's surface patch to the viewer. That is, the outer areas of the disk will be darker than the inner area. Also, the color values will not be the same when looking at the same ball from different directions, because the lighting (e.g. ceiling lights) will look different from a different point of view. If there are windows in your lab, the ball's color values will change during the day because of the movement of the sun. So there are a number of problems to be aware of, and this is not even taking into account imperfections on the ball itself, like the manufacturer's name printed on it, etc.

Many image sources return color values as RGB (red, green, blue). Because of the problems mentioned before, these RGB values will vary a lot for the same object, although its basic color has not changed. Therefore it is a good idea to convert all color values to HSV (hue, saturation, value) before processing and then mainly work with the more stable hue of a pixel.

The idea is to detect an area of hue values similar to the specified object hue that should be detected. It is important to analyze the image for a color "blob", or a group of matching hue values in a neighborhood area. This can be achieved by the following steps:

a. Convert RGB input image to HSV.

b. Generate binary image by checking whether each pixel's hue value is within a certain range to the desired object hue:
$$binary_{i,j} = |\, hue_{i,j} - hue_{obj} \,| < \varepsilon$$

c. For each row, calculate the matching binary pixels.

d. For each column, calculate the matching binary pixels.

e. The row and column counter form a basic histogram. Assuming there is only one object to detect, we can use these values directly:

> Search the row number with the maximum count value.
> Search the column number with the maximum count value.

   f.   These two values are the object's image coordinates.

EXPERIMENT 23 Object Tracking

Extending the previous experiment, we want the robot to follow the detected object. For this task, we should extend the detection process to also return the size of the detected object, which we can translate into an object distance, provided we know the size of the object.

Once an object has been detected, the robot should "lock onto" the object and drive toward it, trying to maintain the object's center in the center of its viewing field.

A nice application of this technique is having a robot detect and track either a golf ball or a tennis ball. This application can be extended by introducing a ball kicking motion and can finally lead to robot soccer.

You can think of a number of techniques of how the robot can search for an object once it has lost it.

Lab 11 Robot Groups

Now we have a number of robots interacting with each other

EXPERIMENT 24 Following a Leading Robot

Program a robot to drive along a path made of random curves, but still avoiding obstacles.

Program a second robot to follow the first robot. Detecting the leading robot can be done by using either infrared sensors or the camera, assuming the leading robot is the only moving object in the following robot's field of view.

EXPERIMENT 25 Foraging

A group of robots has to search for food items, collect them, and bring them home. This experiment combines the object detection task with self-localization and object avoidance.

Food items are uniquely colored cubes or balls to simplify the detection task. The robot's home area can be marked either by a second unique color or by other features that can be easily detected.

This experiment can be conducted by:

   a.   A single robot

   b.   A group of cooperating robots

   c.   Two competing groups of robots

EXPERIMENT 26 Can Collecting

A variation of the previous experiment is to use magnetic cans instead of balls or cubes. This requires a different detection task and the use of a magnetic actuator, added to the robot hardware.

This experiment can be conducted by:

    a.   A single robot

    b.   A group of cooperating robots

    c.   Two competing groups of robots

EXPERIMENT 27 Robot Soccer

Robot soccer is of course a whole field in its own right. There are lots of publications available and of course two independent yearly world championships, as well as numerous local tournaments for robot soccer. Have a look at the web pages of the two world organizations, FIRA and Robocup:

-   `http://www.fira.net/`
-   `http://www.robocup.org/`

SOLUTIONS

F

Lab 1 Controller

EXPERIMENT 1 Etch-a-Sketch

```
1   /* ------------------------------------------------------
2   | Filename:     etch.c
3   | Authors:      Thomas Braunl
4   | Description:  pixel operations resembl. "etch a sketch"
5   | ------------------------------------------------------ */
6   #include <eyebot.h>
7
8   void main()
9   { int k;
10    int x=0, y=0, xd=1, yd=1;
11
12    LCDMenu("Y","X","+/-","END");
13    while(KEY4 != (k=KEYRead())) {
14      LCDSetPixel(y,x, 1);
15      switch (k) {
16        case KEY1: y = (y + yd +  64) %  64; break;
17        case KEY2: x = (x + xd + 128) % 128; break;
18        case KEY3: xd = -xd; yd = -yd; break;
19      }
20      LCDSetPrintf(1,5);
21      LCDPrintf("y%3d:x%3d", y,x);
22    }
23  }
```

EXPERIMENT 2 Reaction Test Game

```
1   /* -----------------------------------------------------
2   | Filename:      react.c
3   | Authors:       Thomas Braunl
4   | Description:   reaction test
5   | ----------------------------------------------------- */
6   #include "eyebot.h"
7   #define MAX_RAND  32767
8
9   void main()
10  { int time, old,new;
11
12    LCDPrintf(" Reaction Test\n");
13    LCDMenu("GO"," "," "," ");
14    KEYWait(ANYKEY);
15    time = 100 + 700 * rand() / MAX_RAND; /* 1..8 s */
16    LCDMenu(" "," "," "," ");
17
18    OSWait(time);
19    LCDMenu("HIT","HIT","HIT","HIT");
20    if (KEYRead()) printf("no cheating !!\n");
21     else
22     { old = OSGetCount();
23       KEYWait(ANYKEY);
24       new = OSGetCount();
25       LCDPrintf("time: %1.2f\n", (float)(new-old) / 100.0);
26     }
27
28    LCDMenu(" "," "," ","END");
29    KEYWait(KEY4);
30  }
```

EXPERIMENT 3 Analog Input and Graphics Output

```
1   /* --------------------------------------------------------
2   | Filename:     micro.c
3   | Authors:      Klaus Schmitt
4   | Description:  Displays microphone input graphically
5   |               and numerically
6   | ------------------------------------------------------- */
7   #include "eyebot.h"
8
9   void main ()
10  {    int disttab[32];
11       int pointer=0;
12       int i,j;
13       int val;
14
15       /* clear the graphic-array */
16       for(i=0; i<32; i++)
17         disttab[i]=0;
18
19       LCDSetPos(0,3);
20       LCDPrintf("MIC-Demo");
21       LCDMenu("","","","END");
22
23       while (KEYRead() != KEY4)
24       { /* get actual data and scale it for the LCD */
25         disttab[pointer] = 64 - ((val=AUCaptureMic(0))>>4);
26
27         /* draw graphics */
28         for(i=0; i<32; i++)
29         { j = (i+pointer)%32;
30           LCDLine(i,disttab[j], i+4,disttab[(j+1)%32], 1);
31         }
32
33         /* print actual distance and raw-data */
34         LCDSetPos(7,0);
35         LCDPrintf("AD0:%3X",val);
36
37         /* clear LCD */
38         for(i=0; i<32; i++)
39         { j = (i+pointer)%32;
40           LCDLine(i,disttab[j], i+4,disttab[(j+1)%32], 0);
41         }
42
43         /* scroll the graphics */
44         pointer = (pointer+1)%32;
45       }
46  }
```

Lab 2 Simple Driving

EXPERIMENT 4 Drive a Fixed Distance and Return

```
1   /* --------------------------------------------------------
2   | Filename:     drive.c
3   | Authors:      Thomas Braunl
4   | Description:  Drive a fixed distance, then come back
5   | -------------------------------------------------------- */
6   #include "eyebot.h"
7   #define DIST   0.4
8   #define SPEED  0.1
9   #define TSPEED 1.0
10
11  void main()
12  { VWHandle     vw;
13    PositionType pos;
14    int          i;
15
16    LCDPutString("Drive Demo\n");
17    vw = VWInit(VW_DRIVE,1); /* init v-omega interface */
18    if(vw == 0)
19      {
20        LCDPutString("VWInit Error!\n\a");
21        OSWait(200); return;
22      }
23    VWStartControl(vw,7,0.3,7,0.1);
24    OSSleep(100); /* delay before starting */
25
26    for (i=0;i<4; i++) /* do 2 drives + 2 turns twice */
27    { if (i%2==0) { LCDSetString(2,0,"Drive");
28                    VWDriveStraight(vw,DIST,SPEED);
29                  }
30         else    { LCDSetString(2,0,"Turn ");
31                    VWDriveTurn(vw,M_PI,TSPEED);
32                  }
33      while (!VWDriveDone(vw))
34      { OSWait(33);
35        VWGetPosition(vw,&pos);
36        LCDSetPrintf(3,0,"Pos: %4.2f x %4.2f",pos.x,pos.y);
37        LCDSetPrintf(4,0,"Heading:%5.1f",
38                          pos.phi*180.0/M_PI);
39      }
40    }
41    OSWait(200);
42    VWRelease(vw);
43  }
```

INDEX

· ·

Index